1st International Conference on Mechanical Power Transmission 2019

IOP Conference Series: Materials Science and Engineering
Volume 624

Chennai, India
11 – 13 July 2019

ISBN: 978-1-5108-9828-8
ISSN: 1757-8981

Printed from e-media with permission by:

Curran Associates, Inc.
57 Morehouse Lane
Red Hook, NY 12571

CURRAN ASSOCIATES INC.
proceedings
.com

Some format issues inherent in the e-media version may also appear in this print version.

Institute of Physics
Dirac House, Temple Back
Bristol BS1 6BE UK

Phone: 44 1 17 929 7481
Fax: 44 1 17 920 0979

techtracking@iop.org

Additional copies of this publication are available from:

Curran Associates, Inc.
57 Morehouse Lane
Red Hook, NY 12571 USA
Phone: 845-758-0400
Fax: 845-758-2633
Email: curran@proceedings.com
Web: www.proceedings.com

Table of contents

Volume 624

1st International Conference on Mechanical Power Transmission

11–13 July 2019, Chennai, India

Accepted papers received: 23 August 2019
Published online: 7 October 2019

Preface

Papers

IOP Conf. Series: Materials Science and Engineering **624** (2019) 011001 doi:10.1088/1757-899X/624/1/011001

Preface

The Year 2019 is historical for the Indian Institute of Technology Madras, India as it celebrates its Diamond Jubilee, and also received National Rank No.1 for the best engineering institution in the country for the fourth consecutive year. The faculty of the Department of Mechanical Engineering, IIT Madras, ranked in the group 51 – 75 in the Shanghai World Ranking, carries out cutting edge research work in the broad areas of Design, Manufacturing and Thermal streams. Founded in 1968, the Machine Design Section, formerly Machine Elements and Mechanical Handling Laboratory, is well known in India and abroad for decades for its contribution in the advancement of Machine Element Design, Design Optimization, Advanced Materials and Design, Health Monitoring, Tribology, and Sound, Machine Dynamic and vibration.

The research on design and development of mechanical power transmission elements in the modern era has moved to a new dimension with the availability of modern materials such as customizable composites, light weight metals, lubricating polymers, nano-materials, etc, design tools and methods for efficient product and system design, and integrated design and manufacturing methods such as additive manufacturing, surface modification, etc., The increasing demand of the society for reliable and comfortable modern mobility engineering involving hybrid and electrical drive systems where mechanical power transmission elements play a significant role directs the researchers and designers to look for newer materials, manufacturing routes, design tools and testing for better performance.

This conference, first in the series, is organized by the Machine Design Section of the Department of Mechanical Engineering, IIT Madras as part of the Diamond Jubilee celebrations of the IIT Madras, from where significant contributions have been made in this area. As the conference theme is highly specialized and limited research is happening all over the World, it is proposed to have a mixture of eminent senior researchers delivering invited lectures and young researchers presenting contributory papers. The conference is mainly intended to facilitate interaction with lead researchers and expose emerging emerging areas to the young researchers in the following themes:

- Gears, Bearings and Other Drives – Design, Manufacturing and Performance
- Design Optimisation
- Advanced Materials & Treatments
- Life Testing, Reliability, Failure mechanisms, Fault Diagnostics, and Prognostics
- Lubrication Engineering, Noise, Vibration, and Harshness

The reviewing team have contributed a lot in looking in to the quality of the contributory papers and accepted only one third of the abstracts received.

We acknowledge the support provided by the reviewers, administration and volunteers and thank all the participants who have come from many reputed institutes in India and abroad.

We wish everyone an enjoyable and enlightening stay at IIT Madras.

Prof. R Gnanamoorthy
Prof. Shankar Krishnapillai
Dr. Piyush Shakya

IOP Conf. Series: Materials Science and Engineering **624** (2019) 011002 doi:10.1088/1757-899X/624/1/011002

Peer review statement

All papers published in this volume of *IOP Conference Series: Materials Science and Engineering* have been peer reviewed through processes administered by the proceedings Editors. Reviews were conducted by expert referees to the professional and scientific standards expected of a proceedings journal published by IOP Publishing.

IOP Conf. Series: Materials Science and Engineering **624** (2019) 012001 doi:10.1088/1757-899X/624/1/012001

Investigation on modal behaviour of FGM annular plate under hygrothermal effect

Pankaj Sharma[1], Rahul Singh[2]

[1] Assistant Professor,University Department of Mechanical Engineering, Rajasthan Technical University, Kota 324010, India

[2] Research Scholar, University Department of Mechanical Engineering, Rajasthan Technical University, Kota 324010, India

Email: [1]psharma@rtu.ac.in, [2]drengineer92@gmail.com

Abstract. This paper investigates modal characteristics of FGM annular plate under hygrothermal effect. The mechanical properties are supposed to have a continuous variation along the thickness of the plate. The power law or sigmoid law functions are used to govern the resultant mechanical properties. The MUMPS eigen solver which is exists in COMSOL Multiphysics® (version 5.2) software is employed to obtain the eigenfrequencies and corresponding mode shapes. The accuracy of the method is validated by comparing the results with the results published work for particular case. Effects of functionally graded index, radius-to-thickness ratio, and temperature and moisture changes on the natural frequencies are discussed in detail. An excessive stress due to drastic moisture and temperature gradients make the engineering structures susceptible to failure. Due to the fact that, continuously identified natural frequencies of vibration under temperature and moisture effect can provide unique information during both initial fabrication and service life of different mechanical structures. Hence, this study can be helpful directly or indirectly in development of highly efficient environmental sensors that are very necessary in a structural health monitoring perspective. It can be observed that the fundamental frequency decreases with increase in functionally graded index and the moisture content rise under the power law model whereas under sigmoid law model the frequencies remain almost constant with increase in functionally graded index.

1. Introduction

Functionally graded materials (FGMs) are the new form of composite, which provides the facility of smooth material variation across any desired direction. Potential use of FGMs can be found in automotive structural components modelled as circular plates (e.g. thrust bearing plates, the driven plate of a friction clutch etc.) [1]. Annular plates are widely used parts in mechanical power transmission to minimize the weight of the whole structure [2]. Reduction of gear weight also reduce the natural frequency of the gears, thereby weakening the mesh noise [3]. In braking system also, FGM is a suitable material for automotive disc brake and pads [4].

Moisture absorption may be practically occurred during the fabrication of FGM structures (beam, plate and shell etc.) and subsequent variation in temperature and moisture reduces the elastic moduli and degrades the strength of material. Due to this reason there are many FGM structures which are failed because of vibrations under hygrothermal effect. Also, the Eigenfrequencies and corresponding mode shapes of any annular plate is affected by the presence of a hole in it.

2. Literature survey

There are many modelling techniques and solution methods exist in the literature to analyze the vibration characteristics of FGM plate under thermal effect [5] only. Also, it may be seen that in most of the cases the material properties were governed by the power law function.

Hygrothermal effect was first investigated by Whitney and Ashton [6] for laminated composite structures. Using Newton-Raphson method, Lee and Kim [7] calculated the Eigenfrequencies of simply supported FGM plates in hygrothermal environment. Using different plate theories, Zenkour [8,9] studied the hygrothermal effects on FGM plates.

In most of the above papers only power law has been considered to study the hygrothermal effects on modal behaviour of FGM plates. It has been stated in the literature [10] that the interfacial stresses arise which is the main cause of failure of composite structures is reduced up to an extent if the mechanical constants are varied according to sigmoid law distribution. To the author's best knowledge, modal behaviour of FGM annular plate under hygrothermal effect where mechanical constants are varying according to sigmoid law has not been reported in the literature before. Herein, the results for an additional boundary condition clamped-free (C-F) is also reported. The finite element analysis software COMSOL Multiphysics® (version 5.2) is used for computation. The results are validated by comparing with the previous studies.

3. Problem formulation

A Functionally graded annular plate of thickness h which is composed of 100% ceramic (ZrO_2) at the upper and 100% metal (Ti-6Al-4V) at the lower surface is shown in figure 1. It is assumed that the mechanical properties [11] are graded smoothly along x_3 direction from lower surface (metal rich) to upper surface (ceramic rich) according to power law or sigmoid law formulation. For meshing sequence in COMSOL®, physics-controlled mesh shown in figure 1 is used and extra fine element size (35 mm) is applied on the plate.

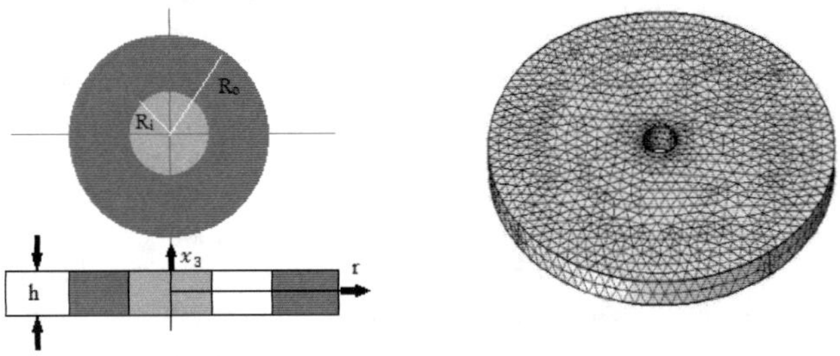

Figure 1. FGM annular plate with physics – controlled mesh (35 mm).

The formulation of power law function can be expressed as [12, 13]

$$P(x_3) = P_{ul}v(x_3) + P_l \tag{1}$$

The formulation of sigmoid law is given by combining the two power law functions as [12, 13]

$$P(x_3) = (P_{ul})v_1(x_3) + P_l \tag{2}$$

$$P(x_3) = (P_{ul})v_2(x_3) + P_l \tag{3}$$

here $v_1(x_3) = 1 - \frac{1}{2}\left\{\frac{\left(\frac{h}{2} - x_3\right)}{\frac{h}{2}}\right\}^n$ for $0 \le x_3 \le \frac{h}{2}$

and $v_2(x_3) = \frac{1}{2}\left\{\frac{\left(\frac{h}{2}+x_3\right)}{\frac{h}{2}}\right\}^n$ for $-\frac{h}{2} \leq x_3 \leq 0$

Where, $P(x_3)$ is effective material property and $v(x_3) = \left(\frac{2x_3+h}{2h}\right)^n$ is volume fraction index. In which, $P_{ul} = P_u - P_l$ and x_3 is coordinate along the thickness of the plate. The P_u and P_l are effective mechanical properties at upper and lower part, h is the thickness and n is the volume fraction index of the FGM annular plate.

Here, uniform hygrothermal load along the plate thickness is considered. The present values of uniform moisture and temperature are:

$$T = T_0 + \Delta T \tag{4}$$

$$C = C_0 + \Delta C \tag{5}$$

where $T_0 = 300$ (k) and $C_0 = 0\%$ are reference values, which are considered at the lower part of the plate. Where ΔT, ΔC are the change in temperature and moisture respectively.

$$W(T) = W_0(W_{-1}T^{-1} + 1 + W_1T + W_2T^2 + W_3T^3) \tag{6}$$

where W_0, W_{-1}, W_1, W_2, W_3 are temperature dependent constants for ceramic and metal materials [11].

4. Convergence test

FGM annular plate with following dimension ($R_o = 1$ m; $R_i = 0.1$ m; h = 0.1 m) is taken for convergence test. All mechanical constants are supposed to be graded by simple power law from ZrO_2 to Ti-6AI-4V. Power law index n = 1 and ΔT = 100 is considered and dimensionless fundamental frequencies for first three modes are obtained with different physics-controlled meshing for clamped - clamped (C-C) case. Results are obtained using COMSOL and presented in table 1. It may be notice that the results are agree well with the previous published work [11] with extra fine meshing (mesh size = 35 mm), so this meshing is used for further computation.

TABLE 1. First three dimensionless frequencies for FGM annular plate with CPU time and different physics-controlled mesh using COMSOL (n = 1, ΔT = 100).

Mesh Size (mm)	Number of elements	CPU time (sec)	Mode (0,0)	Mode (1,0)	Mode (2,0)
DQM [11]	-	-	18.9389	19.9067	25.3866
20	42915	42	18.9386	19.9066	25.3865
35	8883	16	18.9389	19.9067	25.3866
55	2670	13	19.0968	20.0734	25.6044
80	1551	13	19.3278	20.3047	25.9683
100	993	13	19.5338	20.5236	26.2997
150	558	12	19.9712	20.9680	26.9376
190	387	12	20.9544	22.0710	28.6523
300	288	13	22.7261	24.1683	32.0253
500	135	12	27.1954	29.3258	38.7479

5. Validation study under thermal effect

The material constants can be refereed from [11]. The FGM annular plate is made of 100% ceramic (ZrO_2) at the upper and 100% metal (Ti-6AI-4V) at the lower surface. In between these two, the mechanical properties are supposed to be graded with the power law formulation along the thickness of the plate. The Eigenfrequencies are calculated by following relation: $\lambda R_o^2\sqrt{\frac{\rho_c h}{D_{oc}}}$ in which $D_{oc} = \frac{E_{oc}h^2}{12(1-v_{oc}^2)}$ which is used for all computations in this study. Here, λ is the natural frequency, ρ_c, v_{oc}^2, and E_{oc} are the density, Poisson's ratio and Young's modulus for ceramic (ZrO_2) material. Tables 2 and 3 display fundamental Eigenfrequencies for first three modes (0,0), (1,0) and (2,0) under thermal effect for clamped-clamped (C-C) end condition. Here, the first digit in the bracket shows number of nodal diameter while second digit shows number of nodal circles. It may be notice that, COMSOL which is used for current study gives near accurate result and also shows good agreement with the results obtained by differential quadrature method (DQM) [11]. The % difference is calculated using following relation $\left[\frac{(Reference)-(COMSOL)}{(Reference)}\right] \times 100$.

Table 2. Accuracy of the present work (COMSOL), comparison of first dimensionless frequency (λ_1) under clamped-clamped (C-C) boundary condition ($h/R_o = 0.1, \Delta T = 800, n = 1$) with DQM [11].

	DQM [11]	COMSOL	Diff. (%)	DQM [11]	COMSOL	Diff. (%)
Mode		$R_i/R_o = 0.1$			$R_i/R_o = 0.3$	
(0,0)	18.438	18.9389	-2.71	30.006	30.5071	-1.66
(1,0)	19.230	19.9067	-3.51	30.695	31.2401	-1.77
(2,0)	24.528	25.3866	-3.50	33.331	33.9722	-1.92

Table 3. Accuracy of the present work (COMSOL), comparison of first dimensionless frequency (λ_1) under clamped-clamped (C-C) boundary condition ($h/R_o = 0.2, \Delta T = 800, n = 1$) with DQM [11].

	DQM [11]	COMSOL	Diff. (%)	DQM [11]	COMSOL	Diff. (%)
Mode		$R_i/R_o = 0.1$			$R_i/R_o = 0.3$	
(0,0)	15.470	15.3575	0.72	23.603	23.3830	0.93
(1,0)	16.417	16.3527	0.39	24.145	23.9287	0.89
(2,0)	21.215	21.1230	0.43	26.333	26.1110	0.84

6. Parametric study under hygrothermal effect

Modal analysis of FGM annular plate under hygrothermal effect where mechanical constants are varying along plate thickness with power law and sigmoid law for clamped-clamped (C-C) and clamped-free (C-F) end conditions is discussed briefly. Table 4 displays the dimensionless frequency for (ΔC = 1%, 2% and ΔT = 100) with different volume fraction index using sigmoid law formulation.

Table 4. Dimensionless frequencies (λ) of FGM annular plate with hygrothermal effect for C - C & C - F end conditions under sigmoid law using COMSOL ($h/R_o = 0.2, R_i/R_o = 0.3, \Delta T = 100$).

Mode	BCs	ΔC	n = 0	n = 1	n = 2	n = 3	n = 4	n = 5	n = 7	n = 9	n = 10
(1,0)	C-C	1%	26.9433	26.9433	26.9433	26.9433	26.9433	26.9433	26.9433	26.9433	26.9433
		2%	26.8889	26.8889	26.8889	26.8889	26.8889	26.8889	26.8889	26.8889	26.8889
	C-F	1%	13.6069	13.6069	13.6069	13.6069	13.6069	13.6069	13.6069	13.6069	13.6069
		2%	13.5524	13.5524	13.5524	13.5524	13.5524	13.5524	13.5524	13.5524	13.5524
(2,0)	C-C	1%	20.026	20.026	20.026	20.026	20.026	20.026	20.026	20.026	20.026
		2%	20.0209	20.0209	20.0209	20.0209	20.0209	20.0209	20.0209	20.0209	20.0209
	C-F	1%	16.5365	16.5365	16.5365	16.5365	16.5365	16.5365	16.5365	16.5365	16.5365
		2%	16.5325	16.5325	16.5325	16.5325	16.5325	16.5325	16.5325	16.5325	16.5325

IOP Conf. Series: Materials Science and Engineering **624** (2019) 012001 doi:10.1088/1757-899X/624/1/012001

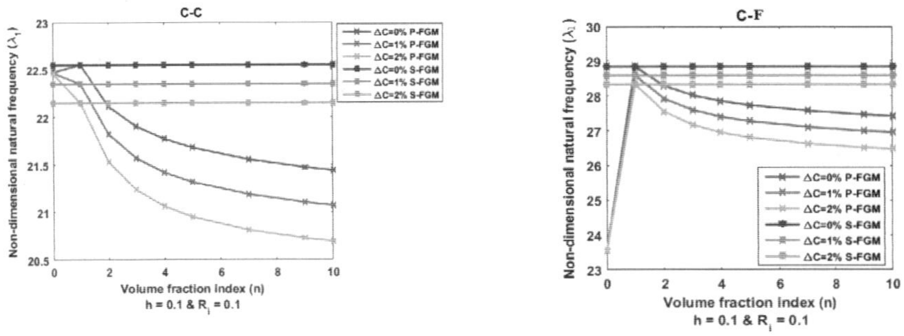

Figure 2. First dimensionless frequency with moisture content rise and volume fraction index using power law and sigmoid law formulation ($\Delta T = 100$) (a) clamped-clamped (b) clamped-free.

Eigenfrequency = 480.38 Hz Eigenfrequency = 489.04 Hz Eigenfrequency = 625.02 Hz

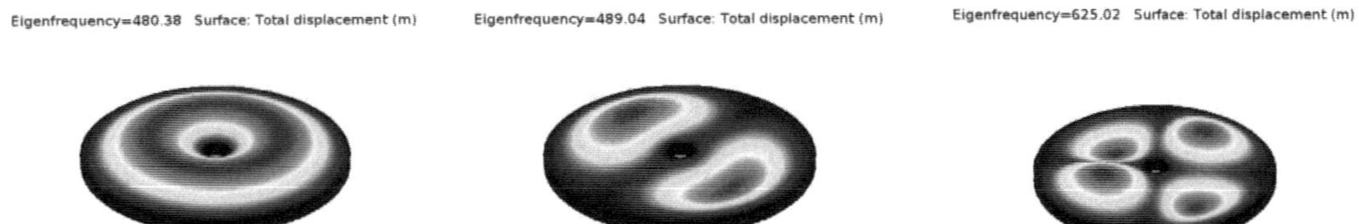

Mode (0,0) Mode (1,0) Mode (2,0)

Figure 3. First three mode shapes for clamped-clamped (C-C) boundary condition under hygrothermal effect using sigmoid law formulation ($\Delta T = 100, \Delta C = 2\%, R_i/R_o = 0.1, h/R_o = 0.1, n = 1$).

Eigenfrequency = 82.539 Hz Eigenfrequency = 331.04 Hz Eigenfrequency = 589.54 Hz

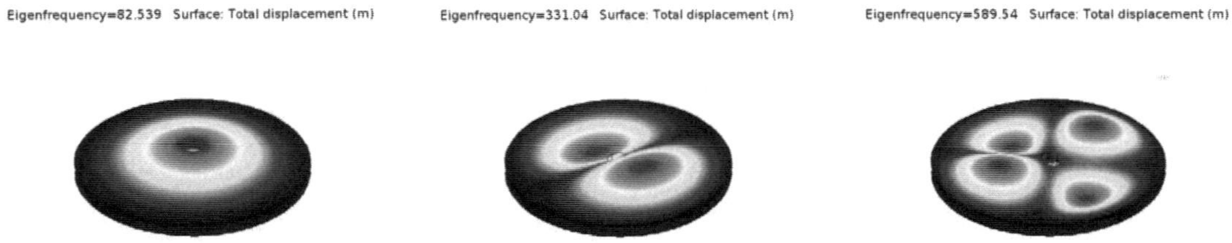

Mode (0,0) Mode (1,0) Mode (2,0)

Figure 4. First three mode shapes for clamped-free (C-F) boundary condition under hygrothermal effect using sigmoid law formulation ($\Delta T = 100, \Delta C = 2\%, R_i/R_o = 0.1, h/R_o = 0.1, n = 1$).

7. Results and discussions

7.1 Effect of moisture concentration rise

Table 4 presents the values of dimensionless frequency for $h/R_o = 0.2$, $R_i/R_o = 0.3$ with $\Delta T = 100$ using sigmoid law distribution. Figure 2 displays the first dimensionless frequency under hygrothermal ($\Delta C = 0\%, 1\%, 2\%$ *and* $\Delta T = 100$) effect with material properties graded using power law and sigmoid law distribution for modes (1,0) and (2,0). With the rise in moisture content, the fundamental frequency decreases with the increase in power law index and the moisture content rise. The frequency remains almost constant when properties are graded by sigmoid law with varying volume fraction index. The frequency of the FGM plate is smaller for high value of moisture content in comparison of FGM plate that without moisture content rise.

7.2 Effect of volume fraction index

It can be observed that, the value of dimensionless frequency decreases sharply as power law index increases from 0 to 1, and decreases smoothly for mid values and then it tends to remain constant for end values of power law index when the power law is used. While the frequencies remain almost constant when sigmoid law is used. Also, it is observed that with moisture content rise, the frequency decreases with the increase of volume fraction index. In case of clamped-free boundary condition, same phenomena can be observed besides of moisture effect because the effect of moisture content becomes insignificant. Figures 3 and 4 display the first three mode shapes (0,0), (1,0) and (2,0) for clamped-clamped (C-C) and clamped-free (C-F) end condition under hygrothermal effect.

8. Conclusion

The modal behaviour of FGM annular plate under hygrothermal effect is studied. Eigenfrequencies and corresponding mode shapes has been obtained for clamped-clamped and clamped-free boundary conditions. The power law and sigmoid law formulation is used to grade the mechanical constants along the plate thickness direction. The finite element analysis software COMSOL Multiphysics® (version 5.2) is used for computation. It can be observed that the frequency decreases with increase in functionally graded index and the moisture content rise under the power law formulation whereas under sigmoid law formulation the frequencies remain almost constant with increase in functionally graded index. Hence, this study can be helpful in making of hygrothermal sensors to provide unique information during both initial fabrication and service life of different FGM structures used in power transmission system.

Acknowledgement

The author(s) received no financial support for the research, authorship, and/or publication of this article.

References

[1] Yevtushenko AA, Grzes P, Adamowicz A. Numerical analysis of thermal stresses in disk brakes and clutches (a review). *Numer. Heat. Tra. A Appl.* 2015; 67: 170-88.
[2] Fan Q, Ikejo K, Nagamura K, Kawada M, Hashimoto M. Application of statistical parameters and discrete wavelet transform to gear damage diagnosis. *J. Adv. Mech. Des. Syst.* 2014; 8: JAMDSM0013.
[3] Redda DT, Nakanishi T, Deng G. Softening s of hardness and surface fatigue of rolling-sliding contact in the case of developed alloy steels. *J. Adv. Mech. Des. Syst.* 2009; 3: 85-92.
[4] Polajnar M, Kalin M, Thorbjornsson I, Thorgrimsson JT, Valle N, Botor-Probierz A. Friction and wear performance of functionally graded ductile iron for brake pads. *Wear.* 2017; 382: 85-94.
[5] Swaminathan K, Sangeetha DM. Thermal analysis of FGM plates–A critical review of various modeling techniques and solution methods. *Compos Struct* 2017; 160: 43-60.
[6] Whitney JM, Ashton JE. Effect of environment on the elastic response of layered composite plates. *AIAA Journal* 1971; 9: 1708-13.
[7] Lee CY, Kim JH. Hygrothermal postbuckling behavior of functionally graded plates. *Compos Struct* 2013; 95: 278-82.
[8] Zenkour AM. Hygro-thermo-mechanical effects on FGM plates resting on elastic foundations. *Compos Struct* 2010; 93: 234-8.
[9] Zenkour AM, Allam MN, Radwan AF. Effects of transverse shear and normal strains on FG plates resting on elastic foundations under hygro-thermo-mechanical loading. *Int J Appl Mech* 2014; 6: 1450063.
[10] Pandey VB, Parashar SK. Static bending and dynamic analysis of functionally graded piezoelectric beam subjected to electromechanical loads. *Proc Inst Mech Eng Part C: J Mech Eng Sci* 2016; 230: 3457-69.
[11] Malekzadeh P, Shahpari SA, Ziaee HR. Three-dimensional free vibration of thick functionally graded annular plates in thermal environment. *J Sound Vib* 2010; 329: 425-42.
[12] Sharma P. Efficacy of Harmonic Differential Quadrature method to vibration analysis of FGPM beam. *Compos Struct.* 2018; 189: 107-16.
[13] Sharma P. Vibration Analysis of Functionally Graded Piezoelectric Actuators. 2019; *Springer.*

IOP Conf. Series: Materials Science and Engineering **624** (2019) 012002 doi:10.1088/1757-899X/624/1/012002

Development of Self-Propelled Material Handling System

G A Walke[1], S N Rivankar[1] and S P Verlekar[2]

[1] Mechanical Engineering Department, Agnel Institute of technology & Design, Assagao, Mapusa, Goa, India, 403-507

[2] Fabricated Engineering Group, Jacobs Engineering India Private Limited, Mumbai, India,400-708

E-mail: gaurishwalke88@gmail.com

Abstract. Material handling system is the vital pillar of manufacturing industries like food, textile, automobile, wood, and other batch type manufacturing system. Research carried out in past shows that material handling accounts for 36% of production cost. A novel material handling system is developed to cut the cost of material transfer and risk of injury. The proposed system consists a mechanism which can self-propel itself and effect the transfer of raw material from one place to another on a shop floor. Since the system is self-propelled, it eliminates external power source in contrast to other material handling systems. The newly developed material handling system utilizes gravitational potential energy of the material loaded on the system to get self-propelled along a straight line. The system returns to original position by retracing the path after unloading the material at the destination. The system not only reduces the energy consumption for material handling but also reduces risk of injury by reducing human interference. It can be used in small and medium size enterprise (SME) to cause improvement in productivity of plant which has limitation on utilization of resources like labour, energy, land and capital.

1. Introduction

Material handling systems are used in industry to safely transfer large volume of heavy materials. A study shows that in most of small and medium sized enterprises (SME) about 2/5th of manufacturing cycle time is spent on material handling. About 20–80% of the total cost is spent on transportation of material within the industry. Also, 30–40% of industrial accidents are cause during material transfer [1] [2]. Most of SME work with constraints like minimum number of technically skilled manpower, limited capital investment on advanced machineries and plant area. Study of existing material handling system reveals that, for employing such advanced systems high initial investment is necessary. Also, operation of automated systems incurs high running and maintenance cost. Dependent nature of SME on manual work, reduces plant productivity. Therefore, to improve the productivity of plant it is necessary to reduce time and effort required for transfer of material. Further, efficient energy consumption will maintain higher plant productivity. This need demands a low-cost automatic energy efficient material handling system suitable for SME. Innovative efforts are required to develop material handling system which improves labour productivity, energy productivity and capital productivity. Improving these partial productivities will cause rise in the value of products [3][4].

2. Concept generation

IOP Conf. Series: Materials Science and Engineering **624** (2019) 012002 doi:10.1088/1757-899X/624/1/012002

Conceptualized material handling system is shown in Figure 1. It has carriage mounted on 4 wheels. The carriage consists load plate resting on spring supports. When the material is loaded gradually on carriage, spring supports under the load plate gets compressed under the gravitational potential energy of load and slides the load plate downward. The downward motion of the load plate is used to drive a mechanism which propels the carriage in forward direction to reach destination point. The material is unloaded at destination point. Upon unloading, elastic potential energy stored in the spring supports during compression is released. The spring supports start extending towards original position causing the load plate to move up. During the extension of the spring support, same mechanism utilises elastic potential energy of spring to retrace the path and reach the start point. The material handling system is held in place by applying brakes to the wheels during loading and unloading process as indicated in the Figure 1. The loading and unloading process is gradual. The transfer of material does not involve any labour input. The energy to drive the mechanism is gravitation potential energy of load during forward motion and elastic potential energy in spring supports during backward motion.

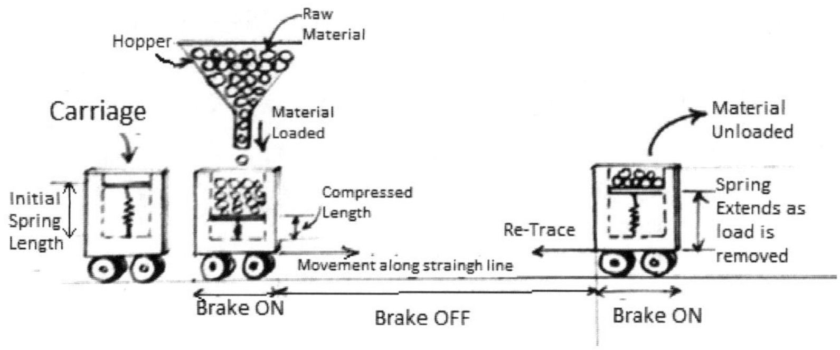

Figure 1. Concept of Self-Propelled Material Handling System.

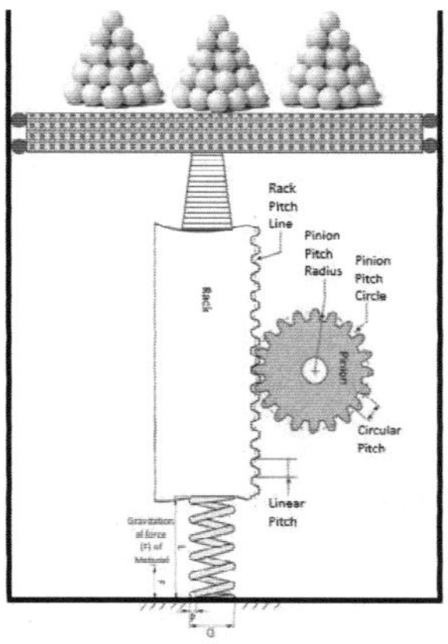

Figure 2. Conversion of gravitation potential energy to elastic potential energy.

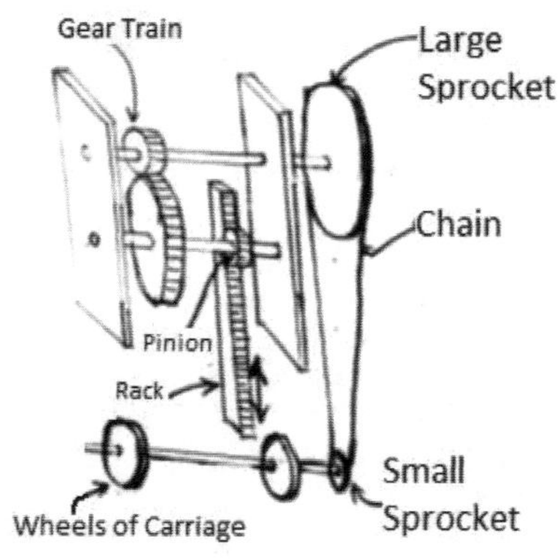

Figure 3. Transmission of motion from rack to the wheels of carriage.

3. Development of virtual prototype

The Rack and pinion mechanism was developed and synthesized which can convert the gravitational potential energy of the material into elastic potential energy, as shown in Figure 2. Top end of the rack is rigidly fixed to bottom of load plate. Load plate is free to slid up and down along walls of the carriage. The spring attached to the bottom of rack compresses during loading operation. This compression is utilized for forward motion of carriage. Compression causes downward motion of rack and proportional rotation of the pinion attached to the rack. Rotation of the pinion is transferred to a large sprocket by using a gear train, as shown in Figure 3. The large sprocket is connected to the small sprocket using a chain. The small sprocket is connected to the wheels of the carriage by mounting them on a common shaft. The rotation of pinion rotates larger sprocket through gear train and rotation of larger sprocket is transmitted to smaller sprocket by a chain. The rotation of small sprocket causes rotation of wheels moving the carriage towards destination point in straight line. Unloading at the destination releases elastic potential energy stored in spring support. Release of elastic potential energy causes extension of the spring. The extension of spring is transmitted to wheels of the carriage by reversing the motion of gear train and chain drive mechanisms. Because of reversal of driving mechanisms, carriage returns to start point by retracing the same path.

The sketches were put together to develop a digital prototype as shown in Figure 4. In virtual prototype load plate slides along the walls of carriage, this concept was updated to new sliding mechanism. The load plate is mounted on top of inner sliding bar. This inner sliding bar slides along the inner walls of outer hollow pipe to compress the spring is contained within the outer hollow pipe. Top end of spring is fixed to inner sliding bar and bottom end is fixed to base frame. Rack is rigidly fixed along the walls of inner sliding bar such that it does not interfere with sliding action of inner bar and outer pipe. Pinion is meshed with the rack. Pinion is mounted on a shaft which is held rigidly a bracket. The bracket also supports gear train and large sprocket. Larger sprocket transmits the motion to smaller sprocket using a chain drive. Smaller sprocket is mounted on driving axel of the carriage wheels.

Figure 4. Digital prototype in SolidWorks software.

4. Design and synthesis of driving mechanisms

Main components of material handling system are rack and pinion, spring support, Gear train and chain drive. These components are designed such that carriage can move 55 Kg load by 5 m along a straight path. Considering ergonomic needs during loading and unloading of material at the start and destination points, height of the carriage was calculated as 1 m [5] [6]. Four plastic wheels of 100 mm diameter were selected to guide and support the carriage. Sixteen revolutions of 100 mm diameter are required to travel 5m along straight line. This calculation was used to synthesize dimensions of smaller and larger sprocket of chain drive. Linear movement of rack is equal to deflection of spring under compressive load. The movement of rack is input for synthesis of gear train. The spring parameters were determined such that spring was strong enough to carry load of 55 kg during loading operation [7] [8]. Elastic potential energy stored in the spring during loading operation should be enough to overcome frictional forces during retrace; therefore, stiffness value was recalculated by considering sliding friction at various joints.

It was necessary to verify the analytical design of gear train by selecting 20MnCr5 as gear material [9] [10]. Finite Element Analysis was carried out using ANSYS R16.1 (Academic). The results of the FEA gave maximum total deformation at the tooth engagement interface as 0.0014 mm and maximum von-mises stress at the tooth engagement interface as 7.6176 Mpa. Deformation and stresses in gear train components were below failure limits as shown in Figure 5.

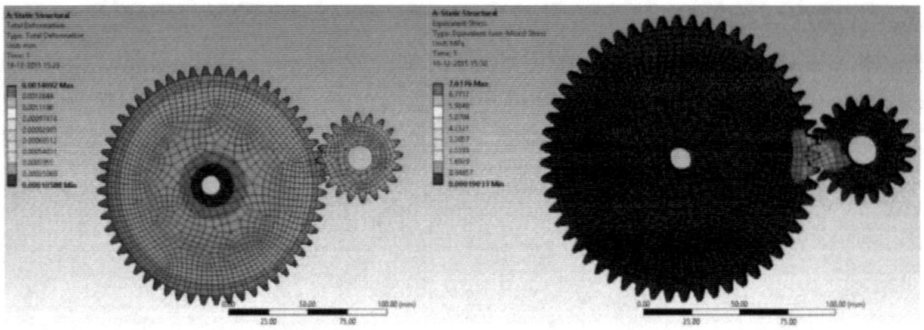

Figure 5. Deformation and stress analysis of gear train.

Verification of rack and pinion design using ANSYS R16.1 (Academic) as shown in Figure 6, gave maximum von-mises stress at the tooth engagement interface as 25.354 Mpa and maximum total deformation at the tooth engagement interface as 0.0035mm. These values were within permissible limits for 20MnCr5 material.

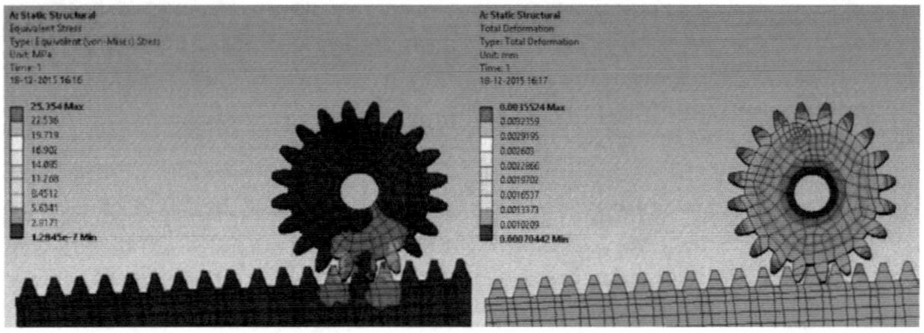

Figure 6. Deformation and stress analysis of rack and pinion.

All other components such as base frame, bearing for shafts, shafts of gear train which transmit the motion were analytically designed [9] [10] [11]. Later the designs were verified in FEA software. Necessary modifications in the designs were made to make manufacturing and assembly operations feasible.

5. Fabrication of physical prototype

Part drawing process diagrams of all components were prepared and were fabricated. The final appearance of fully assembled material handling system is shown in Figure 7. Gear train, chain drive and all other moving components were lubricated to reduce friction.

Figure 7. Physical prototype of Material Handling System.

IOP Conf. Series: Materials Science and Engineering **624** (2019) 012002 doi:10.1088/1757-899X/624/1/012002

6. Results and discussion

Testing of material handling system was carried out by placing the weights on load plate. Observation sheet used is shown in Table 1. The distance travelled by the carriage and the time taken to travel that distance for each load was recorded. The material was then unloaded at the point where carriage stops. Time and distance during retrace of carriage is recorded. To understand the relationship between load placed on load plate, and the distance travelled under that load, a graph of Load (Kg) versus Distance travelled (m) was plotted as shown in Figure 8.

Table 1. Observation Sheet for Testing

Sr No	Load (kg)	Distance travelled (m)	Time taken by equipment (sec)	
			Forward motion	Backward motion
1	18	0.2	41	43
2	20	0.5	38	40
3	25	0.7	34	36
4	30	1	32	32
5	35	1.6	27	28
6	40	2.3	24	24
7	45	2.9	21	20
8	50	3.5	19	17
9	60	4.1	17	16
10	70	5	15	13

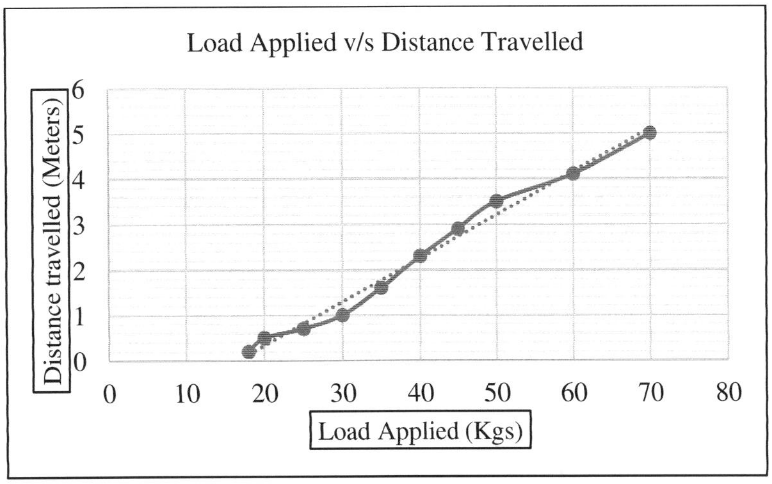

Figure 8. Response of the Machine under various Loads

The graph of load verses linear distance travelled shown in Figure 8, indicates a linear relationship with an intercept on x-axis. Intercept represent the minimum load required to move the carriage. The value of intercept is found as 18 Kg which is required to overcome frictional forces. Further, Load of 70 Kg on load plate displaces the carriage by 5 meters. Time for forward motion and retrace was recorded and it was observed that for all loads, time for forward motion was lesser than that for retrace along same path, shown in Table 1.

7. Conclusion and future scope

A Working prototype of energy efficient self-propelled material handling system was designed and tested. It was intended for an operating range of 35 Kg to 55 Kg, whereas the operating range can be extended based on application of the product. Apart from using this system to transfer heavy raw materials between loading and unloading points, it can find its potential application in assembly lines in automotive industry. it can also find application in assembly lines of automobile industry to transport chassis of vehicle from one station to other station placed along nonlinear paths. Current material handling system is developed for straight line motion between two fixed stations. Further, the system can be modified to travel along desired curved path by incorporating suitable guiding mechanism, thus widening its scope of application in industry.

Acknowledgments

We sincerely thank the workshop instructors of Agnel Institute of Technology & Design, Mechanical Engineering Department, for their valuable support during the manufacturing and assembly of prototype. We thank proprietors of Mardolkar Marine Engineering Works, Thivim Industrial Estate, Goa for providing technical support during development phase of this research work.

References

[1] Deshmukh L M 2005 *Safe handling and storage, Industrial Safety Management: Hazard Identification and Risk Control* (India: Tata McGraw-Hill Professional) p 147

[2] Bolz H A and Hagemann G E 1958 *Analysing Materials Handling Problems, Materials Handling Handbook* (New York: Ronald Press) pp 1-2

[3] Erbe H H 2002 Low cost intelligent automation in manufacturing *IFAC Proceedings* Vol. 35(1) pp 373-378

[4] Birkhofer H 2005 *Design process improvement* (London: Springer) Chapter 23 pp 478-48

[5] Chaudhuri B and Panigrahi AK 2013 Gender Bias in Indian Industry *J. Industrial Statistics*, Vol 2 (1) pp 108-127

[6] Mamidi R S, Kulkarni B and Singh A 2011 Secular Trends in Height in Different States of India in Relation to Socioeconomic Characteristics and Dietary Intakes *Food and Nutrition Bulletin* Vol 32(1) pp 23-34

[7] Valsange P S 2012 "Design of Helical Coil Compression Spring" A Review *Int. J. Engineering Research and Applications* Vol 2 (6) pp 513-522

[8] Patil R V, Reddy P R and Laxminarayana P 2014 Comparison of Cylindrical and Conical Helical Springs for their Buckling Load and Deflection *Int. J. Advanced Science and Technology* Vol 73 pp 33-50

[9] Bhandari V B 2010 *Design of Machine Elements* (India: Tata McGraw-Hill Education) Chapter 9 pp 330-392

[10] Budynas R G, Nisbett J K & Shigley J E 2011 *Shigley's mechanical engineering design* (New York: McGraw-Hill) Chapter 10 pp 509-560

[11] Gujar R A and Bhaskar S V 2013 Shaft Design under Fatigue Loading by Using Modified Goodman Method *Int. J. Engineering Research and Applications* Vol 3 pp 1061-1066

Nonlinear parameter estimation in damping with volterra series through harmonic probing

Hari Prasad Chintha[1] and Animesh Chatterjee[1]

Department of Mechanical Engineering, Visvesvaraya National Institute of Technology, Nagpur, 440011, India
Email: hari.creater@gmail.com, achatterjee@mec.vnit.ac.in

Abstract. In this work nonlinear system identification procedure, based on Volterra series, is presented to distinguish a damping nonlinearity from stiffness nonlinearity, using measured first and third harmonic amplitude characteristics. First and higher order volterra kernel synthesis formulations and frequency response functions (FRFs) have been developed in damping and stiffness nonlinearity. The characteristics of first and higher order harmonic amplitudes at various excitation level is studied for the both the nonlinearity. This paper shows comparison study between the Volterra series, Runge-Kutta fourth order in analysis of damping and stiffness nonlinearity in the mechanical systems. Nonlinearity can lead to various system behaviour, like jump phenomenon, stable and unstable region, super harmonic resonances. Using first and higher order harmonic amplitudes, formulated from Volterra series response representation, nonlinear parameters are estimated through the recursive iteration method.

Keywords: Volterra series, System Identification, Cubic nonlinear stiffness, cubic nonlinear damping

1. Introduction

Most of the engineering systems such as rotor bearing elements [1, 2], nonlinear suspension and isolation system [3, 4], aerospace structures, Nano Electro Mechanical systems, Acoustic nonlinearity of an orifice, cochlear amplifier [5], are inherently nonlinear in nature than linear. If the system exhibits linear or nonlinear behaviour cannot be answered easily unless we know the specified range over which system was operated. Nonlinearity can leads to extensive system behaviour[6, 7], such as, jumps, higher harmonic resonances.

Volterra series represents the input output mapping of physical system, in the form of functional series, which yields a structured and convenient mathematical platform for study of nonlinear systems. Nonlinear system identification and characteristics of first and higher order harmonic amplitudes at different excitation levels for different nonlinear systems are investigated using Volterra series[8]. Recursive iteration algorithm to estimate nonlinear parameter for nonlinear stiffness equation, convergence and error analysis, measurability of higher order harmonics have been presented [9, 10]. Nonlinear bearing stiffness has been studied experimentally and theoretically for rolling element bearings [10]. Mostly the system identification research [1, 8-10] has been limited to, system with stiffness nonlinearity and only few researchers have been studied nonlinear damping [3, 6, 7]. In the present work, Volterra series response is synthesised up to 'k' number of terms in terms of Frequency response functions (FRFs). Response amplitude for damping nonlinearity in terms of lower order kernel transform is presented. Characteristics of dynamic behaviour of nonlinear damping are compared with nonlinear stiffness oscillator [10]. Behaviour of nonlinear system has been studied. Nonlinear parameter is estimated based on recursive iteration method

2. Volterra series representation

Volterra series with input excitation *f(t)* and output response *x(t)*, in a form of functional series given by [9]

$$x(t) = x_1(t) + x_2(t) + x_3(t) + x_4(t) + x_5(t)........., \tag{1}$$

$$x_n(t) = \int_{-\infty}^{\infty} .....\int_{-\infty}^{\infty} h_n(\tau_1,....,\tau_n) f(t-\tau_1)...f(t-\tau_n)\, d\tau_1...d\tau_n \tag{2}$$

$h_n(\tau_1, \tau_2, \tau_3, ..., \tau_n)$ are n^{th} order Volterra kernels. Multidimensional Fourier transforms of Volterra kernel transform with higher order as

$$H_n(w_1, w_2, w_3....w_n) = \int_{-\infty}^{\infty}\int_{-\infty}^{\infty} .....\int_{-\infty}^{\infty} h_1(\tau_1, \tau_2, \tau_3, \tau_4...\tau_n) \prod_{i=1}^{n} e^{-jw\tau_1} d\tau_1 d\tau_2 d\tau_3 d\tau_4 ...d\tau_n \tag{3}$$

Equation of motion for nonlinear stiffness and damping oscillator

$$m\ddot{x}(t) + c_1\dot{x}(t) + kx(t) + k_3x^3(t) = f(t) \tag{4}$$

$$m\ddot{x}(t) + c_1\dot{x}(t) + c_3\dot{x}^3(t) + k_1x(t) = f(t) \tag{5}$$

For a single input harmonic excitation

$$f(t) = A\cos wt = \frac{A}{2}e^{jwt} + \frac{A}{2}e^{-jwt}, \tag{6}$$

Using Volterra, response from eqn. (2) as

$$x(t) = \sum_{n=1}^{\infty} \left(\frac{A}{2}\right)^n \sum_{p+s=n} C_q H_n^{p,s}(w) e^{jw_{p,s}t} \tag{7}$$

Where $H_n^{p,q}(w)$ the higher are order FRFs given by

$$H_n^{p,s}(w) = H_n(\underbrace{w,..........,w}_{p\ times}, \underbrace{-w,.......,-w}_{s\ times}) \qquad w_{p,s} = (p-s)w \tag{8}$$

Volterra series response is synthesised up to '*k*' number of terms. Volterra kernels of higher order are reduced in to lower order volterra kernels and response amplitude for the first and third harmonics as

$$X(w) = AH_1(w) + \frac{3}{4}A^3H_3(w,w,-w) + \frac{5}{8}A^5H_5(w,w,w,-w,-w) + higher\ order\ terms \tag{9}$$

$$\begin{aligned} X(w) = {} & AH_1(w) - \frac{3}{4}A^3c_3jw^3H_1^{\,3}(w)H_1(-w) \\ & + A^5\left(\begin{aligned} & \frac{9}{16}c_3^{\,2}w^6H_1^{\,4}(w)H_1^{\,3}(-w) - \frac{9}{16}c_3^{\,2}w^6H_1^{\,4}(w)H_1^{\,2}(-w)H_1(3w) \\ & -\frac{9}{8}c_3^{\,2}w^6H_1^{\,5}(w)H_1^{\,2}(-w) \end{aligned}\right) \end{aligned} \tag{10}$$

$$X(3w) = \frac{A^3}{4}H_3(w,w,w) + \frac{5}{16}A^5H_5(w,w,w,w,-w) + higher\ order\ terms \tag{11}$$

$$\begin{aligned} X(3w) = {} & \frac{A^3}{4}c_3jw^3H_1^{\,3}(w)H_1(3w) \\ & + A^5\left(\frac{9}{16}c_3^{\,2}w^6H_1^{\,5}(w)H_1(-w)H_1(3w) + \frac{9}{8}c_3^{\,2}w^6H_1^{\,4}(w)H_1(-w)H_1^{\,2}(3w)\right) \end{aligned} \tag{12}$$

3. Nonlinearity structure identification

One can predict whether the system is linear or nonlinear based on response spectrum. Presence of odd and even harmonics in the response spectrum will indicate that, the system is nonlinear in nature. Response spectrum of stiffness (eqn. (4)) and damping nonlinearity (eqn. (5)), for first and third harmonic amplitudes, with different excitation level and constant excitation frequency is shown in Figure 1. For analysis of single degree of freedom nonlinear system parameters are considered as m=1kg, c_1=0.1Ns/m, c_3=0.1Ns/m, k_1 =1 N/m.

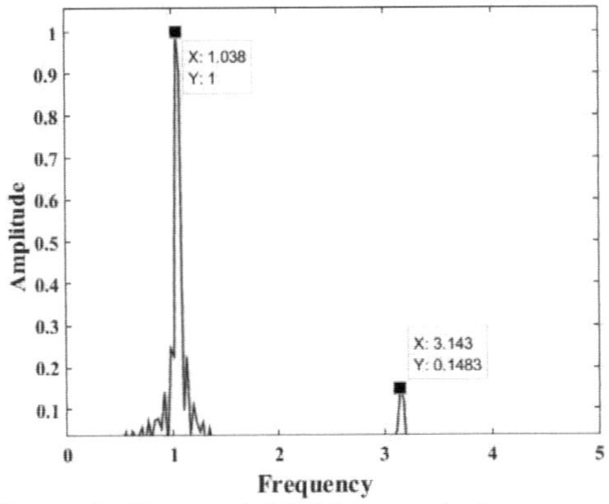

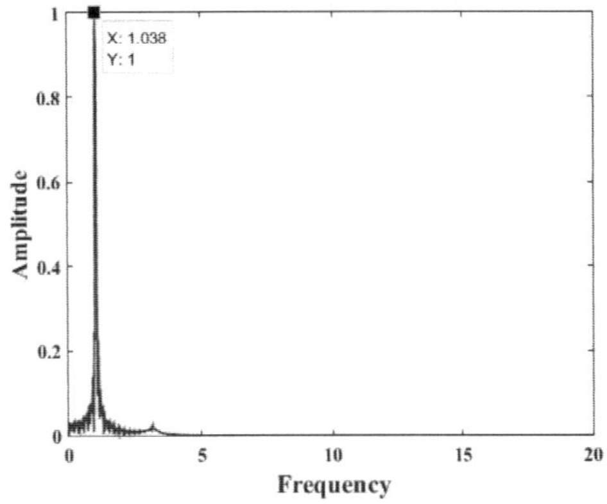

Figure 1. First and third harmonic in response spectrum (a)System with cubic stiffness nonlinearity, (b) System with cubic damping nonlinearity

3.1. Comparison of dynamic behaviour of stiffness and damping nonlinearity

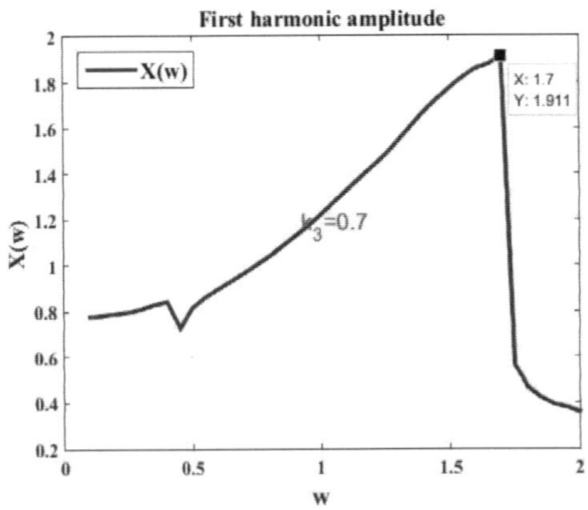

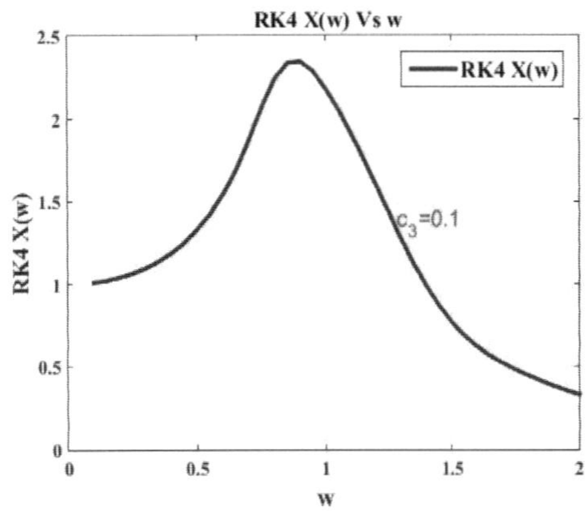

Figure 2. Runge-Kutta method First harmonic amplitude (a)System with cubic stiffness nonlinearity, b) System with cubic damping nonlinearity

Figure 2. Shows that the first harmonic amplitude X(ω), in cubic stiffness and cubic damping nonlinearity by Runge-Kutta method for given values c_1=0.1, c_3=0.1, ω =0.33. It is observed that in the Figure 2(a), amplitude undergoes sudden jump near the resonance, this occur with increasing frequency of excitation. Amplitude gradually increases until certain point and then suddenly jumps to a large value and diminishes along the curve to its straight. Amplitude frequency plot is unstable. Higher harmonic peak occurs at one third of fundamental frequency (ω =1/3 ω_n). Figure 2(b) Show cubic damping nonlinearity in the first harmonic amplitude, there is no jump phenomenon, no unstable region and no super harmonic resonance.

4. Parameter estimation in system with nonlinear cubic damping

4.1. Convergence error analysis

The relative error between the finite term response series approximation and the exact amplitude of the harmonics is [10].

$$ {}_{k}^{n}e = \left| \left[X(nr) - {}_{k}X(nr) \right] / X(nr) \right| \tag{13}$$

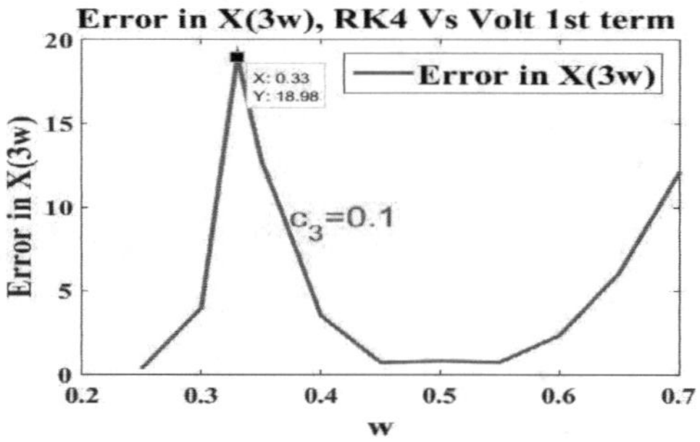

Figure 3. Variation of series approximation error in third harmonic amplitude X (3ω)

For an eqn. (5), Figure 3 show the Volterra approximation error variation in response of third harmonic amplitude X(3ω) in comparison with Runge-Kutta fourth order method under constant excitation amplitude for a single term Volterra series approximation over a frequency range (ω =0.25 to 0.7). It can be seen that the approximation error is very high at natural frequency (ω =0.33), that is 1/3 rd of its fundamental frequency (ω = ω/3=0.33), which is termed as a super harmonic excitation. Error is relatively high at frequency ω =0.33.

4.2. Measurability of higher order harmonics

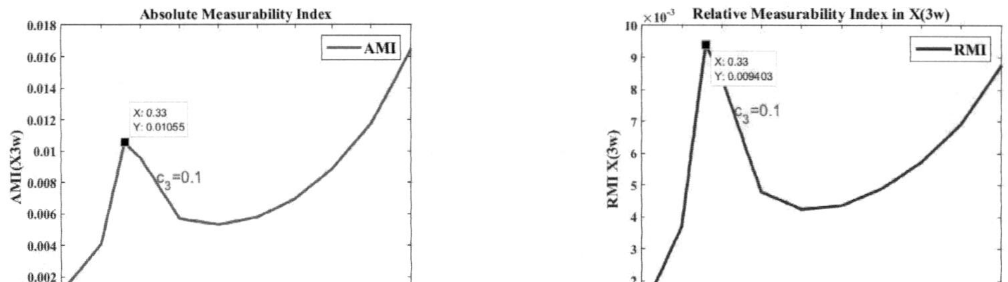

Figure 4. Excitation level corresponding to non-linear damping coefficient for a rage frequency (a) Absolute measurability index, AMI in X(3 ω), (b) Relative measurability index, RMI X(3 ω)

Higher order harmonic amplitudes are used to estimate the nonlinear parameters. For weakly nonlinear systems, compared to first harmonic, higher order harmonics are very much smaller, because of vibration and noise. Therefore, it is essential to measure the response spectrum with suitable higher order harmonics. Figure 4(a) show the Absolute Measurability index (AMI) =X(3ω) of third order harmonic amplitude for a range of frequency, with constant excitation at c_3=0.1. This Figure has been plotted for a given nonlinear eqn. (5) by Runge-Kutta fourth order method.

Relative measurability index (RMI) is defined in terms of strength of relative signal, which is the ratio of harmonics of higher order with respect to the first harmonic amplitude [10].

$$RMI\ (mw) = X(mw)/X(w) \quad where \ \ m = 1,2,3.. \tag{14}$$

Figure 4(b). Show the RMI variation of third harmonic amplitude X(3ω) for a range of frequency. It is shown that peak is max, at ω /3=0.33 frequency, is named as peak measurability. This can be suggested that, X(3ω) must be measured nearer to the 1/3rd frequency of a given system. RMI in X(3ω) Shows in Figure 5 for a different excitation level considering different non-linear parameter c_3. It can be illustrated that higher order harmonics for weakly nonlinear systems are of much lower amplitude, compare to the fundamental harmonic. It can be seen that the response strength is significantly smaller at one third and near resonant frequency. Better signal/noise ratio is required to measure the third harmonic amplitude.

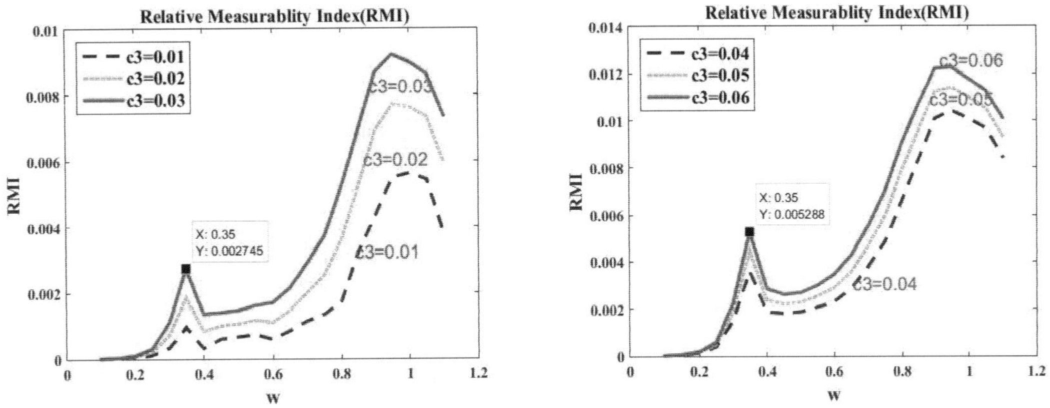

Figure 5. Relative measurability index, RMI in X(3ω) for a different excitation level with corresponding to non-linear damping coefficient (a) c_3=0.01,0.02,0.03, (b) c_3=0.04,0.05,0.06

4.3. Recursive iteration method

Recursive iteration method is used to estimate non-linear parameters c_3 and linear parameters m, c, k_1. The response harmonic series *X (n ω)* is truncated after a finite number of terms, k, and re arranged for first and third harmonic response [10].

$$H_1(w) \approx \frac{1}{A}\left[X(w) - \sum_{i=2}^{k} \sigma_i(w)\right] \tag{15}$$

$$H_3(w,w,w) \approx \frac{4}{A^3}\left[X(3w) - \sum_{i=2}^{k} \sigma_i(3w)\right] \tag{16}$$

Where Response harmonic amplitude, $X(nw)$

$$X(nw) = \sum_{i=2}^{\infty} \sigma_i(nw) \tag{17}$$

$$\sigma_i(nw) = 2\left(\frac{A}{2}\right)^{n+2i-2} {}^{n+2i-2}c_{i-1} H_{n+2i-2}^{n+i-1,i-1}(w) \tag{18}$$

4.4. Error estimation in non-linear parameter c_3

The non-linear parameter c_3 will be estimated by single term Volterra series approximation for different excitation level. Error is calculated by comparing estimated value with the obtained value with single term Volterra approximation. Figure 6 Show that the error in c_3 will be decreases up to excitation frequency (ω=0.7) and then error will increase after the excitation frequency (ω =0.7).

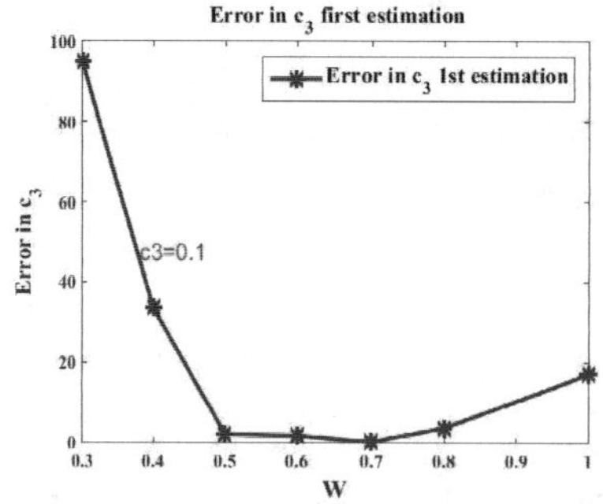

Figure 6. Error in First estimation of c_3 for a different excitation level

5. Conclusion

Using Volterra series, parameter estimation procedure has been developed for polynomial form of damping and stiffness nonlinearity, considering single degree of freedom system. Analysis show jump phenomenon in nonlinear stiffness at the higher harmonic resonance, and there is no such type of behaviour present in damping nonlinearity. Near the resonant frequency results are not converging for nonlinear damping by Volterra and Runge-Kutta methods, and even there is no jump phenomenon at this point. By measurement of harmonics of higher order, nonlinear parameters are estimated. It is observed that higher order harmonics for weakly nonlinear systems are of much lower amplitude, compare to the fundamental harmonic because of background vibration and noise. Response strength is significantly very small. In order to measure the response amplitude better signal/noise ratio is required. The recursive method is used to estimate nonlinear parameter.

References

[1] Chatterjee A and Vyas N S 2003 Nonlinear parameter estimation in rotor-bearing system using Volterra series and method of harmonic probing *J. vibr. acoust* **125** 299-306.
[2] Cheng C, Peng Z, Zhang W and Meng G 2017 Volterra-series-based nonlinear system modeling and its engineering applications: A state-of-the-art review *Mech Syst. Signal.* **87** 340-364.
[3] Jing X J and Lang Z Q 2009 Frequency domain analysis of a dimensionless cubic nonlinear damping system subject to harmonic input *Nonlinear. Dyn* **58** 469.
[4] Peng Z, Meng G, Lang Z, Zhang W, and Chu F 2012 Study of the effects of cubic nonlinear damping on vibration isolations using harmonic balance method *Int. J. Non Linear. Mech* **47** 1073-1080.
[5] Elliott S, Tehrani M. G, and Langley R 2015Nonlinear damping and quasi-linear modelling *Phil.Tran.R.Soc.A* **373** 20140402.
[6] Jones J. P and Yaser K 2018 Recent advances and comparisons between harmonic balance and Volterra-based nonlinear frequency response analysis methods *Nonlinear. Dyn* **91** 131-145.
[7] Peng Z, Lang Z, Billings S, and Tomlinson G 2008 Comparisons between harmonic balance and nonlinear output frequency response function in nonlinear system analysis *J. Sound. Vib* **311** 56-73.
[8] Chatterjee A 2009 Identification of System Nonlinearity Structure by Harmonic Probing *Soc. Exp. Mech. Inc Proceedings of the IMACXXVII* 9-12 (Orlando, Florida USA).
[9] Chatterjee A and Vyas N S 2000 Convergence analysis of Volterra series response of nonlinear systems subjected to harmonic excitation *J. Sound. Vib* **236** 339-358,.
[10] Chatterjee A and Vyas N S 2003 Non-linear parameter estimation with Volterra series using the method of recursive iteration through harmonic probing *J. Sound. Vib* **268** 657-678.

IOP Conf. Series: Materials Science and Engineering **624** (2019) 012004 doi:10.1088/1757-899X/624/1/012004

Power loss prediction in asymmetric spur gear considering gear tooth dynamic load

Benny Thomas[1], K Sankaranarayanasamy[2], S Ramachandra[1] and Suresh Kumar SP[1]

[1] Gas Turbine Research Establishment, DRDO, Bangalore,India-560093
[2] National Institute of Technology, Tiruchirapalli, India- 620015

Email ID: bennythomas@gtre.drdo.in

Abstract. In many gear drives, one side of the flank is subjected to relatively higher load for a longer duration than the other side. Asymmetric spur gears with drive side pressure angle higher than the coast side reflects this functional difference. Conventional design criteria and procedures followed for symmetric spur gears are suitably modified and applied to predict the gear tooth bending, contact stress and power loss in asymmetric spur gears. Quasi-static gear tooth load and empirical friction coefficient formulae were applied in the past to predict the sliding power loss in asymmetric spur gears. In the present work, Finite element method is used to determine the time varying mesh stiffness of the normal contact ratio asymmetric spur gear tooth. Computed gear tooth stiffness is used to predict the dynamic load at two different speeds under non-extended contact condition. Sliding power loss during the course of meshing is analytically calculated under quasi static and dynamic load conditions. Study demonstrates the difference in sliding power loss computed based on friction formulae and empirical friction coefficient formulae under static and dynamic load.

1. Introduction

Asymmetric gears are those in which drive flanks are designed with a higher pressure angle than the coast flank to enhance drive side load carrying capacity. Helicopter main drive gears, turboprop gearbox and gear pumps are some of the applications that demand unidirectional or relatively higher load on one side of the gear flank. Reduction in bending and contact stress, longer life and independent design of coast and drive side profile that helps in managing the gear tooth stiffness are some of the advantages of asymmetric gears. Relatively higher bearing load, non-standard gear tooth parameters for optimal performance and the need for special tool for manufacture are the disadvantages of asymmetric gears.

Performance of gear drives are mainly assessed based on life, dynamic behaviour and power loss. Conventional practice is to compute the gear tooth bending stress, contact stress and power loss under quasi static load. Research work to predict the power loss in transmission systems is given weightage with an intention to maximise the efficiency of the system and thereby reduce fuel burn and associated environmental effects. In geared systems sliding power loss is the major component that contributes to gear mesh loss and hence sliding power loss in asymmetric gears is the focus of present study.

Bernard et al [1] compared the empirical laws of friction with experimental law of friction and presented the variation in friction along the path of contact. Sekar et al [2] predicted the quasi static load in non-standard spur gears using Finite element method (FEM) and analytically computed the power loss. Enhancement in efficiency and load carrying capacity by using non-standard gears was demonstrated in their work. Yenti et al [3] found sliding power loss predicted based on friction coefficient formula proposed by ISO TC 60 to be closer to experimental results. Load and speed dependent losses were analysed by Heingartner et al [4] and found that sliding power losses increase with load. Hai Xu [5] introduced a new expression to determine friction coefficient based on elasto-hydrodynamic model for a specific grade of gear oil and validated it experimentally in his research work. Detail design procedures for asymmetric spur gears and direct gear design approach are presented by Kapelevich [6]. Thomas et al [7] introduced a novel approach to analytically predict the gear tooth bending stress in normal contact ratio asymmetric spur gears.

Researchers have predicted the power loss in asymmetric gears based on standard empirical friction laws considering quasi-static load. Successful performance of many geared systems designed based on conventional approach have proven the adequacy of the procedures. However, these methodologies do not consider the gear tooth dynamic behaviour and its effect on the life and performance of gears. This drawback in the general procedures gives space for possible performance enhancement based on the inferences from more accurate stress and dynamic behaviour predictions. Present work is focused on a mixed FEM and analytical method to compute the sliding power loss in normal contact ratio (NCR) asymmetric spur gear under quasi static and dynamic load. Two different coefficient of friction models are used in this comparative study.

2. Gear tooth stiffness

Asymmetric gear with parameters as given in Table 1 is selected in the present work to investigate the influence of dynamic load on the sliding power loss. Figure 1a shows the model of three teeth of a 40 teeth 20°/31° drive to coast side pressure angle asymmetric spur gear used to generate a 2D meshed model to determine the gear tooth stiffness. Element size and the aspect ratio relations [8] is maintained in the selection of element size when a concentrated load is used to represent the gear tooth contact to determine the deflection within the interval of definition.

Table 1. Gear tooth parameters

Number of tooth z	40
Gear ratio u	1
Module m (mm)	2
Tool dedendum coefficient hf_{dc}	1
Tool addendum coefficient hf_{ac}	1.25
Face width b_{fw} (mm)	25
Tool tip radius $TR_{d,c}$ (mm)	0.1
Coast side pressure angle α_c (degree)	20°
Drive side pressure angle α_d (degree)	31°
Backlash	0
Young's Modulus (GPa)	210
Mass (kg)	1
Applied normal tooth load (N)	2916

Deflection along the line of action corresponding to the point of load application is used to calculate the gear tooth stiffness k_q (N/m) at that point as per equation (1). Single tooth and double tooth time varying mesh stiffness (TVMS) is calculated based on equation (2)

$$k_{qp,qg} = \frac{F_n}{\delta_{qp,qg}} \tag{1}$$

where $\delta_{qp,qg}$ is the pinion and gear tooth deflection (m) respectively along the line of action at qth point due to the application of normal tooth load F_n(N) at that point.
Mesh stiffness of the jth mesh pair at the qth point of contact (k_{qj}) is given as

$$k_{qj} = \frac{k_{qgj}k_{qpj}}{k_{qgj} + k_{qgj}} \tag{2}$$

Double tooth mesh stiffness is the direct summation of mesh stiffness of the two pairs of contact. Figure 1b shows the stiffness of the gear pair with parameters as per Table 1 during the course of meshing. Mesh stiffness calculated as per equation (2) is also presented in the figure.

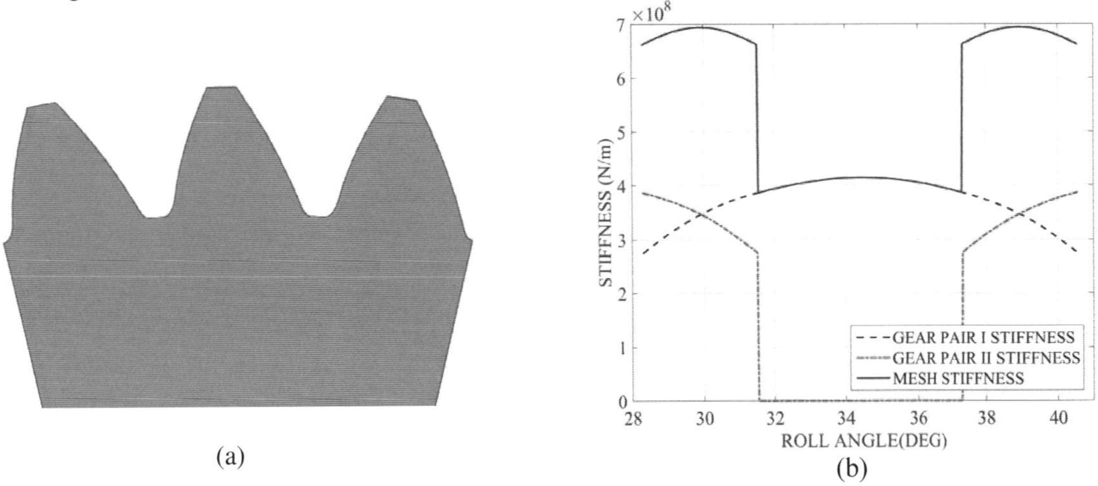

Figure 1. Model and mesh stiffness of asymmetric spur gear with parameters as per Table 1 (a) three teeth of a 40 teeth 20°/31° drive to coast side pressure angle (b) Gear pair stiffness and mesh stiffness computed based on gear tooth deflection determined through FE method

3. Dynamic tooth load

Lumped parameter model is used in this study to analytically calculate the dynamic gear tooth load due to gear tooth stiffness. Rigid wheels representing the gear wheels are connected by spring elements with time varying mesh stiffness ($\mathbf{k_I}$ and $\mathbf{k_{II}}$ in N/m); and damper of damping coefficient $\mathbf{c_m}$ (N-m/sec). Y axis of the rectangular coordinate system in this model is chosen along the line of action.

$$I_{Jg}\,\ddot{\theta}_g + R_{bdg}F_{ST} = R_{bdg}(F_{DYI} + F_{DYII}) \pm \rho_{gI}\mu_I F_{DYI} \pm \rho_{gII}\mu_{II}F_{DYII} \tag{3}$$

$$I_{Jp}\,\ddot{\theta}_p + R_{bdp}(F_{DI} + F_{DII}) = R_{bdp}F_{ST} \pm \rho_{pI}\mu_I F_{DYI} \pm \rho_{pII}\mu_{II}F_{DYII} \tag{4}$$

$$F_{ST} = \frac{T_g}{R_{bdg}} = \frac{T_p}{R_{bdp}} \tag{5}$$

where I_J is the mass moment of inertia (Kg-m^2), $\ddot{\theta}$ is the angular acceleration, ρ is the instantaneous radius (m) at the point of contact, R_{bd} is the drive side base circle diameter (m), μ is the coefficient of friction between gear tooth surface in mesh, F_{DY} is the dynamic load (N), F_{ST} is the static gear tooth load (N) and I, II denotes gear mesh pair I and II respectively.

Elasto-hydrodynamic lubrication (EHL) based coefficient of friction derived by Xu et al [5] is given by

$$\mu = e^{f(SR,P_h,\nu_0,S)}{P_h}^{b_2}|SR|^{b_3}{V_e}^{b_6}{\nu_0}^{b_7}{R_R}^{b_8} \tag{6}$$

where $f(SR, P_h, \nu_0, S) = b_1 + b_4|SR|P_h\log_{10}(\nu_0) + b_5e^{-|SR|P_h\log_{10}(\nu_0)} + b_9e^S$

b_j=-8.916465,1.03303,1.036077,-0.354068,2.812084,-0.100601,0.752755,-0.390958, 0.620305 for j=1 to 9 respectively. SR is the slide to roll ratio, V_e is the entraining velocity (m/sec), $P_h$is the maximum Hertzian pressure (GPa), S is the surface finish (microns), ν_0 is the dynamic viscosity in centipoise, R_R is the relative radius of curvature(m).

Mass moment of inertia of the gear and pinion wheels can be expressed as

$$I_{Jg} = M_g\left(R_{bdg}\right)^2 \tag{7}$$

$$I_{Jp} = M_p\left(R_{bdp}\right)^2 \tag{8}$$

where M is the mass (kg)
Substituting the expression for mass moment of inertia from (7) and (8) and coefficient of friction μ from (6) in equation (3) and (4), we get

$$\ddot{x}_d + 2\omega\zeta\dot{x}_d + \omega^2 x_d = \omega^2 x_s \tag{9}$$

where $\ddot{x}_d$ is the relative acceleration along the line of action (m/sec^2), $\dot{x}_d$ is the relative velocity along the line of action (m/sec), x_d is the relative displacement along the line of action (m), x_s is the static transmission error (m) and ω is the frequency (Hz).

Equation (9) is numerically solved using fourth order Runge-Kutta Method. In the numerical solution damping ratio (ζ) of 0.17 and kinematic viscosity (ν) of 44.5cSt at 60°C is assumed. Length Length of action is divided into sufficiently large number of divisions, typically 500, to accurately capture the transition phase from double tooth contact to single tooth contact and vice versa. Iterative Iterative solution for $\dot{x}_d$ and x_d is continued till a predefined accuracy is reached. Dynamic gear tooth tooth load can be approximated as the product of the gear mesh stiffness and x_d. Figure 2 (a) shows the gear tooth load under quasi static and dynamic condition. Oscillation seen in dynamic gear tooth tooth load is the response of the gear of definite mass, mesh stiffness and damping to sudden load application during transmission.

Dynamic factor is the ratio of the maximum dynamic gear tooth load to the quasi-static load during the course of meshing. Figure 2(b) shows the variation in dynamic factor due to the influence of varying mesh stiffness and damping.

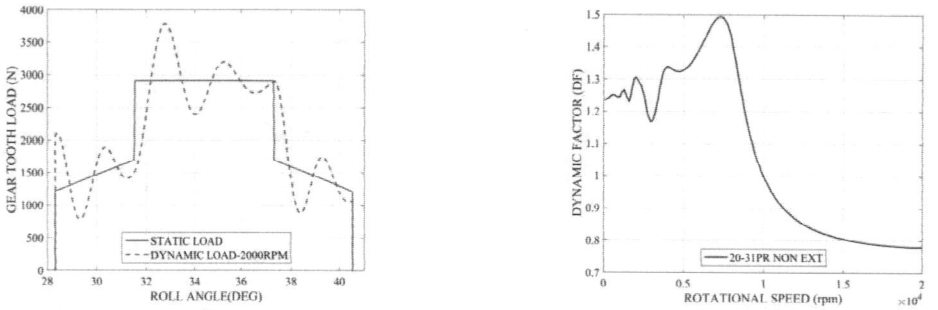

(a) (b)

Figure 2. Quasi static load, dynamic tooth load at a given speed and dynamic factor for specified speed range for a 40 teeth asymmetric spur gear with parameters as per Table 1 (a) Quasi static load and dynamic tooth load computed under non-extended tooth contact at 2000 rpm (b) Variation in dynamic factor with speed

4. Sliding power loss

Total gear tooth mesh loss is the sum of rolling and sliding friction loss. Present investigation is on the power loss due to sliding as the sliding friction gear mesh loss is relatively higher than rolling power loss. Magnitude of the instantaneous sliding friction power loss at a given point of contact between two gears in mesh is proportional to the coefficient of friction, normal tooth load shared by the gear pair and the sliding velocity.

Two models for predicting the coefficient of friction, first by Dowson and Higginson and second by Hai Xu [5] et al, is used in the present study to calculate the gear tooth sliding power loss under quasi-static and dynamic gear tooth load.

Coefficient of friction proposed by Dowson and Higginson is given by

$$\mu = 18.175\, \nu^{-0.15} \left|V_{e/s}\right|^{-0.15} |v_s|^{-0.5} (R_r)^{-0.5} \tag{10}$$

where ν is the kinematic viscosity in centi-stokes, $V_{e/s}$ is the entraining to slide velocity, v_s is the sliding velocity (mm/sec), R_r is the relative radius of curvature at the contact point (mm). Variation in in coefficient of friction, along the course of action defined with respect to roll angle, computed using using the two models for two different speeds is depicted in Figure 3(a) and (b). Maximum value of μ μ predicted using equation (10) is limited to 0.2. Both the models show drop in coefficient of friction friction with increase in speed. However, the two models widely differ primarily in the single tooth contact zone.

Sliding velocity during the mesh cycle calculated for 2000rpm and 7300rpm is shown in Figure4(a) and (b). As seen from the figure, sliding velocity is zero at pitch point and at all other points sliding velocity increases with increase in speed.

Instantaneous sliding power loss P_s at the qth point of contact is given by

$$(P_s)_q = \mu_q (v_s)_q F_q \tag{11}$$

Sliding power loss during the course of meshing due to quasi static and dynamic load for two different speeds are shown in Figure 5(a) and (b). As mentioned earlier, sliding velocity at the pitch point is zero and hence the sliding power loss at this point is zero.

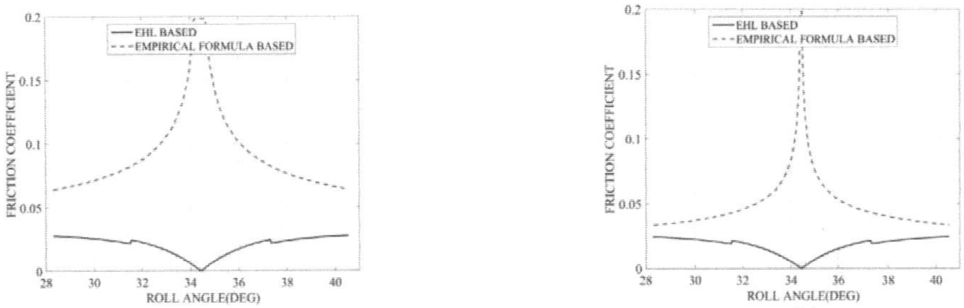

(a) (b)

Figure 3. Variation in coefficient of friction along the course of action for gear with parameters as per Table1 computed as per equation (6) and (10) (a) At 2000 rpm (b) at 7300rpm

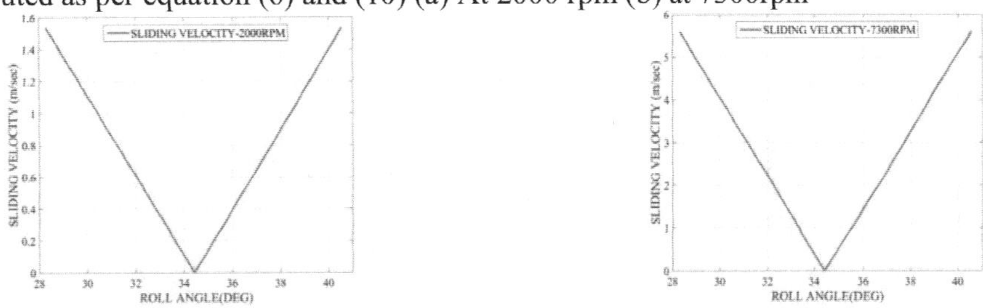

(a) (b)

Figure 4. Variation in sliding velocity during the mesh cycle for gear with parameters as per Table1 (a) At 2000 rpm (b) at 7300rpm

(a) (b)

Figure 5. Variation in sliding power loss during meshing computed for the gear as per Table 1under quasi static and dynamic load (a) At 2000 rpm (b) at 7300rpm

5. Conclusion

Difference in coefficient of friction, computed based on two models, along the course of meshing of a NCR asymmetric spur gear is relatively higher in the single tooth contact zone. Both the models predict lower friction coefficient at higher speed and EHL model predicts relatively lower value of friction coefficient. Mesh sliding power losses computed based on EHL model is relatively lower than that based on empirical formula. Difference in power loss computed under quasi static and dynamic load is found to be relatively higher as the speed approaches resonance speed.

References

[1] Mushirabwoba Bernard, Lahcen Belfals, Najji Brahim, Lasri Abdelilah, A Comparative Study of Friction Laws Used in Spur Gear Power Losses Estimation, Contemporary Engineering Sciences, Vol. 9, 2016, no. 6, 279 – 288

[2] R Prabhu Sekar, V Edwin Geo and Leenus Jesu Martin, A mixed finite element and analytical method to predict load, mechanical power loss and improved efficiency in non-standard spur gear drives, Proc IMechE Part C: Journal of Mechanical Engineering Science February 2017, Volume: 231 issue: 11, page(s): 1408-1424, DOI: 10.1177/1350650117697594

[3] Chakrit Yenti, Surin Phongsupasamit, and Chanat Ratanasumawong, Analytical and Experimental Investigation of Parameters Affecting Sliding Loss in a Spur Gear Pair, Engineering Journal, Volume 17 Issue 1,pp.79-94, January 2013, DOI:10.4186/ej.2013.17.1.79

[4] Heingartner and David Determining power losses in the helical gear mesh; Case study, Proceedings of DETC'03 ASME 2003 Design Engineering Technical Conferences and Computers and Information in Engineering Conference Chicago, September 2-6, 2003 Paper No. DETC2003/PTG-48118, pp. 965-970; 6 pages, DOI:10.1115/DETC2003/PTG-48118

[5] Hai Xu, Development of a generalized mechanical efficiency prediction methodology for gear pairs, Dissertation, The Ohio State University, 2005

[6] Alexander L. Kapelevich, Direct gear design, CRC Press, Taylor & Francis Group, International Standard Book Number-13: 978-1-4398-7619-0

[7] Benny Thomas, K Sankaranarayanasamy, S Ramachandra, SP Suresh Kumar. Search method applied for gear tooth bending stress prediction in normal contact ratio asymmetric spur gears, Proc IMechE Part C: Journal of Mechanical Engineering Science, January 2018, Volume: **232** issue: 24, page(s): 4647-4663, DOI:10.1177/0954406217753235

[8] John J. Coy, Charles Hu-Chih Chao, A method of selecting grid size to account for Hertz deformation in Finite Element analysis of spur gears, NASA-TM-82623, September, 1981

IOP Conf. Series: Materials Science and Engineering **624** (2019) 012005 doi:10.1088/1757-899X/624/1/012005

Assessment of structural behavior of torque converter dampener

Das, Ujjwal Kumar[1], Vinogradov Victoria[2], Ch Bhaskara[3]
[1,3]GM Tech Center India, India.
[2]General Motors, United States.

E-mail id: ujjwal.k.das@gm.com

Abstract. A Torque Converter Dampener is a primary component for transmittal of power between the Engine and the Transmission. It comprises of three major components – driving cage plates, driven cage plates and set of springs. Being a spring based system, it provides a smooth conversion of torque from engine to the driveline. When the vehicle performs a gearshift, these springs damp the driveline oscillations and they get locked when the vehicle runs at a higher speed. A Torque Converter Dampener assembly is typically subjected to high torque and centrifugal loading and sometimes, there is a chance of breaking of the springs under operating loading conditions. This paper explains detailed method of analyzing Torque Converter Dampener using Finite Element Analysis. Cage plates and springs are modeled accurately with interactions and boundary conditions properly defined similar to actual hardware, in order to capture the physics of the system. As the problem is highly complex in nature, material non linearity, temperature effect and effect of dynamic fluid pressure are not taken into consideration. The analysis process takes care of the spring's installation followed by applying centrifugal force and torque in alternating sequence. Similar method is repeated for different spring orientation to find out the worst case model and loading sequence. High stress locations on springs and Driving/Driven cage plates are identified and correlated with the actual hardware test results.

1. Introduction

Torque converters are used in automatic transmissions as torque transfer devices that perform the same purpose as a clutch in a manual transmission vehicle. Without this device, the vehicle would be unable to start or stop without preventing the engine from turning over. A torque converter is a form of fluid coupling that transmits torque while allowing the engine to spin almost independently of the transmission.

The torque converter has a low efficiency because of friction between the working fluid (oil) and mechanical components (impeller, turbine and stator). In order to improve the efficiency of the torque converter, when the slip between the impeller and the turbine is relatively small, the torque converter is locked. This is possible by using a lock-up clutch, which links mechanically the impeller with the turbine. This way there is no more friction between the oil and components, and engine power is mechanically transferred to the gearbox. The lock-up clutch has a vibration dampener, which damps the oscillations during the torque converter lock-up phase. Conventional torque converter dampeners use coil springs to isolate vibration; the energy of the combustion forces is absorbed by the springs and then returned during the compression cycle of the engine.

The Dampeners are typically subjected to cyclic loading resulting in a high probability of spring failure. In this study, static analysis has been performed through Finite Element Analysis to understand the system level behavior of a Dampener under specific loading conditions and predict the springs' failure by identifying the critical locations.

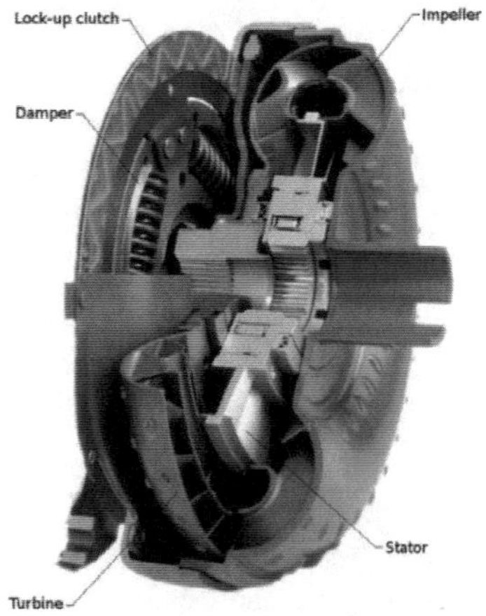

Figure 1. Torque Converter of an Automatic Transmission

Figure 2. Typical failures in Dampener

2. Methodology of torque converter dampener modeling

The torque converter spring dampener to be studied is shown in figure 3 schematically. A dampener primarily is of two types, single stage & 2-stage. [1] In single stage dampener, there are four arc/primary springs which are parallel between TC primary inertia and intermediate retainer. The arc springs are sliding against the holding plate outer wall. In 2-stage dampener, there is another set of linear springs between intermediate retainer and turbine flange and the arc spring set and linear spring set are in series.

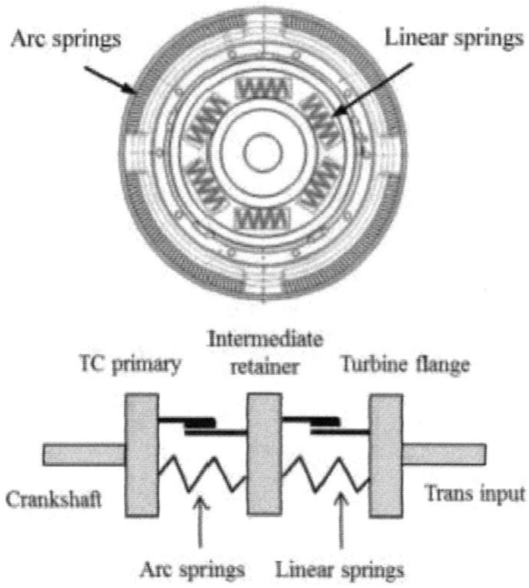

Figure 3. The schematic diagram of the torque converter dampener [1]

3. Spring design parameters

The number of spring segments will determine the dampener structure and dampener maximum travel limit [1]. Because springs are arranged in parallel, more segments will usually produce bigger dampener torque capacity but less dampener travel. There are four spring segments in the system to be studied referring to figure 3. The number of coils in each spring is related with mass distribution and also spring travel.

IOP Conf. Series: Materials Science and Engineering **624** (2019) 012005 doi:10.1088/1757-899X/624/1/012005

4. FE analysis of torque converter dampener

A Dampener springs' orientation in the hardware setup is totally random and it has been observed that failure of the springs are highly dependent on its orientation and the sequence of applied torque and rpm loads. A couple of different spring's orientation along with different combinations of loading are evaluated to understand the sensitivity of these parameters on springs' failure. This work is based on a single stage Dampener where four sets of two coaxial springs are placed in parallel. Few hardware failures were observed in the outer coils of the springs, hence static analyses in Abaqus/Standard are performed to understand the structural behavior of the Dampener and identify the critical areas on the spring coils to predict the failures.

In this model, Driving & Driven Plates and springs are considered for the analysis. Modeling software Unigraphics is used to develop 3D geometry model which is meshed using preprocessing tool Hypermesh. Abaqus is used as a solver and postprocessor. The model comprises of first order hexahedral (C3D8I) elements and second order tetrahedral elements (C3D10I). Several surface based and edge based contact interaction have been defined between all interacting interfaces.

4.1 Analysis assumptions

FEA of Torque Converter Dampener becomes complicated because of its high deformation pattern and due to having too many contacts in the model. The computational time is quite high due to higher model size and more number of iterations required to solve each increment. Hence few assumptions have been made to simplify the model, Figurewhich in turn improves the convergence rate and takes lesser time to solve. For this problem, quarter model with cyclic symmetry has been considered for the analysis and few features which are non-cyclic and would not affect the analysis results, remain ignored. Metal plasticity or post yielding behavior has not been considered as it adds more convergence issues. As failures were observed at the outer spring coils, hence they are modeled accurately with 3D elements whereas the inner spring has been modeled with 1D connector element with equivalent stiffness. Few sharp edges at the contact interfaces have been replaced with small fillets to avoid convergence difficulties. Rivets are not also not modeled explicitly and replaced by 1D rigid elements.

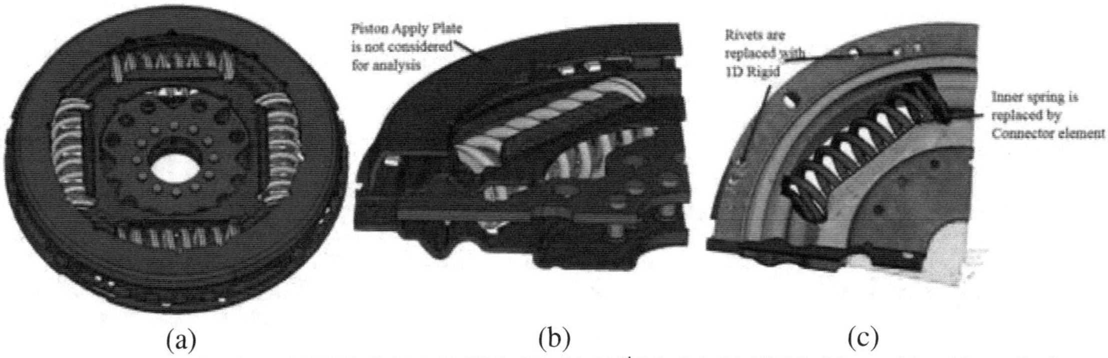

(a) (b) (c)

Figure 4. Simplification of FE Model (a) Full Model, (b) 1/4th Model, (c) FE Model considered in analysis

4.2 Loads and boundary conditions

As MPC (Multi point constrains) have been created between the center node and the ID nodes for all the plates, boundary conditions applied on these center nodes will control the behaviors of these plates. Being in uninstalled condition in the CAD model, springs need to be installed and preloaded through analysis before applying all the external loads.

In order to do that, spring is compressed by 2 mm from each side by keeping all the plates constrained in all six degrees of freedom followed by unloading the spring and get it installed by its own stiffness. After spring installation, torque & rpm load are applied in two different load steps. Torque in

IOP Conf. Series: Materials Science and Engineering **624** (2019) 012005 doi:10.1088/1757-899X/624/1/012005

the form of angular rotation has been applied to the other driving cage plate.

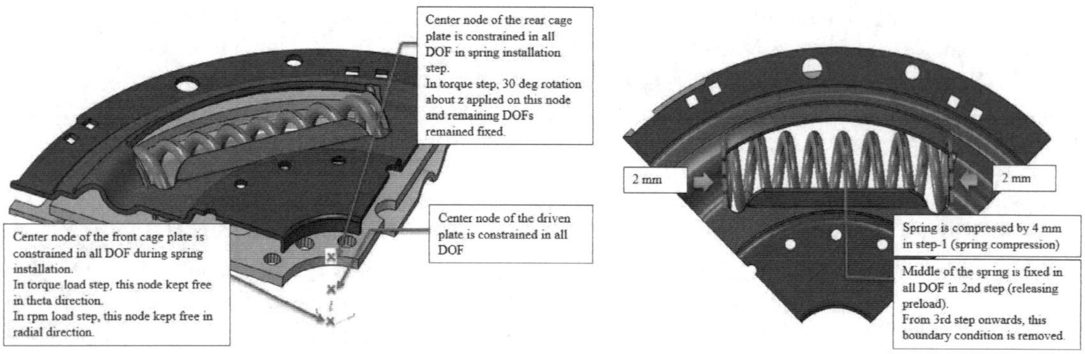

Figure 5. Load and Boundary Conditions

4.3 Analysis iterations

There are 2 different spring orientations considered for analysis. They are termed as A and B shown in Figure 6. Orientation A has spring's free end orientated towards ID and Orientation B has the spring's free end oriented towards OD of the Dampener, as shown in figure 8. Four different load cases are the various combinations of torque and rpm.

Table1. Shows Four load cases performed on orientation A

Cases	Orientation	First Load	Following Load
1	A	Max Torque	-
2	A	Max rpm	-
3	A	Max Torque	Max rpm
4	A	Max rpm	Max Torque

Above table shows 4 load cases performed on orientation A. Similar load cases are evaluated for orientation B as well. After analyzing all these 8 cases, the worst case combination needs to be identified and corresponding stress, deflection, contact force results to be reported.

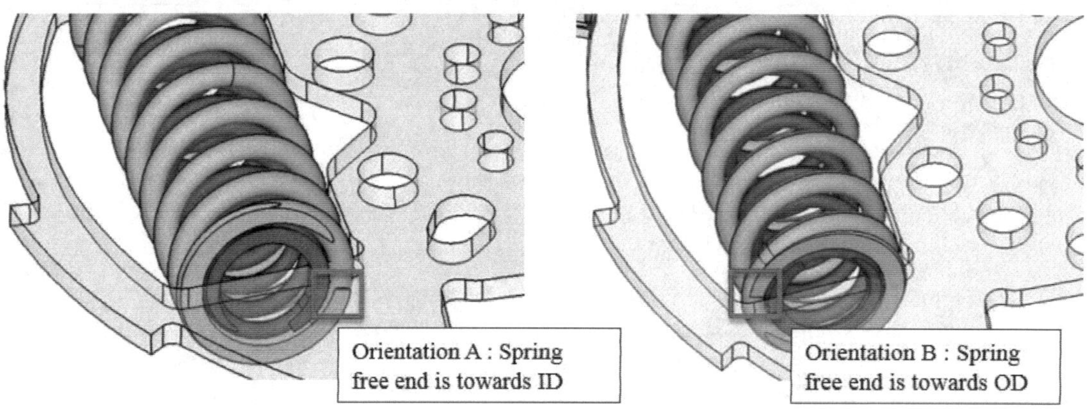

(a) (b)

Figure 6. Details of Orientation (a) Spring free end is towards ID, (b) Spring free end is towards OD

IOP Conf. Series: Materials Science and Engineering **624** (2019) 012005 doi:10.1088/1757-899X/624/1/012005

5. Results

Instead of applying torque on the system, angular rotation has been given to the Driving plate and reaction torque has been measured and compared with the Dampener torque-travel curve. From the FE analysis, reaction torque is measured as T2 whereas the actual torque as per the torque-travel curve is T1 for θ_3 degrees of angular rotation. The deviation of T2 from T1 is below 5% and the same may have been caused due to various analysis assumptions and is consistent throughout all the iterations.

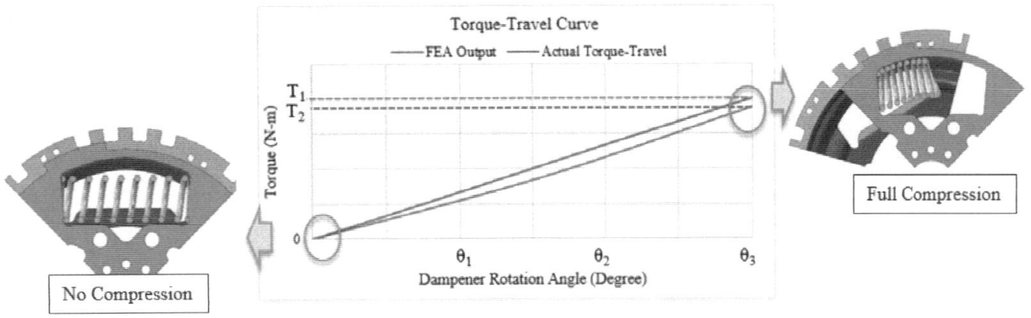

Figure 7. Dampener Torque-Travel curve FEA vs Actual

5.1 Results for Orientation A for Max torque followed by Max rpm

Figure 8 shows some of the stress and contact pressure results for load case 3. As material properties are considered as linear in the analysis, stress values are coming very high. Spring material has a yield limit of 1200 MPa, hence any stress value more than 1200 MPa is damaging for the springs. As shown in Figure 7, location "a" which is 180° from the spring's free end, has shown some material failure which is correlated to the very high tensile stress (4000 MPa) at the same location. Similarly, the "break point" which is 270° clockwise from the free end, is showing high tensile stress (1400 MPa) in analysis. Also locations "c" and "d" are showing wear in the physical hardware and the similar locations are showing very high contact pressure and contact forces as per the analysis too.

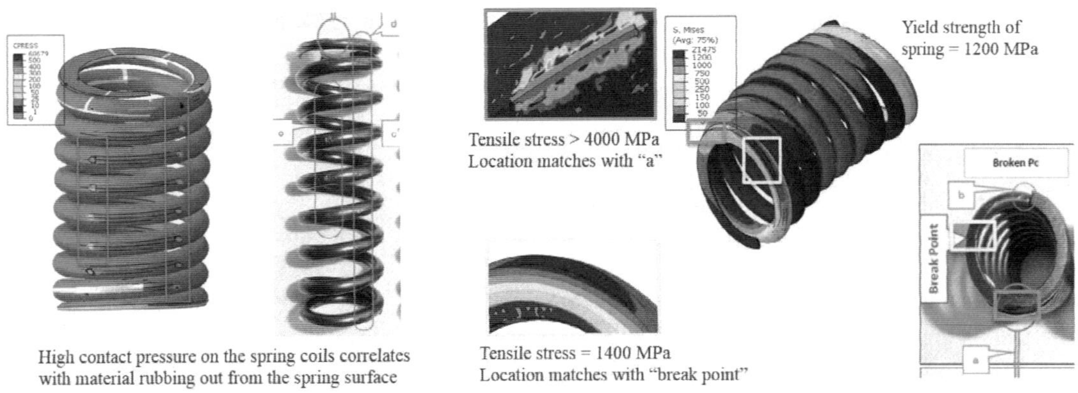

(a) (b)

Figure 8. Results for Orientation A for Max torque followed by Max rpm (a) Contact Pressure (CPRESS) plot, (b) Von misses stress plots

5.2 Results for Orientation B for Max torque followed by Max rpm

Figure 9 shows the analysis results for orientation B which has the spring's free end oriented towards the OD of the dampener. There is a localized high stress location just next to the spring's free end, which has 2600 MPa compressive von Mises stress. The same location matches quite well with location "b" from the hardware. The outer surfaces of the spring coils are subjected to high contact pressure and contact forces, which could significantly contribute to the material wear from locations "c" and "d".

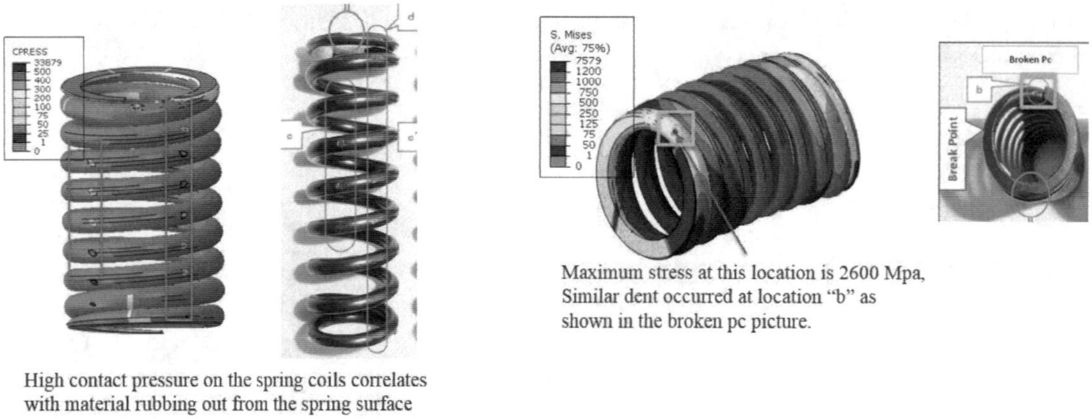

(a) (b)

Figure 9. Results for Orientation B for Max torque followed by Max rpm(a) Contact Pressure (CPRESS) plot, (b) Von misses stress plots

6. Conclusions

In this work, structural behavior of a Torque Converter Dampener has been studied for various spring orientations and loads using static analysis. Results are also correlated with the physical hardware test. Following conclusions can be derived from this work.

1. From static analyses on eight iterations, it is found that the combination of torque and rpm is the most damaging load case for Dampener.
2. As mentioned earlier, breaking of springs are highly sensitive towards the spring orientation and it is observed that orientation A caused the spring to break but orientation B doesn't. Hence orientation A is more critical than B.
3. Irrespective of the spring's orientation, there are significant material wear on the outer surface of the springs and it is occurred due to the spring's rubbing off against the cage plates.
4. Significant material wear happened at about 0.5 coils from the end tip because of very high compression against driven plate and localized material yielding.
5. This method can be used to study the system level behavior of Torque Converter Dampener for any transmission. It is also capable to predict the failure in the spring coils for a given orientation, which may potentially replace the physical test.

References

[1] Li, Z and Sandhu, J. "Transmission Torque Converter Arc Spring Dampener Dynamic Characteristics for Driveline Torsional Vibration Evaluation". SAE Int. J. Passeng. Cars - Mech. Syst. 6(1):2013, doi:10.4271/2013-01-1483

IOP Conf. Series: Materials Science and Engineering **624** (2019) 012006 doi:10.1088/1757-899X/624/1/012006

Study of the lubricating properties of hybrid liquid paraffin with TiO_2 and CuO as nano-additives for engine oil application

Adarsh Kumar Shah[1], K P Vineesh[2] and M L Joy[2]

[1]Corresponding author Department of Mechanical Engineering, National Institute of Technology, Calicut, Kerala 673601, India
[2]Department of Mechanical Engineering, National Institute of Technology, Calicut, Kerala 673601, India

E-mail: shahadarsh1994@gmail.com

Abstract. Mineral based paraffin oil (L14480) lubricant finds it difficult to give optimum characteristics for modern engine application. In this work, possibility of hybrid paraffin oil which consists of paraffin oil (PO) and sesame oil (SESO) in equal proportions to be used for engine oil application is investigated. Properties of hybrid paraffin oil and SAE 20W40 were estimated and compared. Estimation of thermal characteristics of hybrid paraffin oil shows it have comparable properties with SAE 20W40. Low viscosity characteristics of hybrid paraffin oil can be improved by addition of EVA particles. Addition of TiO_2 and CuO nanoparticles shows enough improvement in tribological properties of hybrid paraffin oil. Poor oxidative stability of hybrid paraffin oil can be improved by addition of butylated hydroxytoluene (BHT).

1. Introduction

Nanotechnology is remarked as the most wide ranging technology of the 21st century. It can be preowned in many areas and lead engineering into a new generation. There have been many research on the tribological properties of lubricants with distinct nanoparticles combination. In literature it has been reported that blending of nanoparticles in lubricants exhibit good reduction in wear and friction. The friction-reduction and anti-wear behaviours were relying on the attribute of nanoparticles, like shape, size, and concentration [1–6]. The dimensions of nanoparticles are usually in the limit of 2–120 nm. Use of copper oxide nanoparticles to improve tribological and anti-wear properties are reported [8-9]. TiO_2 nanoparticles displayed satisfactory dispersion capacity and attractive anti-wear behaviour in liquid paraffin [10]. In the present work, properties of hybrid paraffin oil which is a blended mixture of paraffin oil (PO) and sesame oil (SESO) were evaluated and it is compared with the evaluated properties of SAE 20W40. Nanoparticles of copper oxide (CuO) and Titanium di-oxide (TiO_2) are added to the hybrid paraffin oil to improve the tribological properties. Properties of these mixtures were evaluated and compared with SAE 20W40. Similar additions of EVA particles and butylated hydroxytoluene are carried out to improve rheological properties and oxidative stability. EVA particles polymers were selected for improving rheological properties considering their remarkable physical and chemical properties coupled with less cost, excellent process ability and recyclability.

IOP Conf. Series: Materials Science and Engineering **624** (2019) 012006 doi:10.1088/1757-899X/624/1/012006

2. Materials and Methods

2.1 Preparation of modified paraffin oil

The hybrid paraffin oil (PO) is prepared with sesame oil (SESO) in 50/50 weight fractions. The mixing was done using a magnetic stirrer at a speed of 500 rpm at room temperature for 5-6 hours, formulating uniformly diffuse samples. For improving the rheological properties, EVA particles were combined in percentage weight proportions ranging from 2% to 5%. EVA particles were heated with hybrid paraffin oil in a magnetic stirrer at 500 rpm at 70°C for 4-5 hours. The blend was kept overnight and the homogeneity was confirmed. Improvement of tribological properties of hybrid paraffin oil was made by adding nanoparticles of copper oxide (CuO) and Titanium di-oxide (TiO_2). The average size (40nm) and structure of the TiO_2 and CuO nanoparticle were characterized by adopting field emission scanning electron microscope (FESEM). The nanolubricants samples are subjected to the magnetic stirrer for 5 hour and then a water bath sonication was used to dissolve the nanolubricants properly in the base oil for 6 hours. The BHT particles were added in the percentage weight proportion of 1%, 1.25%, 1.5% and 2.25% to improve the oxidative stability. BHT particles were heated with hybrid paraffin oil in a magnetic stirrer at 400 rpm at 80°C for 6-7 hours.

Figure 1. Experimental Setup of Four Ball Tester.

2.2 Evaluation of thermal properties

Thermal properties of hybrid paraffin oil were evaluated by using flash point, fire point, pour point and cloud point. A Cleveland open cup setup was adopted for the determination of Fire and Flash point according to ASTM D92. The pour point was calculated to an exactness of ±3°C as per ASTM D97-96.

2.3 Evaluation of rheological properties

Dynamic viscosities were estimated for a temperature range of 25°C to 100°C. A Modular Compact Rheometer (MCR 102, Anton Paar Austria) having parallel plate tool geometry was used.

2.4 Evaluation of tribological properties

Standard four ball tester equipment (figure 1) was used to evaluate tribological properties of samples. Chrome alloy steel balls with a diameter of 12.7 mm and a hardness of HRC 61-64 were used for the test. Washing of balls and attachment was carried out using acetone in an ultra-sonication bath for 40

IOP Conf. Series: Materials Science and Engineering **624** (2019) 012006 doi:10.1088/1757-899X/624/1/012006

min and then in n-hexane. Wear test was performed with a load of 40 kg at 1200 rpm for a duration of 60 min according to ASTM D 4172-94. Lubricant was maintained at a temperature of 75°C.

2.5 Analysis of oxidative properties
Hot Oil Oxidation Tests (HOOT) were conducted on samples by keeping it in a dark oven at 100°C for 140 hours in a box container. Before and after the tests, viscosities of samples were estimated 40°C.

3. Results and Discussion

3.1 *Result of thermal analysis*
Estimated flash point, fire point, cloud point, and pour point are given in Table 1. Flash and fire point of hybrid paraffin oil are greater than SAE 20W40 indicates that hybrid paraffin oil is thermally more stable. Cloud point and pour points are also comparable with SAE 20W40.

Table 1. Comparison of thermal properties of hybrid paraffin oil and SAE 20W40.

S. NO	SAMPLE	THERMAL PROPERTIES (°C)			
		FLASH POINT	FIRE POINT	CLOUD POINT	POUR POINT
1.	Hybrid paraffin oil	226	239	-10	-17
2.	SAE 20W40	204	209	-15	-20

3.2 *Result of rheology modifiers*
Table 2 shows the viscosities of sample at 40°C and 100°C. The viscosity of hybrid paraffin oil is low compared to SAE 20W40. Table 3 shows the variation of viscosities of hybrid paraffin oil with different weight proportions of EVA particles. It can be seen that addition of 2-3% EVA particles improves the viscosity of hybrid paraffin oil which is comparable with SAE 20W40. Intermolecular forces among the molecules and difference in molecular weight of the polymers are believed to be the reason for this.

Table 2. Comparison of viscosity of hybrid paraffin oil and SAE 20W40.

S. No	SAMPLE	VISCOSITY at 40°C (mPa-s)	VISCOSITY at 100°C (mPa-s)
1.	Hybrid paraffin oil	39.42	8
2.	SAE 20W40	117.86	14.18

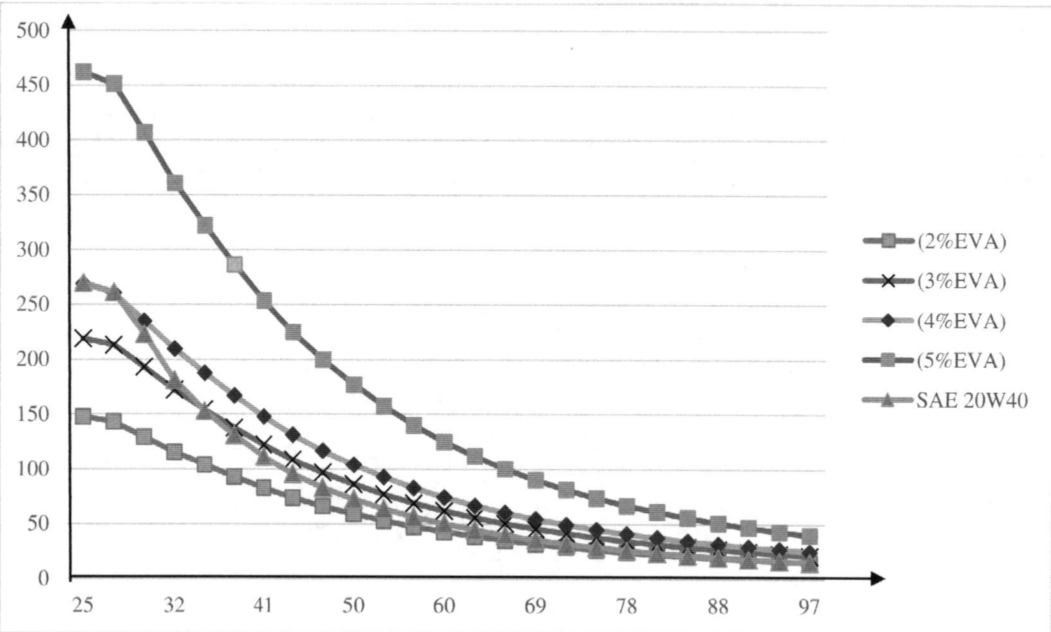

Figure 2. Comparison of various percentages of EVA particles in hybrid paraffin oil and SAE 20W40.

Table 3. Effect of addition of EVA particle on viscosity of hybrid paraffin oil.

TEMPERATURE (°C)	VISCOSITY (mPa-s)				
	SAE 20W40	Hybrid Paraffin oil + EVA			
		2% EVA	3% EVA	4% EVA	5% EVA
40	117.86	85	125	152	260
60	49.37	41	60	72	121
80	24.78	23	32	39	62
100	14.18	14.66	20	23.71	38

3.3 *Additives to improve tribological properties*

Table 4 shows the wear scar diameter and friction coefficient of hybrid paraffin oil SAE 20W40. It can be noted that wear scar diameter of hybrid paraffin oil is 50% higher than SAE 20W40. Table 5 shows the wear scar diameter with addition of nano particles to hybrid paraffin oil. The following points can be noted: (i) Addition of nano-particles decrease the wear scar diameter, and (ii) Addition of 0.15% TiO_2 decrease the wear diameter by about 22%. This can be due to spherical shape of TiO_2 nanoparticles operating as a rolling medium within the contacting surfaces, which eliminates the metal to metal contact between the asperities of the two mating surfaces.

Table 4. Comparison of Wear Scar Diameter of hybrid paraffin oil and SAE 20W40.

S. No	SAMPLE	COEFFICIENT OF FRICTION	WEAR SCAR DIAMETER (mm)
1.	Hybrid paraffin oil	0.0779	0.6577
2.	SAE20W40	0.0823	0.4381

Table 5. Effect of nanoparticles additives on wear scar diameter of hybrid paraffin oil.

S. No	Samples	Wear Scar Diameter (mm)	% Reduction in WSD
1.	Hybrid paraffin oil	0.6577	
2.	Hybrid paraffin oil + 0.5% CuO	0.5939	9.7004%
3.	Hybrid paraffin oil + 1% TiO_2	0.5203	20.890%
4.	Hybrid paraffin oil + 0.15% TiO_2	0.5130	22.009%
5.	SAE 20W40	0.4381	

3.4 Additives to improve the oxidative properties

Table 6 shows the change in viscosity of hybrid paraffin oil and SAE 20W40 after the HOOT (Hot Oil Oxidation Test). Viscosity of hybrid paraffin oil changes by 69.44 mPa-s which is relatively high compared to change in viscosity of SAE 20W40. Table 7 shows the improvement of viscosity of hybrid paraffin oil with blending of antioxidants Butylated hydroxytoluene (BHT). It can be seen that 1% of BHT blending reduce the change in viscosity of about 14 mPa-s. But higher % addition increases the change in viscosity.

Table 6. Comparison of oxidative stability of hybrid paraffin oil and SAE20W40.

SAMPLE	CHANGE IN VISCOSITY (mPa-s)
Hybrid paraffin oil	69.44
SAE 20W40	2.26

Table 7. Effect of oxidative additives on hybrid paraffin oil.

SAMPLE	CHANGE IN VISCOSITY (mPa-s)
Hybrid paraffin oil+1% BHT	14.641
Hybrid paraffin oil +1.25%BHT	28.199
Hybrid paraffin oil +1.5% BHT	18.798
Hybrid paraffin oil +2.25%BHT	24.759
SAE 20W40	2.26

4. Conclusions

Properties of hybrid paraffin oil which consists of equal proportions of paraffin oil and sesame oil were estimated to be used as lubricant oil. Properties are compared with the commercially available SAE 20W40. The following conclusions can be made, (i) Thermal properties of hybrid paraffin oil are comparable with SAE 20W40, (ii) Hybrid paraffin oil have lower viscosity compared with SAE 20W40. Addition of EVA particles can improve the rheological properties, (iii) Tribological tests shows higher wear scar diameter for hybrid paraffin oil. Addition of TiO_2 can improve the tribological properties, (iv) Oxidative stability of hybrid paraffin oil can be improved by adding butylated hydroxytoluene.

References

[1] F. Chinas-Castillo and H.A. Spikes, (2003), *Mechanism of action of colloidal solid dispersions*, Trans, ASME 125, 552–557.

[2] J. Zhou, J. Yang, Z. Zhang, W. Liu and Q. Xue, (1999), *Study on the structure and tribological properties of surface-modified Cu nanoparticles*, Mater. Res. Bull. 34 (9), 1361–1367.

[3] L. Rapoport, Y. Feldman, M. Homyonfer, H. Cohen, J. Sloan and J.L. Hutchison, (1999), *Inorganic fullerene-like material as additives to lubricants: structure–function relationship*, Wear 225–229, 975–982.

[4] S. Chen, W. Liu and L. Yu, (1998), *Preparation of DDP-coated PbS nanoparticles and investigation of the antiwear ability of the prepared nanoparticles as additive in liquid paraffin*, Wear 218, 153–158.

[5] Q. Xue, W. Liu and Z. Zhang, (1997), *Friction and wear properties of a surface modified* TiO_2 *nanoparticle as an additive in liquid paraffin*, Wear 213, 29–32.

[6] W. Liu and S. Chen, (2000), *An investigation of the tribological behavior of surface modified ZnS nanoparticles in liquid paraffin*, Wear 238, 120–124.

[7] Wu Y, Tsuia W and Liub T (2007) *Experimental analysis of tribological properties of lubricating oils with nanoparticle additives*. Wear 262, 819–825.

[8] Thottackkad MV, Perikinalil RK and Kumarapillai PN (2012), *Experimental evaluation on the tribological properties of coconut oil by the addition of CuO nanoparticles*. Int J Precis Eng Manuf 13, 111–116

[9] Qunji, Liu and Zhijun Zhang (1997): '*Friction and wear properties of a surface-modified* TiO_2 *nanoparticle as an additive in liquid paraffin*'.

[10] Z.S. Hu and J.X. Dong, (1998), Wear 216, 92–96

IOP Conf. Series: Materials Science and Engineering **624** (2019) 012007 doi:10.1088/1757-899X/624/1/012007

Dynamic characteristics analysis of a lead screw by considering the variation in thread parameters

Alex S Syriac[1,2], Shital S Chiddarwar[3]

[1,3] Department of Mechanical Engineering, Visvesvaraya National Institute of Technology, Nagpur-440010, India

Email ID: alex_sherjy@yahoo.com

Abstract. A lead screw is a mechanical transmission device which transforms rotary motion into translational movement. The main elements are screw and nut which establishes connection among them by the means of threads on respective element. It plays a major role in transferring the load between screw and nut during the travel stroke. This paper investigates the behavior of the lead screw, inside the withdrawal part of the circuit breaker of switchgear, by analyzing the mechanical response of the lead screw system. The various thread parameters such as thread pitch, nominal diameter etc. play an important role in force transmission as well as in overall efficiency of the mechanical system. This study is done by creating models of various thread profile such a square thread, acme thread, trapezoidal thread and simulating the results in the finite element analysis software ANSYS Workbench. This paper focuses on the effect of the thread pitch, nominal diameter, helix angle, screw length on the stress state and deformation within the screw. Also, the influence of screw pitch, screw length and screw nominal diameter on the withdrawal part of the circuit breaker's natural frequency along its whole stroke is carried out using modal analysis.

1. Introduction

Lead screw is a mechanical device used in power transmission defined by threads that transfer force between the nut and the screw. Lead screw drives are the simplest, most efficient, accurate and highly economical devices used for application that require precise motion and transmission of force. The most common thread profiles used in lead screws are square threads, trapezoidal threads, and acme threads. Square threads poses better efficiency but are weaker at the root. They are also costly to manufacture. From manufacturing perceptive, trapezoidal threads can be manufactured easily and thicker at the root of the thread and hence poses higher loading carrying capacity. Withdrawal devices housed within a medium voltage switchgear framework are periodically disconnected for maintenance, testing or replacement. This disconnection of the withdrawal device from the switchgear is known as "racking out". The connection of the withdrawal device to the switchgear is known as "racking in". For both of the above mentioned operations lead screw plays an important role. In this application, lead screw transfers the torque from the

IOP Conf. Series: Materials Science and Engineering **624** (2019) 012007 doi:10.1088/1757-899X/624/1/012007

motor (rotational motion), coupled to the screw shaft, to the nut, carrying the circuit breaker which moves linearly. The present system employs manual racking of the circuit breaker by the means of a handle. Manual racking is a laborious job and dangerous due to the potential for an arc flash. The operator who does the racking operation must wear protective equipment to prevent sudden arc flash, injuries or death. So an automatic racking mechanism is a better alternative. For this purpose, the design and dynamics of the lead screw, a vital element in the power transmission in the withdrawal part, is studied in the next session for the purpose of automation.

1.1 Background

Enormous research has been done to find the distribution of force in the threads of the lead screws as well the effects of thread parameter on the efficiency of transmission. Varanasi [1] developed a dynamics model of a lead screw system considering the inertia of screw along with the damping and compliance of bearing and nut. Dragoni [2] studied the effect of thread pitch and coefficient of friction on the stress concentration within the screw. Murphy [3] studied the distribution of load on a lead screw undergoing wear subjected to varying operating conditions. Pick and Burns [6] determined load distributions in threaded end closures by finite element method. Miller et al [7] modeled the threaded connection element by axisymmetric rings by springs to evaluate the distribution of load in threaded connections.

2. Design of lead screw

The main element in the withdrawal part is the lead screw which helps in racking in and racking out of the circuit breaker. So, the design of the lead screw is calculated in this section. Even the torque required to drive the mechanism including torque required for overcoming the friction is calculated. The lead screw in the withdrawal part of the withdrawal indoor type circuit breaker is constrained with the help of rotary bearings and rotated by a remote portable driver which houses a motor with a gearbox and coupling. The nut is attached to the carriage whose movement is constrained axially by sliding over fixed supports on either side. The end of the screw nearest to the drive motor is axially constrained to the base by means of a thrust bearing. Analyzing the mechanics including friction and inertia of the load to be positioned is the first step in determining the requirements of a motion control system.

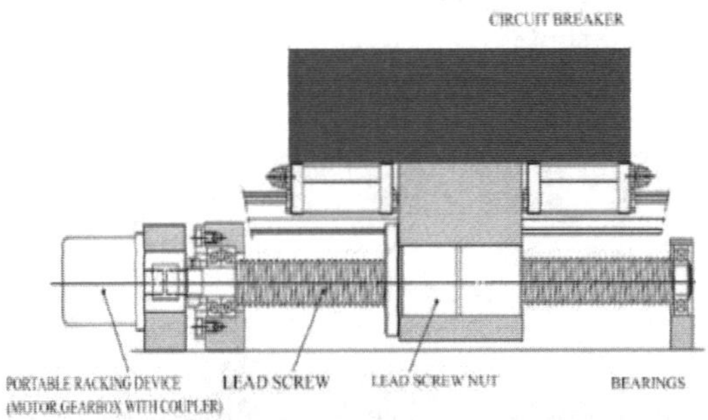

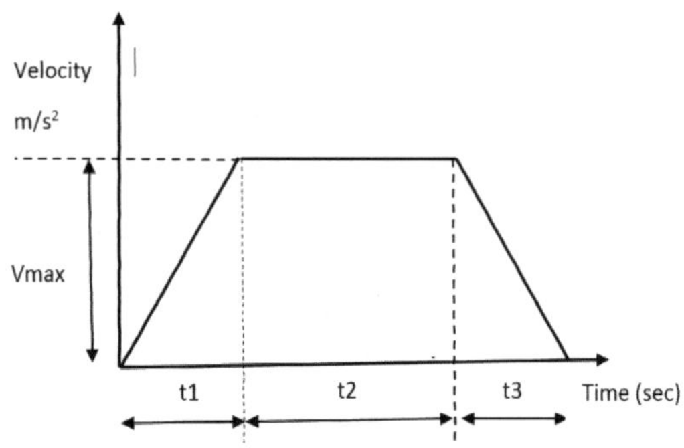

Figure 1. The schematic sketch of the lead screw system. **Figure 2.** Trapezoidal move profile.

According to the loading condition, bending theory and comparing in ANSI Standard table for threads Ac= 519 mm^2 , dc = 25.7062 (26mm approx.) , d = 30mm , p=3.5mm

Table 1. Formulas used in the design.

	Formula	Value
Load torque reflected to motor	$T_r = \frac{F_l}{2\pi Pe} + \frac{\mu F_{pf}}{2\pi P}$	3.442588 Nm
Load torque due to friction	$T_f = \frac{F_f}{2\pi Pe}$	1.996759 Nm
Maximum velocity		$8.2222*10^{-3}$ m/s
Torque due to acceleration	$T_a = (\frac{J_l}{e} + J_{ls} + J_{cp} + J_{motor})\frac{\omega}{t_a}$	0.035506 Nm
Peak Torque	$T = T_r + T_f + T_a$	5.4745 Nm
Rms Torque	$T_{rms} = \sqrt{\frac{T_1^2 t_1 + T_2^2 t_2 + T_3^2 t_3}{t_1 + t_2 + t_3}}$	1.9968 Nm

3. Modelling and Simulation

The withdrawal part which houses the lead screw system is modeled in CATIA V5 as shown in the figure 3 and is analyzed in ANSYS Workbench. The boundary conditions are fixed supports, a load of 2800N acting on the nut vertically downwards and a torque 6 Nm applied at the left end of the lead screw, which is shown in the figure 4.

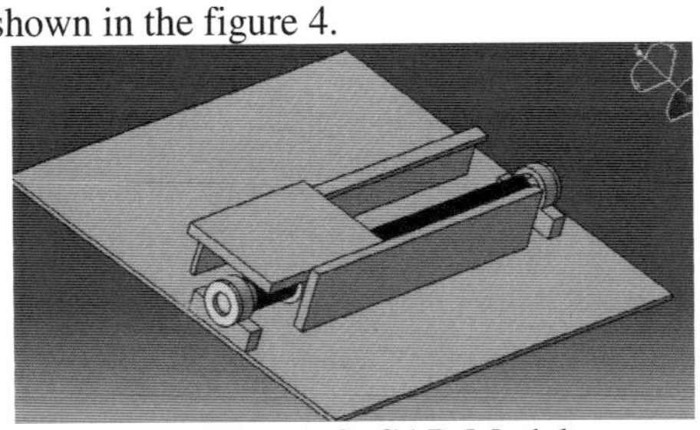

Figure 3. CAD Model.

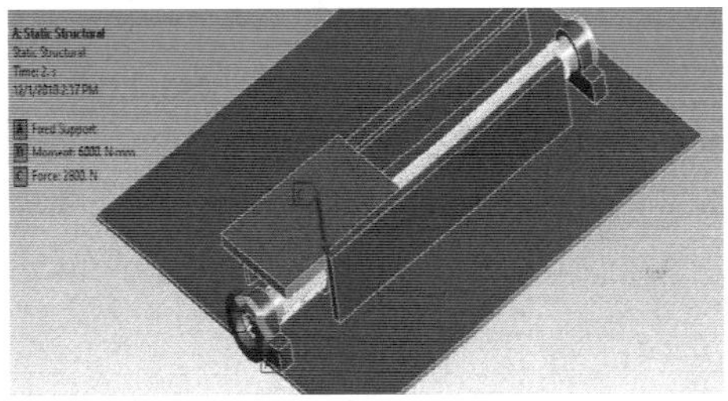

Figure 4. Model with boundary Condition.

Table 2. Parameters of the constructed lead screw drive for simulation.

Parameters of the lead screw drive	Value	Units
Material of the lead screw	Structural steel	-
Density ρ	7850	kg/m^3
Modulus of elasticity E	$2x10^{11}$	Pa
Poisson's Ratio μ	0.3	-
Bulk Modulus	166666666666.667	Pa
Shear Modulus	76923076923.0769	Pa
Root Diameter of screw d_c	26	mm
Pitch Diameter of screw p	3.5	mm
Nominal Diameter of screw d	30	mm
Screw length	370	mm
Inertia of lead screw	0.0354375	$kg\ m^2$
Mass of load m_l	280	kg

4. Results and Discussion

4.1 Influence of the screw length on the natural frequency

The effect of thread screw length on the natural frequency of the system is depicted in figure 5.The thread profile is square of pitch 3.5mm and the position of the nut is at the center of the lead screw. The parameters for simulation are from table 2 with a change to the lead screw length. (300mm/350mm/400mm/450mm). The natural frequency decreases with screw length. For the lower range of screw length, all the frequency mode values indicate relatively high variation in natural frequency value. While on further increasing the screw length, the variation in the magnitude of the natural frequency decreases in case of Mode1, Mode 2and Mode 3.

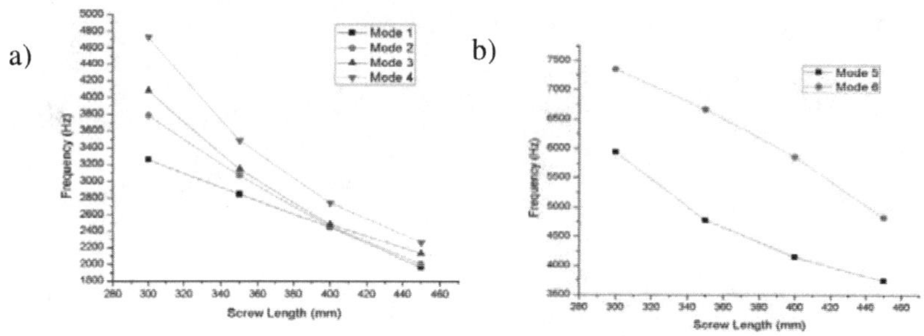

Figure 5. Natural frequency variation for different screw length a)Mode 1 to Mode 4 b)Mode 5,Mode 6.

4.2 Influence of the screw pitch on the natural frequency

The effect of thread pitch on natural frequency is illustrated in figure 6. The position of the nut is at the center of the screw and screw thread profile is squareof different thread pitches, such as 2.5 mm, 3.5 mm, 4.5 mm, 5.5mm and 6.5mm. The other parameters used for the simulation are shown in table 2.

From figure 6 it is observed that the screw pitch has minimum influence on the value of the natural frequency. Moreover, as the pitch of the lead screw changes, the system frequency is not drastically affected. The various modes of natural frequency show the same trend with the lead screw having different pitch that is the natural frequency of the system shows slight decrease with increase in the screw pitch and after reaching a minimum value it again increases.

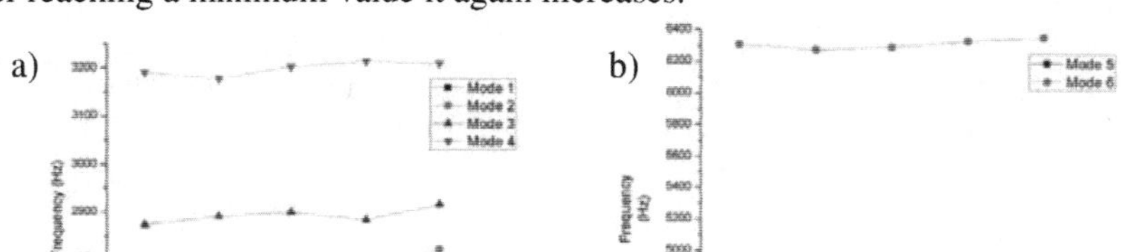

Figure 6. Natural frequency variation for different screw pitcha)Mode 1 to Mode 4 b)Mode 5,Mode 6.

4.3 Influence of the screw nominal diameter on the natural frequency

Figure 7 shows the effect of the nominal diameter of screw of square thread profile, such as 29.1mm, 29.3mm, 29.5mm, 29.7mm, and 29.9mm, on the value of the natural frequency of the system. The screw is of square thread profile with thread pitch 3.5mm and the nut is located at the center of the screw. Other parameters for simulation are kept same as shown in table 2.

4.4 Influence of the position of worktable on the natural frequency

The natural frequency changes with the variation in worktable location as seen in figure 8. The thread profile considered is square of pitch 3.5mm. Other parameters for simulation are as shown in table 2. The trend seen is that the natural frequency increases as the worktable moves from the front support end to the center and decreases when the worktable moves further towards the rear support end.

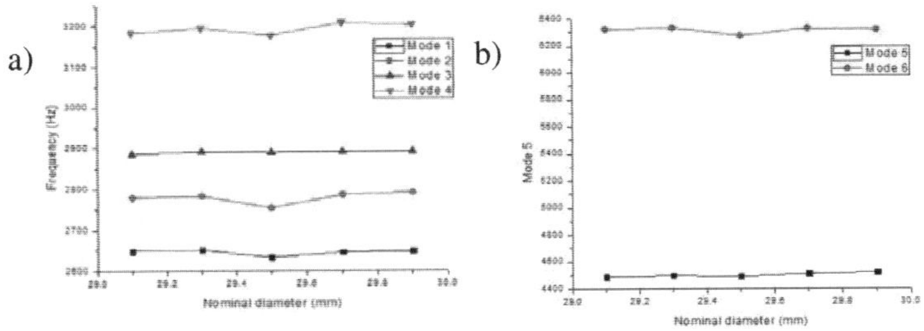

Figure 7. Natural frequency variation for the different screw nominal diametera) Mode 1 to Mode 4 b) Mode 5,Mode 6.

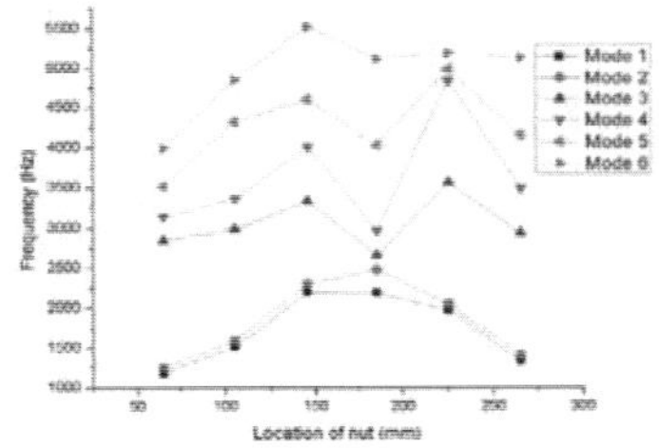

Figure 8. Natural frequency variation for different position of worktable.

4.5 Influence of thread pitch variation on stress distribution

The stress distribution is evaluated for the lead screw for different thread profiles with pitch ranging from 2.5mm to 5.5mm. The effect of pitch variation on stress values at different location of the nut along the screw is depicted in figure 9.The stress value decreases as the nut moves from left support end towards the center. Later it increases as it moves towards the right end support. As in case of all the thread profiles, the trend seen is similar. As the location of nut increases, the variation of stress values is significantly less near the center of the screw in all thread profiles. It is seen in three thread profiles that under the same loading condition, trapezoidal thread stress values are higher compared to square and acme profile at almost all location of the nut.

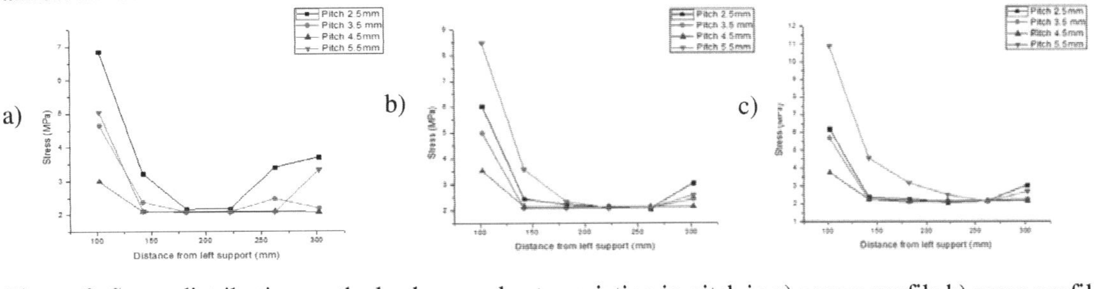

Figure 9. Stress distribution on the lead screw due to variation in pitch in a) square profile b) acme profile c) trapezoidal profile.

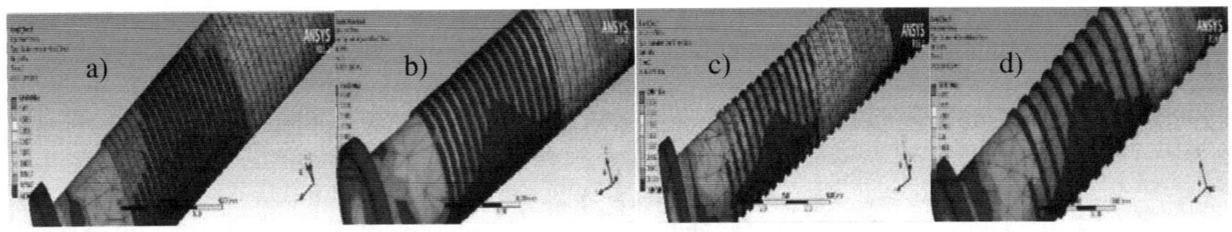

Figure 10. Stress distribution on the lead screw with square profile of various pitch of a) 2.5mm b) 3.5mm c) 4.5mm d) 5.5mm.

4.6 Influence of nominal diameter variation on stress distribution

The stress distribution is evaluated for the lead screw for different thread profiles of pitch 3.5mm with nominal diameter ranging from 29.1mm to 29.7mm. The effect of nominal diameter variation on stress values at different location of the nut along the screw is illustrated in figure 11.Similar trend as in the case of variation of thread pitch is seen here also. The magnitude of stress in all the thread profile decreases towards the center of the shaft and stress values increase as the nut approaches the other end support.

a) b) c)

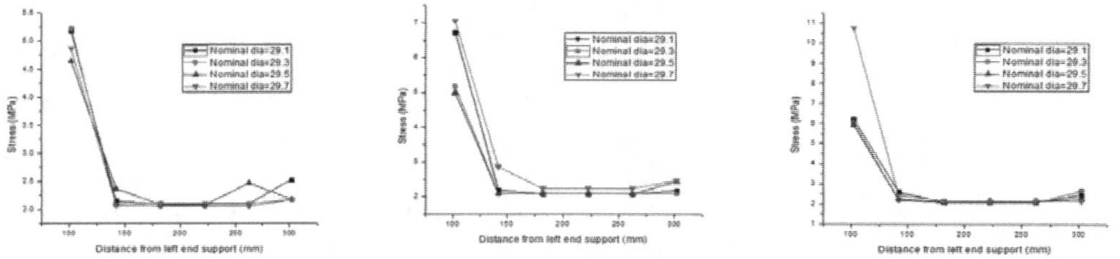

Figure 11. Stress distribution on the lead screw due to variation in nominal diameter in a) square profile b) acme profile c) trapezoidal profile.

5. Conclusion

The present analysis is performed to study the effect of the change in thread parameters on the frequency of the system and stress distribution. From study, it can be concluded that the parameters like screw length, nominal diameter and worktable position remarkably affect the natural frequency of the system. The amplitude of natural frequency of the system decreases with the change in screw length along the stroke length. The change in thread pitch shows less influence on the natural frequency. The natural frequency changes remarkably with the change in worktable position along the stroke length. The amplitude of stress by varying the pitch decreased apparently with increase in movement of the nut towards the center of the screw and further shows an increase as it moves to the rear end.

6. Reference

[1] Varanasi K K and Nayfeh S A 2004 The dynamics of lead-screw drives: low-order modeling and experiments *Journal of Dynamic Systems, Measurement, and Control* Vol. **126** pp 388-396.

[2] Dragoni E 1994 Effect of thread pitch and frictional coefficient on the stress concentration in metric nut-bolt connections *Journal of Offshore Mechanics and Arctic Engineering* Vol. **11**pp 21-27.

[3] Murphy D J and Blanche T A 2008 Load redistribution on lead screw threads wearing under varying operating conditions *Journal of Tribology* Vol. **130** pp 045001-1-13.

[4] Cojocaru Vasile et al 2014 Fatigue analysis of large diameter threaded connections subjected to dynamic axial loads *Applied Mechanics and Materials* Vol. **658** pp 177-182.

[5] Huijie Zhang, et al. 2018 Dynamics analysis of a slender ball-screw feed system considering the changes of the worktable position *Proc. of the Institution of Mechanical Engineers Part C Journal of Mechanical Engineering Science 1989-1996* Vols.**203-210**

[6] Pick R J and Burns D J 1971 Finite element analysis of threaded end closures of thick walled vessels *Engineering Solids Under Pressure 3rd Int. Conf. on High Pressure*, Proceedings of the Institution of Mechanical Engineers, London pp 15–25.

[7] Miller D L, Marshek K M Naji and Mohammad R 1983 Determination of load distribution in a threaded connection Mech. Mach. Theory **18**(6) pp 421–430.

Numerical Simulation and Experimental Validation of Planetary gearbox System Design to Govern Constant Generator Speed in Hydro Power Plant

Bhargav[1], M. A. Parameshwaran[2], Sivaraj S[2]. & Nithin Venkataram[1]

[1]Dept. of MME, Ramaiah University of Applied Sciences Bangalore, India
[2]MAGTORQ Pvt. Ltd., Hosur, Tamil Nadu, India

E-mail: nithin31@gmail.com

Abstract. Micro Hydro Power Plant (MHPP) is a clean and renewable source of harnessing power ranging between 5kW-100kW. The reservoirs are avoided in MHPP and power house is directly placed in the path of the water stream. In India, there exists an abundant resources unutilized due to technical constraints. As the flow of the water stream varies, the input velocity of turbine also varies and so does the generator speed. Hence, the challenge lies in arriving at a gear arrangement to maintain constant generator speed. In the present paper, an auxiliary gearbox with planetary gearbox is designed to maintain constant generator speed for various input turbine speed. Based on these specifications, the planetary gear and auxiliary gearbox is designed through analytical calculations. The complete system is modelled to carry out Multi Body Dynamic simulation to verify kinematic analysis (analytical method) and FEA to analyse stresses developed in planetary gearbox system. The auxiliary unit and planetary gearbox is fabricated and tested for no load condition to maintain the constant generator speed. The test results are found to be close agreement when validated with the analytical method and multi body dynamic simulation results. Hence the designed system is practically feasible for maintaining constant generator speed for fluctuation turbine speed

1. Introduction

Hydro power is a clean, renewable and non-polluting energy source utilized by more than 150 countries around the world. India produces about 26% of its electricity from Hydro Power Plants. Hydro power plants can be of large and small in size. Usually power plants up to 100 kW is called as Micro Hydro Power plant [1]. India has round 6474 potential small hydro sites out of which 1251 sites are operational. India generates 21134 MW of power from small hydro power plant, which accounts for only 2% of total potential for renewable power generation. In India 0.8% of the villages are still not electrified [2]. One of the emerging solutions for village electrification is Micro Hydro Power Plant (MHPP). MHPP are located near small rivers, open canals and streams. MHPP requires little or no reservoir and is generally built on run way of water streams with little arrangement to direct water towards the turbine. Hence it has a very little impact on local ecosystem. In MHPP hydro governors and reservoirs are avoided based on power generation capacity to reduce the overall size and cost. Hence MHPP requires good control system to keep the constant generator speed for fluctuating flow of water in the stream. One of the means to maintain constant generator speed is by modifying the transmission system.

The transmission system used in MHPP is generally speed multiplier, since the turbine speed is less than generator speed. Various types of transmission system in MHPP are belt drive, chain drive, spur/helical drive and planetary gear drive. A planetary chain transmission is developed for MHPP to increase the efficiency. Since the system is complex it is developed for lower transmission ratio [3]. The planetary chain drive is tested for Turgo turbine for a speed ratio of 1:4.8. Such system gives an efficiency of about 67% for speed range of 150-500 rpm. This type of drive is used for lower transmission ratio only [4]. Spur and helical gear drive are used for lower speed ratio. For higher speed ratio multiple stages are required which will increase the overall dimension and cost. In order to have good efficiency, higher speed ratio and compact size, planetary gearbox is used for MHPP. In planetary gear drive the turbine shaft is connected to carriage and ring gear is fixed. The output is drawn from sun gear [5]. The planetary gear box works as speed multiplier but fails to control generator speed. Hence a conceptual design of planetary gearbox system is developed by fixing auxiliary unit to ring gear. The feasibility study of the conceptual design concludes that planetary gearbox system can be used maintaining constant speed. The present paper focuses on multi body dynamic simulation, FEA, and testing of planetary gearbox system for governing constant generator speed for variable turbine speed.

2. Design requirements

The conceptual design includes planetary gearbox which is placed between turbine and generator for speed amplification. The turbine shaft is connected to the carriage of the gearbox and the generator shaft is connected to the sun gear. The auxiliary gearbox unit is attached to ring gear of planetary gearbox as shown in Fig. 1.

The specifications arrived from field survey for designing planetary gearbox system for MHPP is shown in table 1. Based on the specifications from the survey, the planetary gearbox and auxiliary unit are analytical designed [6].

Table 1. Specification for designing planetary gearbox system for MHPP

Sl. No	Particulars	Specifications
1	Turbine speed variation	10%
2	Generator specification	1500rpm, 50Hz
3	MHPP power generation capacity	5kW

3. Simulations of Planetary Gearbox System

The designed planetary gearbox system for maintaining constant generator speed is shown in Fig. 2. The developed model is simulated for Multi Body Dynamic simulation and flexible body simulation. The Multi Body Dynamic simulation is performed for kinematic analysis to verify the analytical design and flexible body analysis (FEA) for determining the stresses developed in the gearbox.

3.1. Multi Body Dynamic Simulations

The developed CAD model is imported to Multi Body Dynamic Simulation environment. The required material properties and boundary conditions are assigned to the components of the imported model and simulated for three conditions based on the working condition of MHPP. When the flow of water (turbine speed) is equal to operation speed, the auxiliary unit will be in fixed condition. Here the gearbox system works as conventional speed amplifier. This condition is considered as operating condition. During simulation, the input shaft to the carriage is provided with turbine speed. The auxiliary unit is fixed by fixing the ring gear. From simulation result

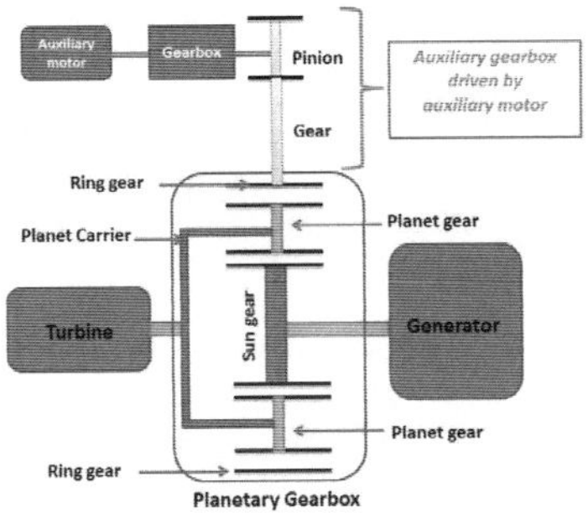

Figure 1. Conceptual design of planetary gearbox system [6]

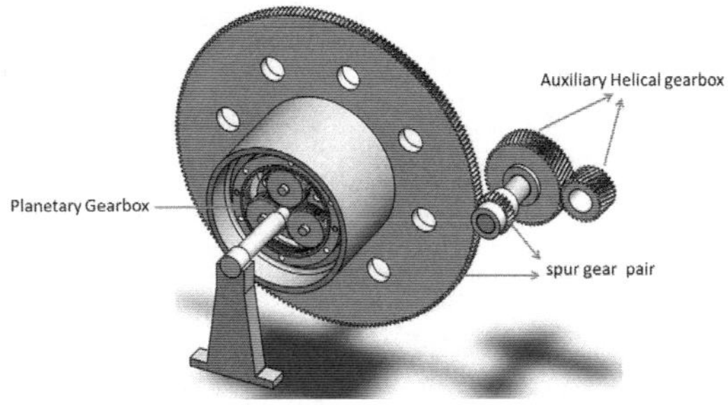

Figure 2. 3D model of the planetary gearbox system

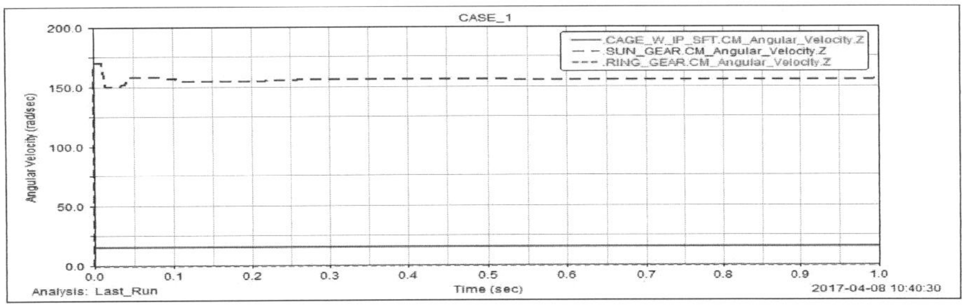

Figure 3. Kinematic simulation for operating condition

shown in Fig. 3. it is observed that the planetary gearbox system acts as convention gearbox by amplifying speed by 9.9 times. The generalized equation is arrived given by Eqn.1

$$outputspeed = inputspeed * 9.9 \tag{1}$$

Second condition, where the turbine speed is more than the operating speed, the auxiliary unit will start to rotate the ring gear and planet gear opposite to each other. This condition leads to speed reduction in the generator shaft (sun gear), is considered as braking condition. During this process the auxiliary unit acts as generator and stores power. While carrying out simulation, the input shaft to the carriage is provided with minimum turbine speed which is less 10% less than operational speed. The auxiliary unit is run at -31.83 rad/s based on analytical design calculations. Negative sign indicates rotation in opposite direction. Simulation output is plotted in Fig. 4. which shows that output speed was equal to generator speed. The generalized equation arrived from simulation result is given by Eqn.2

$$outputspeed = \langle turbinespeed * 9.9\rangle - \langle\frac{aulilaryunitspeed}{2.03}\rangle \tag{2}$$

In third condition, turbine speed is considered less than the operation speed. The stored energy during braking condition is utilized by the auxiliary motor to run auxiliary unit such that ring gear and the planet gear run in the same direction. When the planetary gear and ring gear both run in same direction, extra speed is added to the output (generator shaft) to maintain constant speed. Hence this condition is considered as speed amplification condition.

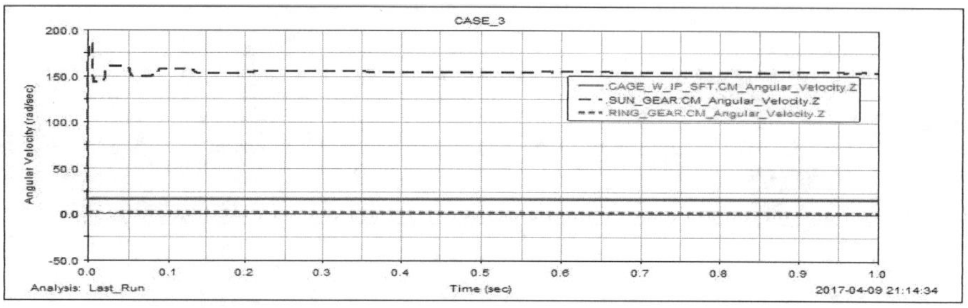

Figure 4. Kinematic simulation for braking condition

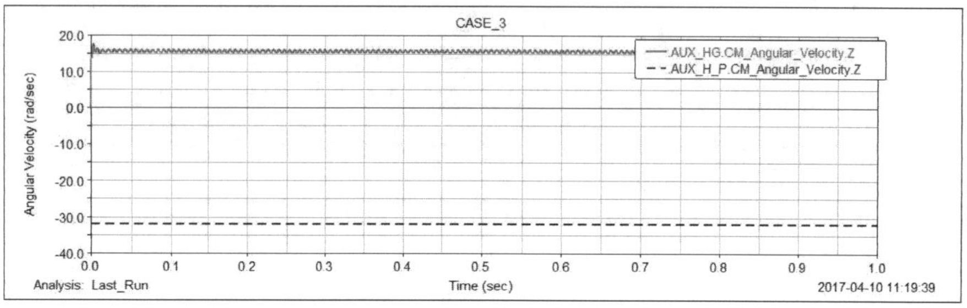

Figure 5. Kinematic simulation for speed amplification condition

During simulation the input shaft to the carriage is provided with maximum of turbine speed which is 10% more than operating speed. The auxiliary unit is run at 31.83 rad/s based on analytical design calculations. Simulation output is plotted in Fig. 5. which shows that output speed was equal to generator speed. The generalized equation is arrived from simulation result is given by Eqn. 3

$$outputspeed = \langle turbinespeed * 9.9 \rangle + \langle \frac{aulilaryunitspeed}{2.03} \rangle \tag{3}$$

3.2. Finite Element Analysis

The results of multi body dynamic simulations are in close agreement with the designed planetary gearbox system. During working of the planetary gearbox, both ring gear and planet gears are in motion, may have higher stresses for the designed load. Hence Finite element Analysis is carried out to analyse the induced stresses. 2D static analysis is carried out instead of 3D Dynamic analysis due to high computation time and memory. Mesh convergence study is carried out to arrive at proper element size. Structured mesh is developed in HYPERMESH and imported in FEA simulation software for simulation and post processing. 4-noded linear quad element with 3 degrees of freedom is used to develop FE model. Table 1, shows the material property assigned during simulation.

The shaft of each gear is modelled as a rigid node to reduce the computation time. The rigid nodes are created at the centre of the each gear and kinematically coupled to provide the boundary condition for the respective gears. Gear contact property is defined between the meshing gears with coefficient of friction as 0.3. Finally, the required boundary condition and loads are applied on the gears for simulating all three conditions.

During FE simulation for operating condition 1280 N-mm and 12720 N-mm are applied

IOP Conf. Series: Materials Science and Engineering **624** (2019) 012008 doi:10.1088/1757-899X/624/1/012008

Table 2. Specification for designing planetary gearbox system for MHPP

Sl. No	Material Properties	Particulars
1	Material Density	$7.85*10^{-}9$ tonne/mm^3
2	Elastic Modulus	$210*10^3$ MPa
3	Poisson's ratio	0.3

on sun gear and carriage respectively. Maximum stress of 307 MPa is observed on all planet gears. Fig. 6 and 7 shows the stress distribution in planetary gear drive for operating condition respectively. 1420 N-mm, -12720 N-mm and 1414 N-mm are applied on sun gear, ring gear and carrier to simulate braking condition. Maximum stress of 475 MPa is observed on planetary gear. Similarly 1160 N-mm, 10410 N-mm and 11570 N-mm are applied on sun gear, ring gear and carrier to simulate speed amplification condition. During simulation maximum stress of 430 MPa is observed on the planetary gear. From the simulation results it is be observed that the maximum stress is induced in braking condition, which is less than the allowable stress of 510 MPa. These simulation results show that the developed system is safe for fabricating and testing.

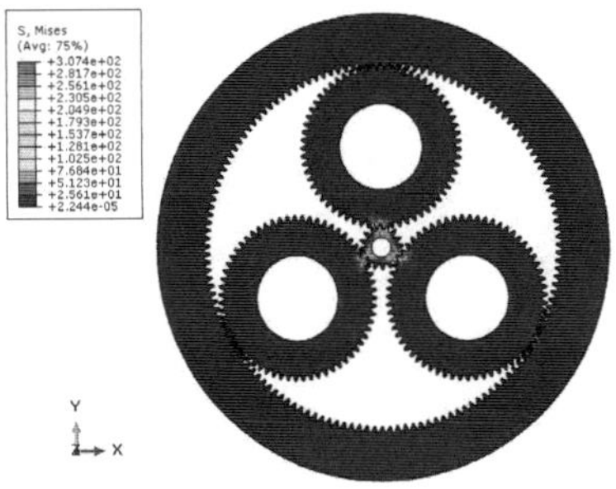

Figure 6. FEA of planetary gearbox

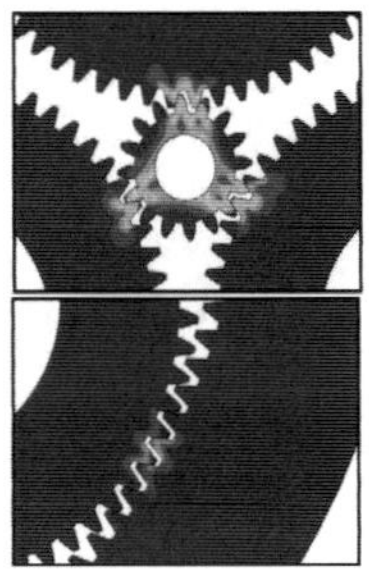

Figure 7. Stress distribution in gears

4. Experimentation

The results from multi body dynamic simulation and FE simulation indicates that the planetary gearbox can be used for maintaining constant generator speed in MHPP. The planetary gearbox system is fabricated and tested for no load condition. The Fig. 8. shows the experimental setup of planetary gearbox system.

The experimental setup is developed to replicate the working conditions. It consists of two variable frequency drives (VFD), planetary gearbox, auxiliary gearbox, dynamometer, electromagnetic brakes and connecting shafts. A 37kW capacity VFD is connected to carrier, which acts as hydro turbine. A 3.7kW capacity VFD is connected to the auxiliary unit to compensate speed during speed amplification and braking condition. Testing are carried out for all the three operating condition with no load at the output shaft. Experiment results found are satisfying the requirement of constant speed. Table 3. shows the testing result for all the three operating conditions. It is observed that the experimental results and analytical results are in close agreement with insignificant error. From the experimental result the compensating speed for auxiliary unit is arrived for fluctuating turbine speed. This compensating speed will be used for automatic control, which is the future scope of work.

IOP Conf. Series: Materials Science and Engineering **624** (2019) 012008 doi:10.1088/1757-899X/624/1/012008

Figure 8. Experimental setup of planetary gearbox system

Table 3. Testing results for no load condition

Condition	Input Speed (rpm)		Generator shaft Speed (rpm)	
	T. Speed	Aux. motor	Exp.	Analytical
Speed Ampl.	-10% of operating speed	-306.9	1505.9	1507.5
Working Condt.	Operating speed	–	1502.4	1500.5
Braking Condt.	+10% of operating speed	305.6	1504.3	1501.5

5. Conclusion

The planetary gearbox system for maintaining constant generator speed is designed using analytical method for 1:9.9 speed ratio and validated using multi body dynamic simulation. Finite element analysis is carried out for all the three operating condition to analyse induced stresses. From simulation it is observed that breaking condition is the worst condition which has maximum stress of 475 MPa in all planet gears. The maximum stresses developed are less than the allowable stress. Later the system is fabricated and tested for no load condition. The testing results were in close agreement with analytical and multi body dynamic simulation result having insignificant error. Hence the design is practically feasible for maintaining constant speed in MHPP.

6. References

[1] Ahmed M.A. Haidar, Mohd F.M. Senan, Abdulhakim Noman and Taha Radman 2012 Utilization of pico hydro generation in domestic and commercial loads *J. Renewable and Sustainable Energy Reviews* **16** 518-524

[2] M.V.S Ranganadham 2018 Energy statistics *Central Statistics office Ministry of statics and programme implementation Govt. of India* **25**

[3] O. Climescu, I. Visa C. Jaliu and R. Saulescu 2010 Comparative analysis of planetary transmissions to be used RES *Bulletin of the Transilvania University of Brasov* **52** 55-62

[4] C. Jaliu D. Diaconescu and R.Saulescu 2008 Speed Multipliers for Renewable Energy Systems-Hydro and Wind *J. Renewable Energy and Power Quality* **06** 650-653

[5] C. Jaliu D.V. Diaconescu M. Neagoe and R. Saulescu 2009 Dynamic Features of Speed Increasers from Mechatronic Wind and Hydro Systems. Part I: Structure Kinematics *Euro. Conf. Mechanical sciences* 351-359

[6] Bhargav, M. A. Parameshwaran Sivaraj S. and Nithin Venkataram 2018 Conceptual design of planetary gearbox system for constant generator speed in hydro power plant *MATEC Web of Conf.* 1-11

IOP Conf. Series: Materials Science and Engineering **624** (2019) 012009 doi:10.1088/1757-899X/624/1/012009

Extraction of vibration behavior in conventional and electric drive two-wheeler using order analysis

Lalsaheb Mulla[1], Piyush Khuba[1], Apeksha Dhere[1,] and Sakthivel Palanivelu[2]
[1]Department of Automotive Engineering, VIT University, Tamil Nadu-632014 India.
[2]Associate Professor, Department of Automotive Engineering, VIT University, Tamil Nadu-632014 India.

E-mail: psakthivel@vit.ac.in

Abstract: Electrification of automobiles results in new and challenging characteristics of vehicle NVH (Noise, Vibration, and Harshness) and ride comfort. Optimized architecture of electric vehicles has been evolved for appropriate placement of electric power drives. This paper presents the experiments performed for order analysis technique to predict the vibrational behavior of conventional and electric scooters which was not used in 2-wheelers earlier and used for 4-wheeler noise predictions and for vibration analysis of rotating machinery. Vibrational analysis of source alone in conventional and electric two-wheelers carried out by order analysis and the result is presented through Campbell Diagrams to observe the critical orders along with the variation of the amplitude of vibration as a function of frequency. Target locations' Vibration Dose Values (VDV) is calculated to assess rider comfort and is compared with the ISO 2631 standard. The higher amplitude of vibration has been observed at 23rd order for electric scooter while for conventional scooter it is noted at 8th and 16th order with reference to the wheel speed. Compared to the conventional scooter, electric scooter exhibits very low overall vibration level and range is away from human perception range resulting in good comfort.

1. Introduction

In recent years, the talk is on 'electric drive' mobility against its fossil fuel counterpart to meet the ecological and economic need. Strong legislation on automotive pollution control by the Environmental Protection Agency (EPA) commanded automotive industries to adopt new and more efficient technologies for vehicles called hybrid electric vehicles and electric vehicles. On one hand, weight reduction of the vehicle aid fuel economy of the vehicle and on the other hand, the vehicle tends to vibrate more and produce unwanted noise and vibration due to the vibration of various components. Higher levels of vibrations are observed in lightly damped structural components. Ride comfort in two-wheelers is characterized by the transmissibility of structural parts that are placed between the source and the rider contact points. Tires and modified suspension systems are available to reduce the vibrations excited by road undulation during tire-road interaction, while the inertial unbalance in power sources like IC engines or electric motors which are mounted on the frame and wheels as the case may be contributing to the vibrations and can easily be transmitted to the prominent locations. This paper deals with the prediction of vibration levels for a conventional and an electric scooter due to their power drives. Though the electric motor led vehicles to perform quieter and smoother than IC Engine vehicles the new NVH issues and challenges due to electrification needs to go for the new methodologies to understand the vibro-acoustic phenomena. Yu et al. [1] performed NVH analysis on the electric drivetrain and stated that, in electric vehicles, the torque fluctuation of motor and the meshing force variations causes the vibration in drive-line. Guo et al. [2] studied the NVH characteristics of the auxiliary power unit and the accessories like power steering pump, electric vacuum pump and AC compressor in a vehicle. The low-frequency torsional resonance of the power unit during start-up is a newly arrived problem which is not generally observed in a conventional engine. The order tracking method is used by Wu et al., [3] to study the interior noise generation in the cab and reported that the reason is the second order excitation of the system with respect to the engine speed. Panza et al., [4] explained different experimental methods to identify source and target responses of the system. There are many order tracking methods such as Fourier Transform based order tracking method, angle domain sampling- based order tracking also known as computed order tracking or digital resampling-based order tracking method, Kalman filter based order tracking method, time variant discrete Fourier transform order tracking method (TVDFT). Blough et al., [5] compared these order tracking methods and stated the advantages and disadvantages of each of these methods and concluded that TVDFT is the most suitable method for order tracking. Order tracking method has been used to identify the rotational behavior and the mode of vibration of the component. Guo et al., [6] worked on the vibration reduction of the engine by using Active Engine Mount (AEM).

It is well known that the vibration is a serious issue pertaining to human discomfort in automobiles, specifically in two-wheelers. There are many literatures available for four-wheelers compared to that of two-wheelers for ride studies, especially of electric power drives. The term, Whole Body Vibration (WBV) is referred to the transmission of vibration to the whole body, through the contact points of the rider. Manzato et al., [7] presented a method for two-wheeler comfort evaluation using simulation in LMS Virtual Lab Motion along with its experimental validation. Referring to vehicle characteristics, it has been stated that two-wheelers are generally harsher vehicles than other automobiles. Chen et al.,[8] conducted an extensive test run measurements for evaluating Root Mean Square (RMS) acceleration, 8-h estimated Vibration Dose Value (VDV) and 8-h estimated daily dose of static compression dose (SCD) for a bike, a scooter and a sedan car on a combination of rural, provincial and urban routes. Further, these estimations were compared with the recommended values by ISO 2631-1 and ISO 2631-5 standards [9]. It has been concluded that the WBV exposure levels of common motorcycle riders are significantly higher than those for sedan cars, even on a normal paved road [10]. This led a motivation for the research project presented in this paper.

2. Experimental Setup

Figure 1 describes the schematic and the actual test setup developed in the laboratory. Three tri-axial accelerometers and a speed sensor are connected to a data acquisition system (Siemens LMS SCADAS) using low impedance cables for data transferring. A laser-based non-contact speed sensor is used to measure the speed of the wheel. Accelerometers are installed at 3 prominent locations one near the source (wheel hub) and the other two at targets (rider's contact points: seat and handlebar) for vibration measurement. LMS Test. Lab. Software is used to define necessary channels and acquisition parameters for measurement and subsequent signal post-processing for calculation of ride metrics.

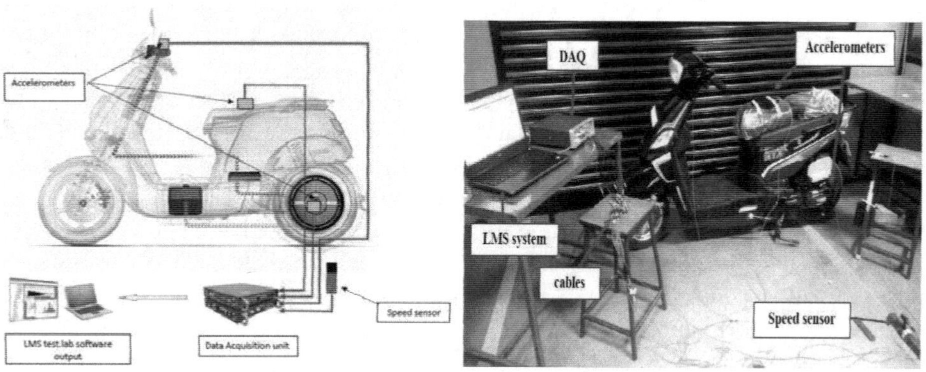

Figure 1. Schematic and the Actual test setup.

Two test scooters are chosen for testing. One for conventional IC engine powered (Hero Pleasure) and the other for electric motor powered (Hero Electric Nyx) respectively. When the test vehicles were put on their stand and powered, the engine runs at idle speed around 1100 rpm, whereas in the electric scooter with a wheel hub motor, the motor rotated only when the throttle was given. In order to match the equivalent scenario for order analysis, the rotating rear wheels of both test scooters are chosen as reference. Wheel speed from 0 rpm to 650 rpm is fixed to compare vibration behavior of both vehicles. The experiment is performed with the rider sitting on the seat to measure the vibration level at the seat and right handlebar location for run up and run down condition. Importantly, these experiments are conducted in the laboratory to account only the power drives as the source and hence the road excitation is eliminated. Though, this adheres to the objective of the current research work in real time riding of a scooter the road excitation also contributes.

3. Measurements and Observations

The time histories of all responses were obtained with reference to ISO right hand co-ordinate axis system. X, Y, Z axes refer to the longitudinal, lateral and vertical direction of the test vehicles. Figure 2 shows the variation of wheel speed and acceleration responses measured in all three directions at the hub, seat, and handlebar as a function of time for conventional and electric scooter respectively.

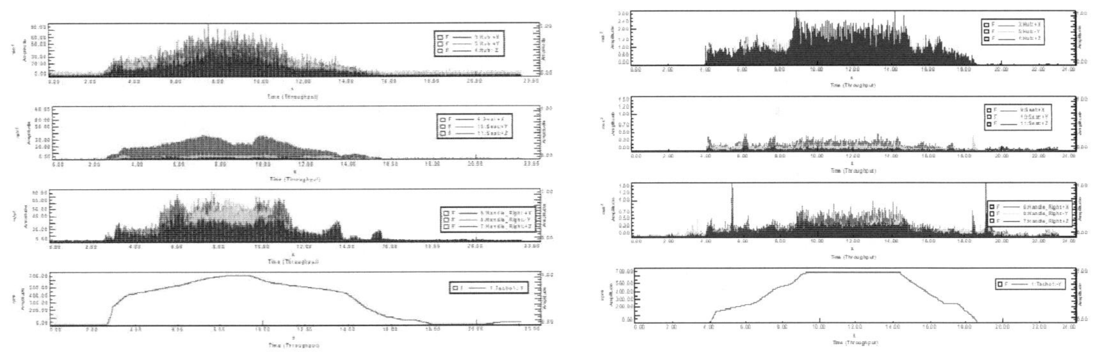

(a) (b)

Figure 2. Observation of wheel speed and acceleration response for (a)Conventional scooter (b) Electric Scooter.

The increase in amplitude followed by a decrease in amplitude represents the responses during the run-up and run-down condition and the same can be seen from the variation of speed increase and decrease between 4 – 18 seconds of measurements. Vibration signatures fora conventional scooter are observed to be of higher amplitudes (up to 80 m/sec^2 in the scale) compared to that of an electric scooter (up to 3 m/sec^2). In both cases, the attenuation of vibration transferred from hub to seat and handlebar location is observed clearly. However, among all, the dominating vibrations are seen in the X direction. Hence considering the vibrations along with X directions at various target locations the comparative study has been done further. Figure 3 compares Campbell diagram for variation of auto power amplitudes along longitudinal direction for the locations wheel hub (source) and seat and right handlebar (targets) respectively between conventional and electric scooters. Wheel hub in the conventional scooter is seen to be vibrating with higher amplitude in 8th order and 16th order of the wheel with the frequency close to

86.72 Hz and 175 Hz respectively. Whereas in the case of the electric motor, the wheel hub is observed to be vibrating with higher amplitudes in 23rd and 46th order of motor speed. Seat contact location the rider feels major vibrations generated by the source. It is evident from the corresponding plots that at the same critical orders as that of the hub, seat location does exhibit the higher vibration levels for both conventional and electric scooter. Next, to the seat, the handlebar is another critical location which influences the rider's comfort. Handlebar in the conventional scooter is seen to be vibrating with higher amplitude at to 550 rpm of the wheel. The observation is seen in the frequency range of 405-450 Hz i.e. 40th and 48th order of wheel at the speed of 550 rpm and the acceleration amplitude of 6 m/s^2. In case

IOP Conf. Series: Materials Science and Engineering **624** (2019) 012009 doi:10.1088/1757-899X/624/1/012009

of an electric motor, the handlebar is observed to have higher amplitudes at 230 rpm with the 46 order frequency of 172 Hz and amplitude of 0.14m/s^2. Also, the considerable peaks are observed in 3rd and 23rd order of the reference hub-motor speed. Electric scooters vibrate with very low amplitudes but the frequency is too high which may cause 'hand-arm vibration syndrome' when the rider is exposed for a very long time.

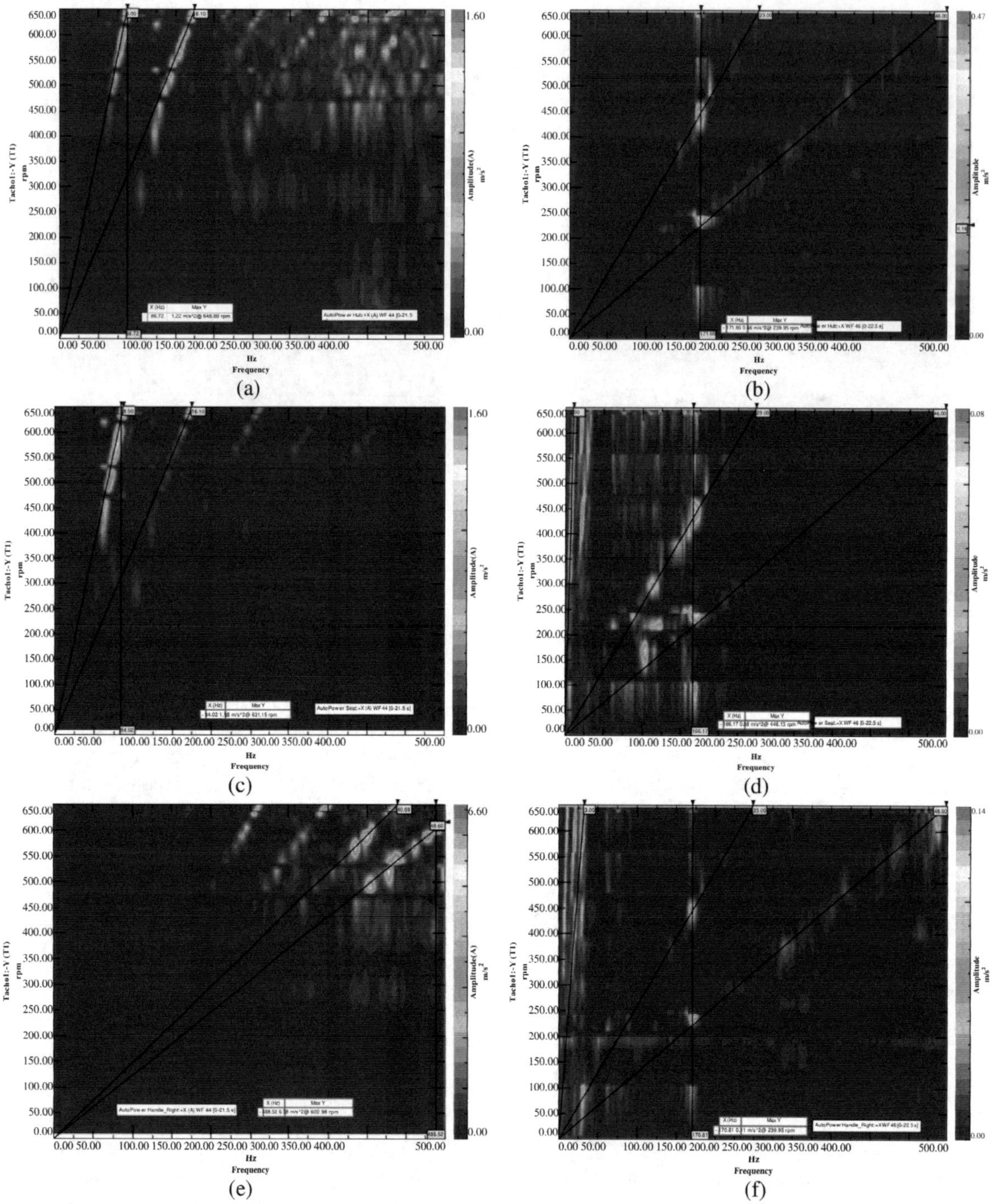

Figure3. Campbell diagram illustrating orders of frequency at different locations. Campbell diagram at hub for, (a) conventional scooter (b) electric scooter. Campbell diagram at seat for, (c) conventional scooter (d) electric scooter. Campbell diagram at handlebar for, (e) conventional scooter (f) electric scooter.

IOP Conf. Series: Materials Science and Engineering **624** (2019) 012009 doi:10.1088/1757-899X/624/1/012009

4. Results and discussions

4.1 Transmissibility

Transmissibility of vibrations from the source is identified by observing the displacement spectrums near the source and at the target. The graphs in figure 4 represent the transmissibility function between target and source on the right vertical axis; on the left vertical axis the response spectrum of the source (hub) and target (seat and handlebar) as a function of frequency. In both conventional and electric scooter, the displacement transmissibility appears to be very high below 100Hz causing amplification in energy transfer to seat with corresponding acceleration amplitudes and beyond 100Hz the vibration isolation is clearly observed. It is interesting to note, on one hand, the vibration response of the handlebar of a conventional scooter is amplified beyond 100 Hz and at the frequency of 178Hz is shows the maximum value. And the same has also been observed in the respective Campbell diagram in 40th and 48th order with respect to wheel speed. On the other hand, handlebar in an electric vehicle is showing amplified vibrations below 100Hz and beyond that good vibrational isolation is observed.

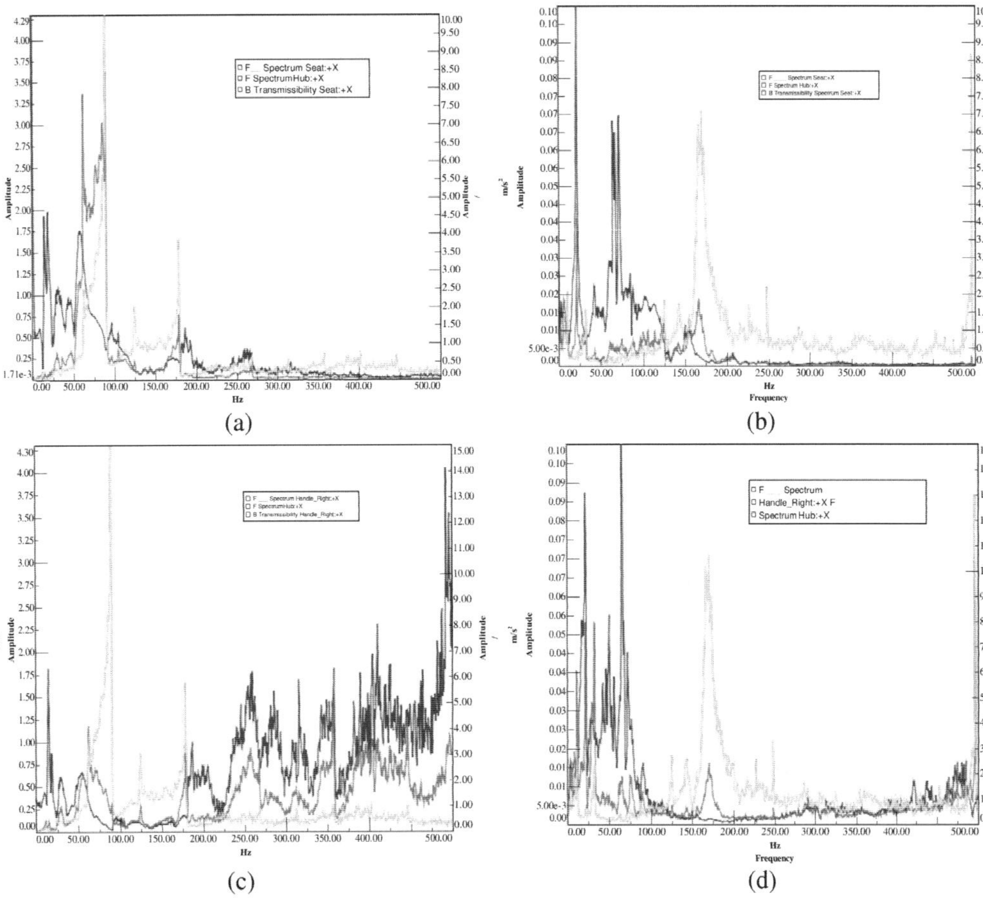

Figure 4. Acceleration responses on the left vertical axis and transmissibility ratio of target locations on the right vertical axis. Transmissibility at the seat for (a) conventional scooter, (b) electric scooter. Transmissibility at handlebar for (c) conventional scooter, (d) electric scooter.

4.2 Vibration Dose Value (VDV)

Vibration Dose value is one of the important metrics to assess the perception of the overall vibration level of the rider. VDV always accumulates for the vibration exposure and does not decay during periods of the low value of vibration magnitude. As recommended by the ISO 2631 standard, daily vibration dose value (VDV) in the region of 15 ($m/s^{1.75}$) usually causes severe discomfort and health-related problems. The assessment of vibration in a vehicle depends upon the method of measurement and evaluation of human responses. ISO 2631 and ANSI S3.18 are mainly used to evaluate the whole-body vibrations of a driver in a vehicle. Table 1 compares the mean VDV for a conventional and electric scooter for the target locations in all three directions. In the conventional case, the mean VDV obtained for seat contact in all the three directions is more than daily vibration dose value and at handlebar only in the longitudinal direction, the value is severe. Whereas in the case of the electric scooter it is observed that the mean VDV value at both target locations is below 1 $m/s^{1.75}$, which are well within the ISO2631 standards for good ride comfort.

Table 1. Comparison of Vibration Dose Value (VDV) (m/s^1.75) at target locations.

Test Vehicle	Conventional Scooter			Electric Scooter		
Location	X direction	Y direction	Z direction	X direction	Y direction	Z direction
Handle bar	26.312	5.656	2.353	0.200	0.116	0.062
Seat	30.561	28.753	18.996	0.451	0.261	0.225

5. Conclusion

The order analysis method is used to extract the critical orders and frequency to understand the vibration behavior in scooters having different excitation sources like engine and hub motor. 8th and 16th orders are observed to be critical for engine operated scooter due to engine excitations and 23rd and 46th orders are observed in motorized wheel scooter having wheel speed as reference speed. It is concluded that the vibration level perceived in electric scooter is far lower than that of its conventional counterpart and higher resonating frequencies are not perceived by driver which improves drivers comfort whereas in conventional scooter the lower resonating frequencies causes high discomfort for rider due to frequencies within human perception range also displacement transmissibility and VDV values are estimated to support the observations. However, this observation is limited by considering power drive sources alone. Further investigation is needed to account another important source due to tire-road interaction. The coupled effect of both power drive and road interaction will bring in the actual rider comfort/discomfort level.

6. References

[1] Yu, P., Zhang, T. and Liu, P.H., 2013. NVH characteristic prediction of an electric vehicle reducer. In *Advanced Materials Research* (Vol. 608, pp. 1656-1659). Trans Tech Publications.
[2] Guo, R., Cao, C., Mi, Y. and Huang, Y., 2016. Experimental investigation of the noise, vibration and harshness performances of a range-extended electric vehicle. *Proceedings of the Institution of Mechanical Engineers, Part D: Journal of Automobile Engineering*, *230*(5), pp.650-663.
[3] Wu, X.B., 2013. Computed order tracking and its application to NVH. In *Applied Mechanics and Materials* (Vol. 321, pp. 692-696). Trans Tech Publications.
[4] Panza, M.A., 2015. A review of experimental techniques for NVH analysis on a commercial vehicle. *Energy Procedia*, *82*, pp.1017-1023.
[5] Blough, J.R. and Gwaltney, G.D., 1999. Summary and characteristics of rotating machinery digital signal processing methods. SAE Technical Paper No. 1999-01-2818.
[6] Guo, R., Gao, J. and Wei, X.K., 2017. *A Method of Acceleration Order Extraction for Active Engine Mount* (No. 2017-01-1059). SAE Technical Paper.
[7] Manzato S, Kirchner M, Erdélyi H, Baglini G, and Pieve M., 2013 Numerical and operational identification and assessment of motorcycle dynamics and comfort. Research Gate Publication:235718877
[8] Chen, H.C., Chen, W.C., Liu, Y.P., Chen, C.Y. and Pan, Y.T., 2009. Whole-body vibration exposure experienced by motorcycle riders–An evaluation according to ISO 2631-1 and ISO 2631-5 standards. *International Journal of Industrial Ergonomics*, *39*(5), pp.708-718.
[9] ISO, I., 1997. 2631-1: Mechanical vibration and shock-evaluation of human exposure to whole-body vibration-Part 1: General requirements. *Geneva, Switzerland: ISO*.
[10] Toward, M.G. and Griffin, M.J., 2011. The transmission of vertical vibration through seats: Influence of the characteristics of the human body. *Journal of Sound and Vibration*, *330*(26), pp.6526-6543.

To Optimise the Performance of Electric Powertrain by Tuning the CVT

S Gawade[1], J Bari[1], N Anagolkar[1] and D Ashok[2]

[1] Department of Automotive Engineering, VIT University, Tamilnadu, India.
[2] Associate Professor, Department of Mechatronics Engineering, VIT University, Tamilnadu, India.

E-mail : gawadeshubham123@gmail.com

Abstract. There is desperate need to find an alternate power source to run the vehicles. This reasons leads to work on electric vehicles. In India, it has been noticed major part of urban air pollution is caused by auto rickshaw emissions as they are not maintained properly. So switching from conventional rickshaw to electric will significantly reduce this problem in urban areas. Continuously variable transmission (CVT) transfers power uninterrupted due to its infinite gear ratios. Moreover, it allows the motor to run at its optimum speed which ultimately reduces the current consumption. CVT available in current market is design to work efficiently with internal combustion engines. This CVT can be made compatible with pure electric power train by CVT tuning. This paper aims to study and analyses the effect on performance of CVT by varying the mass of flyweights. The driver and driven pulley RPM are measured using DAQ system. This data is then use to obtain gear ratio. The iterations are performed using two different mass of flyweights. The gear ratio verses pulley rpm graphs are compared. The result infers that as the flyweight mass is increase the engagement rpm decreases and up-shift rate increases. This enhances performance of CVT.

1. Introduction

One of the major reasons for urban pollution is vehicle emissions. Simultaneously the fossil fuel resources are also depleting at faster rate. This leads to the rise in fuel prices. In addition, Indian government is imposing stringent norms on engine emissions. Hence there is desperate need to find an alternate power source to run the vehicles. In urban areas of India, it has been noticed that major part of air pollution is caused by auto rickshaw emissions as they are a common man's daily commute medium and are not properly maintained. So switching from conventional auto rickshaw to electric will significantly reduce the problem. This gave motivation to optimize the performance of an electrical three wheeler power train.

In this paper a simplified CVT transmission is selected over single reduction transmission for electric vehicle. A simplified CVT transmission improves vehicle dynamic performance, also significantly reduces daily usage cost. The installation cost difference for a single reduction transmission and CVT transmission is not large as a less power motor is used. Motor can work more efficiently and save energy in daily use with the help of larger gear ratio range in CVT. This will benefit the end user.

2. Vehicle specifications

The vehicle selected is BAJAJ RE. The conventional powertrain of Bajaj RE is converted to an electric power train.

Table 1. Conventional auto rickshaw data.

Sr no	Description	Values
1	Mass of empty vehicle	262.32 kg
2	Mass of vehicle with driver	332.32 kg
3	Mass of vehicle with driver and two passengers	400 kg
4	Mass of vehicle with driver and three passengers	542.32 kg
5	Radius of front tyre	0.2 m
6	Radius of rear tyre	0.2 m
7	Frontal area	1.5414 m^2
8	Coefficient of drag	0.385
9	Coefficient of rolling resistance	0.015
10	Transmission efficiency	0.85
11	Maximum gradient angle	5°
12	Density of air	1.2

Figure 1. BAJAJ RE auto rickshaw.

3. Power calculation and motor selection

The crucial component of electric rickshaw is its motor. Unless the proper motor is selected, vehicle will not run at all load conditions and there are chances of failure during extreme loads. So, for selection of motor minimum required power has to be calculated, which is actually the power required to overcome all the dynamic forces acting on vehicle. The calculation of these dynamic forces requires the geometric data and coefficient of drag of vehicle which has to be converted to electric vehicle, which are obtained from the dissertation work of Mogili Babu. [1]. Power requirement at wheel is determined by driving resistance. The calculations are done for one driver plus three passengers (W = 542.32 kg). It consist of 3 main components

Rolling resistance (F_r):

$$Fr = \mu \times W \times g = 0.015 \times 542.32 \times 9.81 = 79.80N \quad (1)$$

Where, μ= coefficient of friction ;W= mass of vehicle (kg);
g= Acceleration due to gravity = 9.81 (m/sec2).

Air resistance (F_d):

$$Fd = 0.5 \times \rho \times Cd \times A \times V^2 = 0.5 \times 1.2 \times 0.38 \times 1.5414 \times 1.92 = 1.1172 \text{ N} \quad (2)$$

Where - density of air; Cd - coefficient of drag; A - frontal Area; V- velocity of vehicle in m/s.

Gradient resistance (F_g):

$$Fg = W \times g \times \sin\alpha = 542.32 \times 9.81 \times \sin 5° = 463.68 \text{ N}. \quad (3)$$

Where, W= mass of vehicle (kg); g= Acceleration due to gravity = 9.81 (m/sec2).

Tractive effort: $$Ft = Fr + Fd + Fg = 79.80 + 1.1172 + 463.68 = 544.59 \text{ N}. \quad (4)$$

Torque at wheels: $$T = Ft \times radius\ of\ wheel = 544.59 \times 0.2 = 108.91\ Nm. \quad (5)$$

Power: $$P = Ft \times V = 544.59 \times 1.9 = 1034 \text{ W} = 1.034 \text{ kW}. \quad (6)$$

The motor available selected has following specification.

Table 2. Motor specification

Make	Compage
Type	BLDC motor
Output Power	1 KW
Required input voltage	48 V_{rms}
Rated rpm	3000
Rated torque	3.2 Nm at 22 Amps
Peak torque	8 Nm at 30 Amps

4. CVT tuning

In order to start vehicle moving, the motor must be engaged at a speed that has sufficient power to start accelerating the vehicle. To obtain low engagement speed the flyweights have to overcome the pretension in the driving clutch and start moving the sheaves together, until enough force is exerted on the belt to start the vehicle moving.[2] As the force increases, the belt pressure will go from 0 to 100%.All shift force is generated by the flyweight system. Flyweight works on the principle of centrifugal force.

$$F = M \times R \times \omega^2 \quad (7)$$

where, M= mass of flyweights in kg

Hence centrifugal force increases proportionally with flyweight mass. The overall effect of increasing the flyweights even slightly will increase centrifugal force. That is the CVT will engage at lower speed than earlier due to increased centrifugal force. This makes it possible to tune the CVT at lower engagement speed

By lowering the engagement speed it is made sure that the shift speed increases during lower ratio acceleration phase.

5. Experimental setup

The complete experimental setup consists of BLDC motor, CVT, motor controller, 48V lead acid battery, LMS SCADA DAQ system, non-contact type laser tachometer and clamp meter. By changing the mass of flyweights multiple iterations of experiments are carried out for tuning CVT for electric vehicle. In this setup, 1kW BLDC motor is connected to CVT by using curved jaw coupling. Both the shaft of CVT and motor are aligned with precision. Motor speed is controlled through throttle which actually varies the input current to motor, keeping voltage constant. RPM of input and output pulley of CVT is measured with the help of non-contact LASER tachometer. Clamp meter with help of shunt resistance is used to measure voltage and current drawn by motor.

5.1. Test Procedure

- Initially flyweight of mass 120 gm is used.
- 25% throttle is given to motor.
- RPM of output and input pulley are measured using tachometer and LMS SCADA DAQ system.
- Voltage and current drawn by motor are noted from clamp meter.
- Similar process is repeated by in increasing the mass of flyweights and varying percentage of throttle input.
- During the process, speed ratios at various throttle position is calculated by measuring the angular speed of primary and secondary pulley.

IOP Conf. Series: Materials Science and Engineering **624** (2019) 012010 doi:10.1088/1757-899X/624/1/012010

Figure 2. Experimental Setup along with DAQ system

6. Results

The mass is varied in two iterations and the results are plotted by measuring the angular velocities of primary and secondary pulley (rpm).

6.1. Iteration 1:

Table 3. Total mass of flyweights = 120 grams

Throttle	Primary pulley rpm	Secondary pulley rpm	Voltage	Current
25%	372	52	48	0.7
50%	1367	453	48	4.9
100%	2460	970	48	6.76

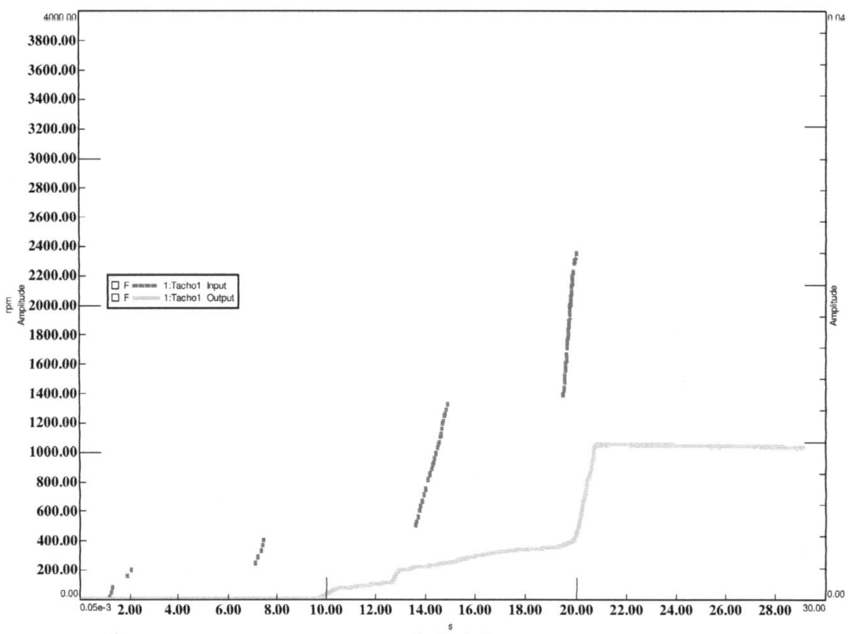

Figure 3. Primary pulley speed (rpm) verses time (seconds) [red-dotted line] and secondary pulley speed (rpm) verses time (seconds)[green-continuous line].

6.2. *Iteration 2:*

Table 4. Total mass of flyweights = 260 grams

Throttle	Primary pulley rpm	Secondary pulley rpm	Voltage	Current
25%	1629	206	48	3.32
50%	2253	2309	48	5.9
100%	2757.85	3726.45	48	8.7

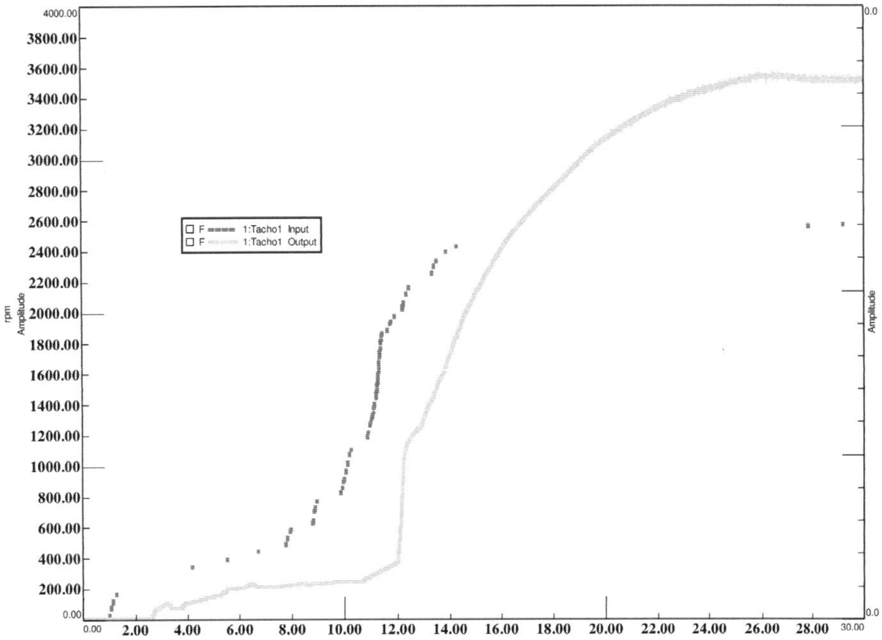

Figure 4. Primary pulley speed (rpm) verses time (seconds) [red-dotted line] and secondary pulley speed (rpm) verses time (seconds)[green-continuous line].

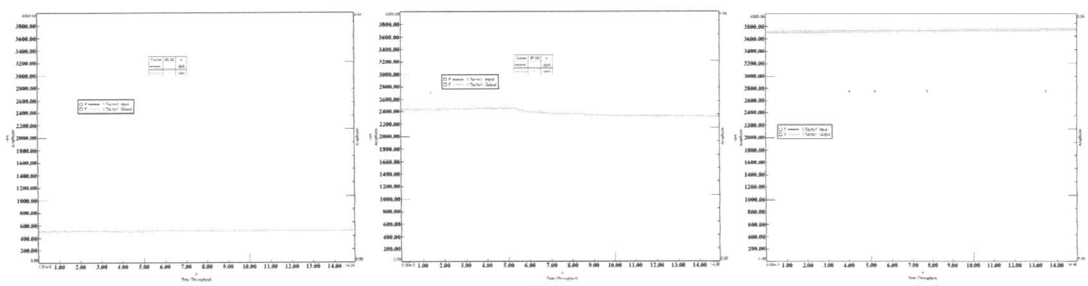

Figure 5. Primarypulley speed (rpm)verses time (seconds) [red-dotted line] and secondary pulley speed (rpm)verses time (seconds)[green-continuous line]. [a] Under drive, [b] speed ratio ≈ 1, [c] Over drive.

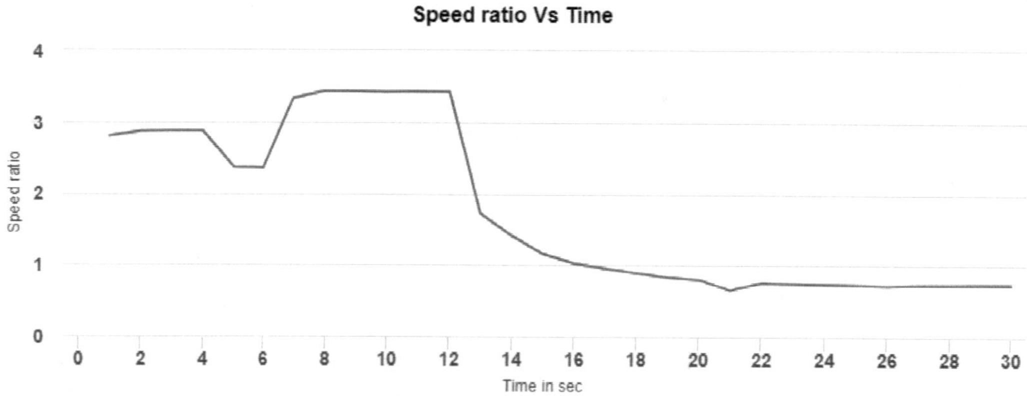

Figure 6. Graph of speed ratio verses time.

The highest speed ratio of 3.433 is obtained at 8seconds which remains constant till 12 seconds. This is the time at which shifting of pulleys starts. Hereafter speed ratio decreases and hence the speed of output pulley increases and eventually reaches a point where speed of both pulleys is equal. This occurs at time equal to 16 seconds and at this point speed ratio reaches to 1. From hear onwards speed ratio decreases and attains over drive.

7. Conclusions

Speed ratio is one of the important parameters considered for designing powertrain of any vehicle. Hence more attention is paid towards obtaining various speed ratio from CVT. Initially when the mass of flyweights was equal to 120 grams, different speed ratios were not obtained. CVT was operating at its highest speed ratio for every throttle position. But when the mass of flyweights was changed to 260 grams, all the speed ratios were obtained ranging from 0.733 to 3.433. It can be concluded that the CVT designed for internal combustion engine can be made compatible with electric motor of 1 kW in a simple and cheaper way. It has been noticed that urban city air pollution can be reduced by electrification of auto rickshaw. The electrification can be made cheaper using a CVT available in market and tune it according to the selected motor power. The results obtained from this study are useful for electrification of auto rickshaw.

References:

[1] B Mogili (1995) Lateral dynamics of bajaj rear engine auto-rickshaw.

[2] Aaen, Olav.(1989) Clutch tuning handbook. *Aaen Performance.*

[3] R Timothy.(2013)The continuously variable transmission: A simulated tuning approach. *Worcester Polytechnic Institute. Project Number: Mqp-Dcpfsae-***E12D13.**

[4] Parker, J Richard, Stuart H,Loewenthal and G Fischer.(1981) Design studies of continuously variable transmissions for electric vehicles. *SAE transactions* 1570-1580.

[5] Lechner, Giesbert, and H Naunheimer.(1999) Automotive transmissions: fundamentals, selection, design and application. *Springer Science & Business.*

[6] Ehsani, Mehrdad, Khwaja M. Rahman, and A Hamid.(1997): Toliyat. Propulsion System design of electric and hybrid vehicles. *IEEE Transactions on industrial electronics* **44.1**:19-27.

[7] Ruan, Jiageng, N Zhang, and P Walker.(2015) Comparing of single reduction and CVT based transmissions on battery electric vehicle. *Proceedings of the 14thIFToMMWorld Congress in Mechanism and Machine Science, Taipei, Taiwan.*

IOP Conf. Series: Materials Science and Engineering **624** (2019) 012011 doi:10.1088/1757-899X/624/1/012011

Degradation analysis of gear oil SAE 90 used in load haul dumper machine

Nagesh Dewangan[1], Anil Kumar Agrawal[1], Mohd A H Siddiqui[2], Somnath Chattopadhyaya[1]

[1] Indian Institute of Technology (Indian School of Mines), Dhanbad, India
[2] Coal India Limited, Western CoalField Limited, Nagpur, India

Email ID: dewangannagesh1993@gmail.com

Abstract: Lubricant degradation is the main cause for low performance of the machine such as heavy engineering machines, mining machinery etc. Lubricants are used for smooth running of gears and they also avoid friction and noise. Its performance is affected by high temperature and shearing, due to high temperature viscosity of oil changes and due to shearing its physical property changed or degraded. This paper analyses the failure behaviour of gear oil used in Load Haul Dumper (Model No: Eimco Elecon – 811MK). Gear oil degradation was measured by oil analysis in two sections: (i) fluid properties and (ii) contamination in the oil. Fluid properties were analysed by Viscometer, Rheometer whereas contamination analysis was done by Fourier Transform Infrared spectrometer (FTIR). All tests were performed carefully and their graphs were plotted. Results show that temperature and shearing were two of the several causes for degradation of oil. Viscosity analysis clearly presents that viscosity of oil exponentially decreases with temperature. Viscosity is also decreasing with shear rate at different temperatures. FTIR analysis confirmed that contamination is increasing with continuous use of oil. Temperature and oxidation are the main reasons for degradation. Different composition forming in used gear oil and those were confirmed by infrared spectroscopy absorption table.

1. Introduction

Lube Oil degradation is one of the major problems for the low performance of the gear. Degradation is taking place in the gear oil due to various causes such as contamination, metal wear particles, and oxidation [1]. Surface contacts occur between running parts, due to that surface roughness took place. Surface roughness generates friction, deformation, ploughing, adhesion, abrasion, delamination in it [2]. All these happen in lube oil due to long use of oil. The thermal breakdown is also occurred in lube oil [3]. Lubricating oil forms a fluid film between the mating surfaces and reduces direct contact of material and it also maintained a proper distance between the surfaces. Lubricating oil helps in reducing friction & wear of contacting surfaces, heat removal as well as removing contaminations and also improve performance and service life of all moving parts [4]. The main properties of lubricating oil are viscosity, anti-wear properties, extreme pressure properties, rust & corrosion inhibiting properties, thermal stability [5]. To increase service life and improve lubrication properties, additives such as anti-wear, anti-oxidant, corrosion inhibitors, extreme pressure additives are added to base oil [4]. All of these additives are added in an appropriate amount so that required property of oil for lubrication could be gained. Modern and advanced techniques are used for extracting health of a machine by performing oil analysis.

During meshing, gears have rolled and sliding contacts except at pitch point. This would generate frictional heating at tooth contacts and increase the temperature of gear assembly & gearbox [6]. This rise in the temperature of oil will accelerate degradation of oil, changes viscosity, oxidation and it also reduces film thickness between gear surfaces appreciably [7]. Viscosity of oil is dependent on the operating temperature and running hours of the machine. Oil viscosity changes with ageing due to contaminants generated, oxidation of oil, consumption of additives added to improve properties [8].

Shearing action of oil is one of the reasons for the degradation of oil [9]. Specific additives are added to the base oil to improve viscosity, generally containing double and triple bonded carbon atoms as poly(methyl methacrylate) (PMMA), olefin copolymer (OCP) and hydrogenated styrene-diene copolymer (SDP) [10]. After shearing of oil, some additives are deformed and formed different types of compound. These compounds change density of lubricant and fluid film thickness, therefore friction or wear took place between contacting surfaces [11].
Gears have elastohydrodynamic lubrication (EHL) due to relative motion between them. It maintains the gap between surfaces by an elastohydrodynamic (EHD) fluid film [12]. It helps in changing nearby surface of gear during EHD film formation [6]. Between EHL gap, small particles are pushed away and moved with the film whereas large particles are easily trapped in the vicinity of lubricant. Large particles accumulate together and generate solid frictional effects. These further produce fluid starvations because of accumulation of debris particles in the area of EHD film, which leads to scuffing [13]. Further, crushing of particles inside an EHD zone induces scratching on the surface during severe sliding [14]. Failures in gear occur due to degraded lube oil in circulation. These lead to occurrence of many types of wear such as micro pitting, pitting, scuffing, squeezing on the surface of gear teeth [7, 15].
This paper presents analysis of gear oil using different ways such as fluid analysis and contamination analysis.

2. Load Haul Dumper Machine

LHD machine is used in mining industries for loading purposes. It uses high torque for movement of machine and transportation of materials. A view is shown in figure 1. It is an electric power-driven machine and requires 550 Voltage and 37.5 KW of power.

Figure 1. Load Haul Dumper Machine

Figure 2. Planet gear of LHD machine

2.1. Specifications of LHD

Bucket capacity of 1.5 m^3 of 1.3 tonne. Payload of 2250 kg. Weight of LHD: 7100 kg. height-1.95-2 m, length: 6.30 m, width: 1.5 m. running speed: 3.8-12.2 kmph.

2.2. Gear System

Differential planetary gear system is used for providing high torque requirement to drive LHD in rough terrain with loads. This gear consists of sun gear, planet gear and gear box for high torque conversion. Planetary gear is shown in figure 2.

3. Material & Methods

3.1 Lubricants:

Gear oil SAE 90 was selected from LHD machine. Oils were collected just after stopping of engine otherwise contaminants may be settling down. Samples were collected in a neat, clean and airtight storage bottle. Fresh oil was also taken as a reference for used oil from the same lot which was added to the gearbox. Two used oils were selected from the machine at an interval of 200 hrs from initial position so, at 200hrs and 400hrs. A picture of all three samples is shown in figure 3. It is clearly visible that colour of the oil is changed and sample 3 is the darkest in colour. A detailed Flowchart (figure 4) is given below which provides complete information of lubricant had been analysed.

3.2 Methods:

Figure 3. Gear Oil Samples

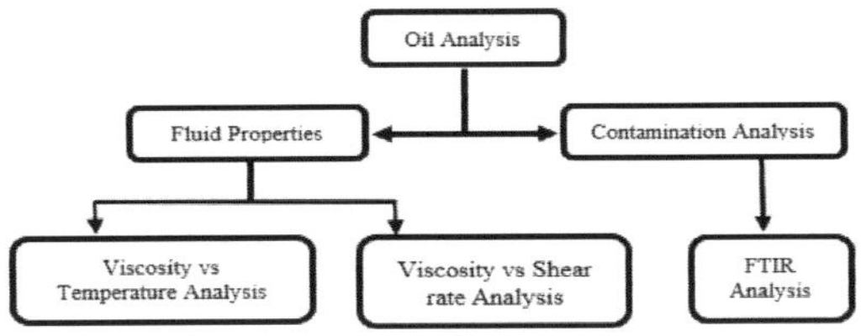

Figure 4. Steps for Oil Analysis

Sample 1: Fresh gear oil. Sample 2: Used gear oil 200 Hrs. Sample 3: Used gear oil 400 Hrs.

3.2.1 Viscometer: Viscosity of samples was measured using Stabinger Viscometer (Model No. - SVM 3000) and it should be according to ASTM D7042 method [16]. First, the instrument was cleaned with toluene/acetone for avoiding any previously tested oil. Then 10ml of a sample was taken in a syringe and put into viscometer. A temperature ranges from 40 °C to 100 °C was fixed for testing. It will provide all data like dynamic viscosity, kinematic viscosity, and density at the different temperature. These processes were repeated for all samples. Viscosity was measured in mPas.\

3.2.2 Air Bearing Rheometer: Shear strain occurring inside the oil and it is measured by rheometer (Model No. - AR 2000) and test is conducted according to ASTM D4440 method. A pressure of 3 bar is supplied to rheometer from air compressor to maintain a minimum distance between bearing and surface. Now, a cup & bob type geometrical measuring instrument (figure 5) is used for measurement for liquid samples with Newtonian & non-Newtonian behaviour [17]. A quantity of 50ml of each sample was taken for the testing purpose. Now, the sample was kept in the measuring geometry and temperature range of 60-100 OC was fixed for testing. Shear strain is varied between 0.01 & 1000 1/s to get viscosity variation at different temperatures. Viscosity was measured in mPas.

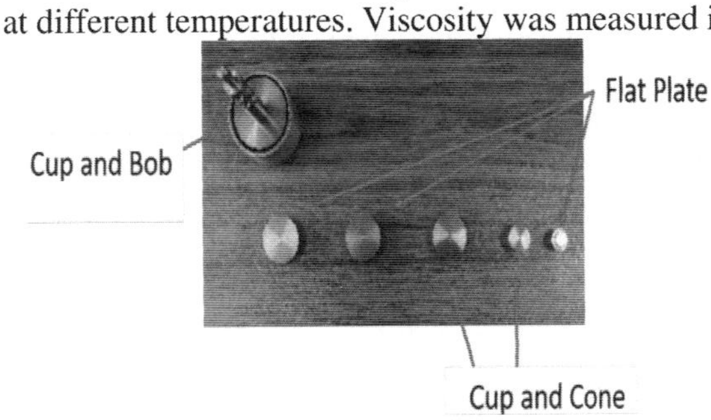

Figure 5. Geometrical Measuring Instrument

3.2.3 Fourier Transform Infrared Spectroscopy: Transmittance change occurring inside the oil molecules and it is measured by FTIR spectrometer (Model No. Cary 600) and test is conducted according to ASTM E168. First, isopropyl alcohol was used to clean the disc surface and background scan was taken. Now, Lube oil is placed on highly reliable, chemically inert diamond ATR fixed on a disc & a metallic needle and pressed to ensure good contact between diamond ATR and sample [18]. This spectrometer for FTIR was run for analysis which uses wavenumber between 4000-400 cm-1. Computer generated a graphical plot between transmittance percentage and wave number.

IOP Conf. Series: Materials Science and Engineering **624** (2019) 012011 doi:10.1088/1757-899X/624/1/012011

4. Results and Discussion

4.1. Variation of Viscosity with Temperature and Time

It can be seen in figure 6 that the viscosity of all samples is varying. It is high for fresh oil (224.82 mPas). It is decreasing exponentially with increase in temperature for all oils and follows Reynold's equation [11]. Also, it is decreasing with running hours. This is taking place due to additives deterioration. Also, there were wear particles formation, oxidation, contaminations occurred in oil. It will lead to overheating, accelerated wear, and ultimately failure of the gear component. In table 1. presents the detailed value of constant, viscosity index and correlation coefficient.

$$\mu = A * e^{-BT} \quad \text{Reynold's equation}$$

where,

μ= Dynamic viscosity in mPas., T= Temperature in °C, A= Constant in mPas, B= Constant in 1/°C.

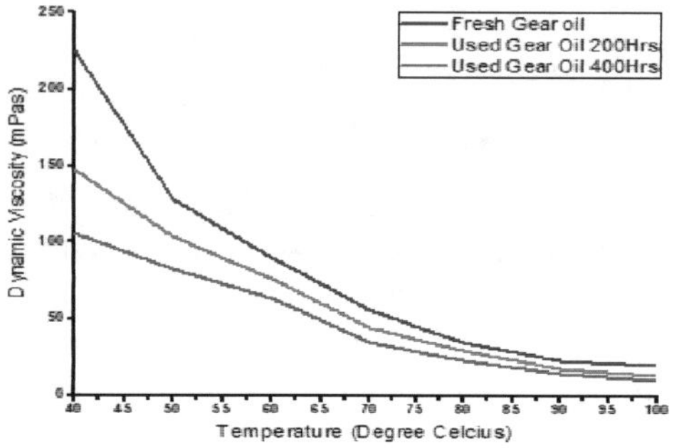

Figure 6. Viscosity variation with temperature

Table 1. Correlation coefficient and VI of Gear Oil

Gear Oil	A (mPas)	B (1/ °C)	Viscosity Index	Correlation Coefficient (R^2 value)
Fresh Oil	1079.4	0.042	100	0.9878
200 Hrs Used Oil	818.33	0.041	87	0.9951
400 Hrs Used Oil	592.99	0.040	84	0.9895

4.2. Shear Rate Test with Temperature and Time.

From the figures (7, 8, 9, 10 and 11) of viscosity vs shear rate, it is clearly visible that viscosity decreases with shear rate increase. For figures 7 & 8, the viscosity of oil at 60 °C and 70 °C almost constant with respect to shear rate. At 80 °C, decrease in viscosity slightly increases (figure 9). From figures 10 & 11, it can be analysed at 90 °C and 100 °C decrease in viscosity with shear rate increases and it is maximum for used gear oil.

From the graph shown, it is clear that shear thinning is occurring in the oil as the increase in shear rate at higher temperature. It is mainly caused by breaking of bonds between molecules of additives added specifically for viscosity improvement. Viscosity at 60 °C is concentrated at about 50 mPas reduced to about 25 mPas at 100 °C. That means the temperature is also affecting viscosity with shear rate. Thus from references, it can be concluded that initially at low temperature fluid behaves as a Newtonian fluid but at later case of higher temperature, it is behaving as non-Newtonian fluid [19].

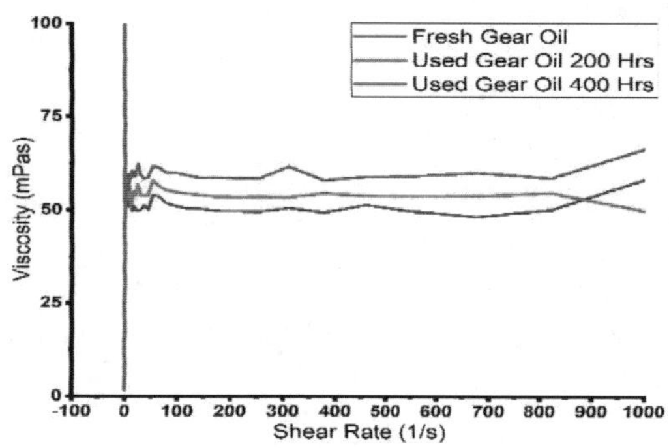

Figure 7. viscosity vs shear rate at 60 °C

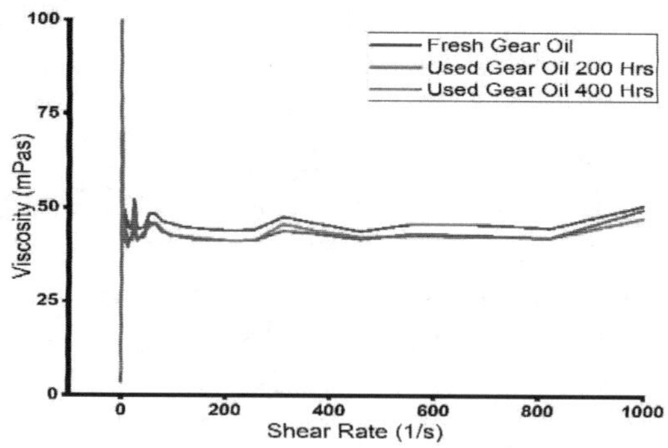

Figure 8. viscosity vs shear rate at 70 °C.

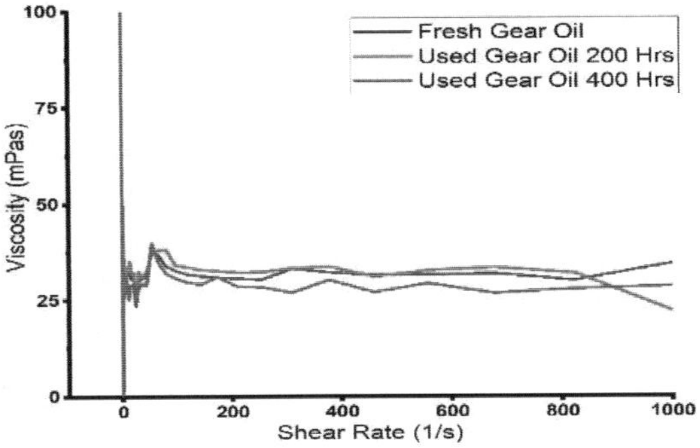

Figure 9. viscosity vs shear rate at 80 °C.

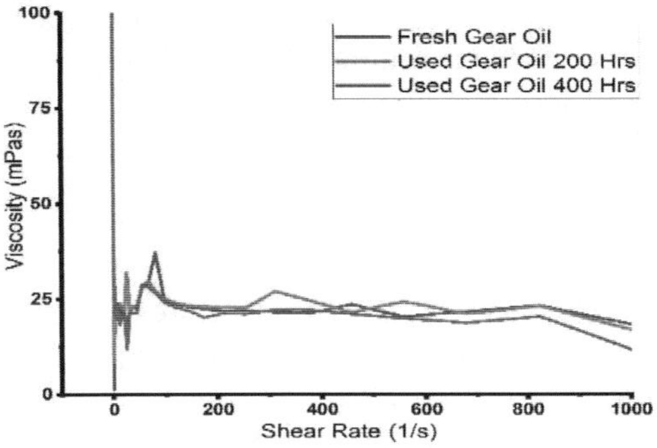

Figure 10. viscosity vs shear rate at 90 °C

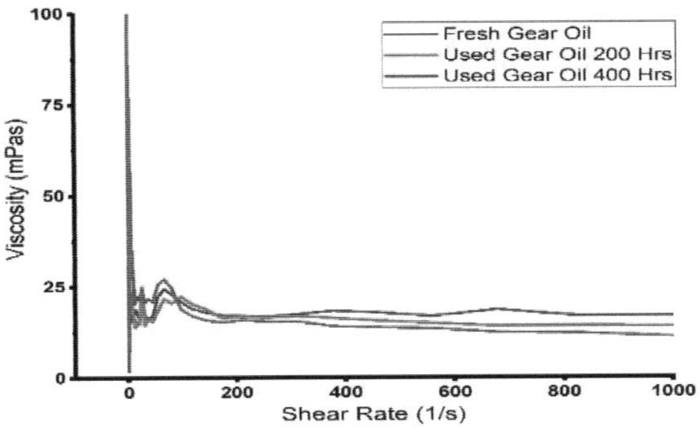

Figure 11. viscosity vs shear rate at 100 °C.

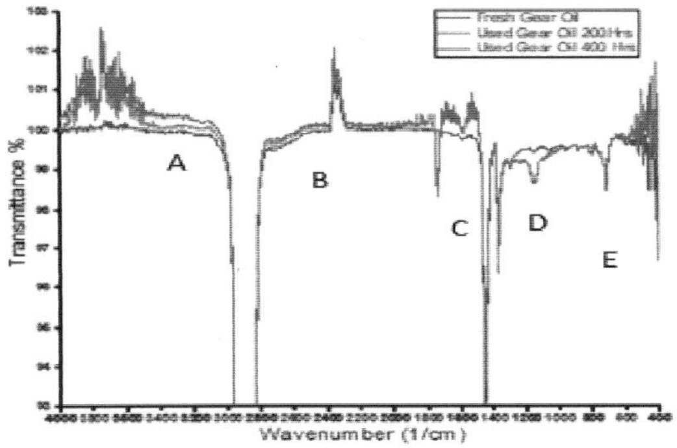

Figure 12. FTIR of gear oil in transmittance band

4.3 Fourier Transform Infrared Radiation Analysis

Gear oil is made of paraffin, aromatic compounds, long chain hydrocarbon, branched chain hydrocarbon and other organo-compounds [5]. Most common degradation processes are oxidation and shearing of oil with increase in temperature. These compounds are degraded and form other by-products [20]. From the figure 12, it can be concluded that.

Region A: In the wavenumber range between 4000-3000 cm^{-1}, alcohols (-OH), amines(-NH), carboxylic acids (-COOH), alkane (-CH) groups have increasing transmittance so there is depletion of additives in the oils. It is suggesting that aging of oil is occurring after long use of oil. Additives having anti-oxidation properties, rust & corrosion inhibiting properties are consumed with aging. **Region B:** In the wavenumber range between 3000-2000 cm^{-1}, Thiols(-SH) and alkynes(-C≡C-) groups have risen transmittance level, results from depletion of additives. These means that these compounds are consumed while changes in thermal and shear properties. **Region C:** In the wavenumber range between 2000-1400 cm^{-1}, aromatic compounds have shown decrease in transmittance level. They are forming after a long duration of use of oil. These shows braking of organo-compounds is happening. **Region D:** In wavenumber range between 1400-1000 cm^{-1}, oxidised sulphur compounds, alkyl ethers, esters present, and oxidation has taken place in the used oil samples. Here, no additives are found. Oxidation of oils forms carbonyls, hydroxy compounds. **Region E:** Similarly, in wavenumber range between 1000-600 cm^{-1}, Substituted carbon chain is decreasing which clearly shows the breaking of bonds of additives.

5. Conclusion

5.1. Viscosity of gear oil is decreasing with increase in temperature. This happens due to contamination in oil, oxidation and various causes. Oxidation is responsible for degradation of gear oil.

5.2. Temperature is not only the cause of viscosity loss of lubricating oil but also due to shearing phenomena. Shearing is occurring in gear oil due to breaking of molecular bonds.

5.3. FTIR results clearly show the formation of oxidation by-products like carbonyl oxidation products, sulphur oxidation products, nitrogen oxidation products and many more.

6. References

[1] Heredia-Cancino J A, Ramezani M and Álvarez-Ramos M E 2018 Effect of degradation on tribological performance of engine lubricants at elevated temperatures *Tribology International 124* pp 230-237.

[2] Bobzin K and Bartels T 2011 *Industrial tribology: tribosystems, friction, wear and surface engineering, lubrication* John Wiley & Sons.

[3] Vishweshwara S C and Al Badi O K H 2015 Density and Thermal Conductivity Changes in Engine Oil During its Life Cycle: An Experimental Study *International Journal of Mutidisciplinary Science and Engineering 6*(10) pp 24-28.

[4] Rohatgi P K, Tabandeh-Khorshid M, Omrani E, Lovell M R and Menezes P L 2013 Tribology of metal matrix composites *Tribology for scientists and engineers* Springer New York NY pp 233-268.

[5] Zeng Q, Dong G, Yang Y and Wu T 2016 Performance deterioration analysis of the used gear oil *Advances in Chemical Engineering and Science 6*(02) p 67.

[6] Olver A V 2002 Gear lubrication—a review *Proceedings of the Institution of Mechanical Engineers Part J: Journal of Engineering Tribology 216*(5) pp 255-267.

[7] Höhn B R and Michaelis K 2004 Influence of oil temperature on gear failures *Tribology International 37*(2) pp 103-109.

[8] Abusaad S, Brethee K, Assaeh M, Zhang R, Gu F and Ball A D 2015 The detection of lubricating oil viscosity changes in gearbox transmission systems driven by sensorless variable speed drives using electrical supply parameters *Journal of Physics: Conference Series* 628 No 1 p.012078 IOP Publishing.

[9] Feng R and Ramesh K T 1993 The rheology of lubricants at high shear rates *Journal of tribology 115*(4) pp 640-647.

[10] Ghosh P, Das T and Nandi D 2011 Shear stability and thickening properties of homo and copolymer of methyl methacrylate *American Journal of Polymer Science 1*(1) pp 1-5.

[11] Sander D E, Allmaier H, Priebsch H H, Reich F M, Witt M, Füllenbach T, Skiadas A, Brouwer L and Schwarze H 2015 Impact of high pressure and shear thinning on journal bearing friction *Tribology International 81* pp 29-37.

[12] VAN GAAL J C, Van der Graaf W T, Rikhof B, VAN HOESEL Q G, Teerenstra S, Suurmeijer A J, Flucke U E, Loeffen J L, Sleijfer S and DE BONT E S 2012 The impact of age on outcome of embryonal and alveolar rhabdomyosarcoma patients. A multicenter study *Anticancer research 32*(10) pp 4485-4497.

[13] Nikas G K, Sayles R S and loannides E 1998 Effects of debris particles in sliding/rolling elastohydrodynamic contacts *Proceedings of the Institution of Mechanical Engineers Part J: Journal of Engineering Tribology 212*(5) pp 333-343.

[14] Wan G T Y and Spikes H A 1988 The behavior of suspended solid particles in rolling and sliding elastohydrodynamic contacts *Stle Transactions 31*(1) pp 12-21.

[15] Ognjanovic M 2004 Progressive gear teeth wear and failure probability modeling. *RESEARCH, Tribology in industry 26*(3&4) pp 44-49.

[16] Kanaveli I P, Atzemi M and Lois E 2017 Predicting the viscosity of diesel/biodiesel blends *Fuel 199* pp 248-263.

[17] Nagy D and Belina K 2018 Measuring viscosity of polyethylene blends using a rotational rheometer *Journal of Physics: Conference Series* 1045(1) p.012030 IOP Publishing.

[18] Pavithra R, Gunasekaran S, Sailatha E and Kamatchi S 2015 Investigations on Paper Making Raw Materials and Determination of Paper Quality by FTIR-UATR and UV-Vis DRS Spectroscopy *Int. J. Curr. Res. Aca. Rev 3* pp.42-59.

[19] Zhou S Q, Ni R and Funfschilling D 2010 Effects of shear rate and temperature on viscosity of alumina polyalphaolefins nanofluids *Journal of Applied Physics 107*(5) p.054317.

[20] Patty D J and Lokollo R R 2016 FTIR spectrum interpretation of lubricants with treatment of variation mileage *Adv. Phys. Theor. Appl, 52* pp 13-20.

IOP Conf. Series: Materials Science and Engineering **624** (2019) 012012 doi:10.1088/1757-899X/624/1/012012

Crack detection in a shaft using finite difference technique

R Nigam[1] and S K Singh[2]

[1]Research Scholar, Department of Mechanical Engineering, Indian Institute of Technology (ISM), Dhanbad, India
[2]Assistant Professor, Department of Mechanical Engineering, Indian Institute of Technology (ISM), Dhanbad, India

E-mail: ravinigam264@gmail.com

Abstract. Fatigue cracks arise in a rotor shaft due to cyclic loading. A growing crack in a machine element may lead to a calamitous failure of the machinery. Identification of crack at an early stage supports in safe and economic running of machinery. The appearance of crack in a shaft reduces local stiffness and causes a slope discontinuity in the shaft elastic line during vibration. Detection of slope discontinuity reveals the presence of crack in a shaft. However, in the presence of measurement noise, identification of slope discontinuity is difficult. In the present work, finite element model of a cracked simply supported shaft is considered for the simulation work. A transverse surface crack is considered for the cracked model. The shaft is assumed to be non-rotating and an external sinusoidal forcing is used to excite the shaft. To imitate actual experimentation, Gaussian noise is added in the simulated cracked shaft response. A dominant peak is observed at the crack position, which is obtained by using the central finite difference approximation on the simulated response of simply supported shaft. The method is used for the localization of the cracks present in the shaft. It is based on detecting the slope discontinuity in the elastic line of the shaft. The finite difference method based evaluation is also implemented for varying crack depth ratios at different signal to noise ratio (SNR).

1. Introduction

Condition monitoring of rotating machines is important for continuously running plant such as power generation industries. Some rotor faults, such as cracks in the shaft need to be detected at early stage. The crack appears in rotating machinery due to the periodic stresses imposed in the shaft. Presence of crack in a shaft affects flexibility and entire dynamic vibrational response of the shaft. Vibration measurements from rotating machines have important information regarding the condition of the machineries. A crack in the shaft has mainly two parameters its depth and position. The crack position can't be seen directly from the shaft response. The investigation of the dynamic behaviour of cracked rotor can lead to proper diagnosis of machinery and prevent the machinery from sudden failure [1-2]. A thorough state-of-the art reviews are presented by [3-4]. In the earlier literature [5-6], the damage identification has been mainly focused on modal analysis of vibration signals. These analyses mainly use frequency and mode shape information.

The appearance of crack in shaft reduces local stiffness which causes slope discontinuity in the shaft elastic line. Resulting slope discontinuity is localized and hence its identification reveals the crack existence in the shaft. A multi crack detection and localization algorithm (MCDLA) is developed [7-8] for the detection of slope discontinuity due to cracks in a shaft. The reduction of stiffness causes abrupt change in the first derivative of deflection i.e. slope and second derivative of deflection i.e. curvature. The observation of these modal derivatives helps to identify the crack in the structure [9]. A Vibration-based Structural Health Monitoring (SHM) method was developed for cantilever beam with the artificial delamination to analyse mode shapes for different crack depth and locations. The analysis was performed on ANSYS software. Curvature mode shapes were calculated by using a central finite difference approximation to identify the crack position [10]. In the present paper, the finite difference method is used to detect the crack position in the rotor shaft.

The purpose of the present paper is to present the potentiality of the proposed method using simulated signals. The present paper is structured as follows: Section (2) introduces the finite difference method and the term signal to noise ratio (SNR). Section (3) presents the mathematical model of the rotor shaft with a transverse open crack which is based on finite element method. The simulated response of the cracked rotor shaft with measurement noise is evaluated in section (4). In this section, we also discussed the finite difference method based crack detection results at different SNR values. Finally, the paper is concluded in Section (5).

2. Theory

2.1 The proposed finite difference method

The finite difference method plays an imperative role for the numerical solution of differential equations, mainly in boundary value problems. In the present paper we are using second order central difference approximation, as defined below

$$y''(z) \approx \frac{d^2y}{d^2z} = \frac{y(i+\Delta z) - 2\times y(i) + y(i-\Delta z)}{(\Delta z^2)} \tag{1}$$

where, $y(i)$ is the point of interest, $y(i+\Delta z)$ is the point of interest plus step size, $y(i-\Delta z)$ is the point of interest minus step size, Δz is the step size.

2.2. Signal to noise ratio (SNR)

Finite difference based evaluation is required to be consistent in the presence of measurement noise for efficient analysis. In order to specify the effect of white Gaussian noise in the signal, the SNR is defined in dB as SNR = 20 $\log_{10}(Y_s/Y_n)$ dB. Here Y_s and Y_n is the standard deviation of signal and white noise respectively.

3. Mathematical model of a cracked rotor shaft

The Timoshenko beam theory is used in the modelling of the rotor shaft. A discretized model of the cracked simply supported shaft is developed by using finite element method. The finite element model of the rotor shaft with a transverse open crack is shown in figure 1. Each node of shaft element has two rotational and two translational degrees of freedom.

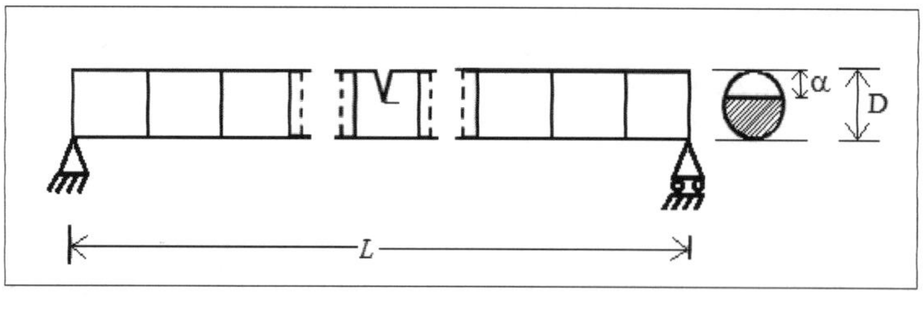

Figure 1. Finite element model of the rotor shaft

These degrees of freedom describes the position and rotation at the ith node of a shaft element, which is collected in the generalized displacement vector X_i .

$$X_i^T = [u_{xi} \quad u_{yi} \quad \theta_{xi} \quad \theta_{yi}]$$

A cracked shaft element and its generalized displacements at the nodes i and i+1 are shown in the figure 2.

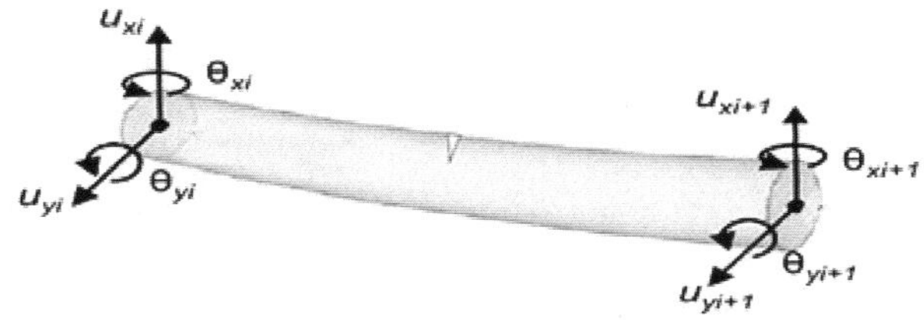

Figure 2. Cracked shaft element with generalized displacement at nodes *i and i*+1

Here u_{xi} and u_{yi} are translational degree of freedom along x and y direction and θ_{xi}, θ_{yi} are rotational degree of freedom along x and y direction respectively.

The governing equation of motion for non-cracked rotor shaft system is

$$[M]\{\ddot{X}\} + [C]\{\dot{X}\} + [K]\{X\} = \{F\} \tag{2}$$

In case of cracked rotor-shaft system, a crack is assumed to be placed within finite element of the model. The presence of crack increases the local flexibility, which affect stiffness. The total flexibility matrix of the cracked shaft element is given by,

$$[S] = [S_O] + [S_c] \tag{3}$$

where $[S_O]$ is the flexibility in an uncracked element and $[S_c]$ is the additional flexibility in the shaft element due to the crack. The term $[S_c]$ is expressed as [2],

$$[S_c] = \begin{bmatrix} S_{22} & 0 & 0 & 0 \\ 0 & S_{33} & 0 & 0 \\ 0 & 0 & S_{44} & S_{45} \\ 0 & 0 & S_{54} & S_{55} \end{bmatrix}$$

where $[S_{22}]$, $[S_{33}]$, $[S_{44}]$, $[S_{55}]$ are the direct and $[S_{45}]$, $[S_{54}]$ are the cross-coupled flexibility coefficients. Now, the stiffness of the cracked shaft element is obtained as [2]

$$[K_c]^e = [T]\,[S]^{-1}\,[T]^T \tag{4}$$

where [T] is a transformation matrix.

The assembled mass matrix of the non-cracked system is considered to be same as that of the cracked system. Now, the governing equation of motion for cracked rotor shaft system can be written as:

$$[M]\{\ddot{X}\} + [C_c]\{\dot{X}\} + [K_c]\{X\} = \{F\} \tag{5}$$

Here, $[M]$ is assembled mass matrix. The term $[K]$ and $[K_c]$ are assembled stiffness matrices of a rotor shaft in the absence and presence of crack respectively. The effect of the proportional damping is also included in the rotor-shaft system, so the assembled damping matrices of a rotor shaft in the absence and presence of crack are $[C] = a_0[M] + a_1[K]$ and $[C_c] = a_0[M] + a_1[K_c]$ respectively, where a_0 and a_1 are the Rayleigh damping factors. The term $\{X\}$ is the nodal displacement vector of the rotor shaft and $\{F\}$ represents the excitation force vector. Dimension and material properties for the rotor shaft system is given in table 1.

Table 1. Details of the rotor shaft parameters

Parameters	Physical dimensions
Shaft length, *L* (m)	1
Shaft diameter, *D* (m)	0.01
Number of elements on the shaft	500
Cracked element position	300
Young's modulus, *E* (N/m^2)	2.06×10^{11}
Poisson's ratio, *v*	0.3
Density, ρ (Kg/m^3)	7800

4. Results and discussions

4.1. Response of a cracked shaft

A simply supported rotor shaft with a transverse open crack is considered for the simulation. The shaft is supposed to be non-rotating and an external forcing is applied in vertical direction to excite the shaft. The cracked shaft is excited at 110 radian/sec of forcing frequency. This excitation frequency is near the first natural frequency of shaft. The vertical response in the space domain is continuous by nature. In the real scenario, the shaft response is contaminated by measurement noise. Hence to imitate real experimentation, white Gaussian noise is added in the simulated shaft response. The response of the cracked rotor shaft is shown in figure 3. The noise level (SNR) is kept at 90 dB. This shaft response is further used an input signal for the finite difference method based analysis.

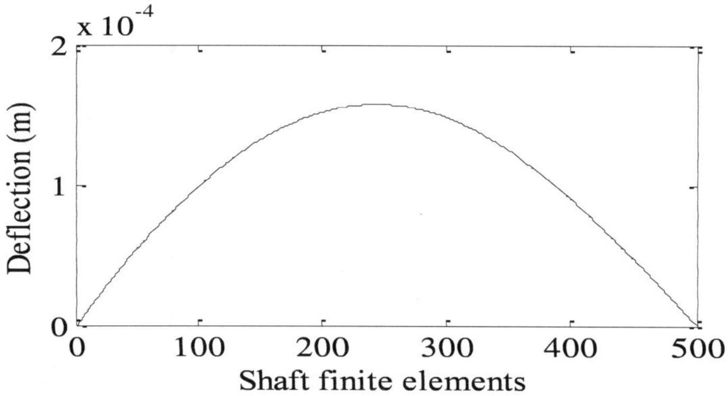

Figure 3. Defection response of the shaft with crack

4.2. Detection of crack position at different SNR level

The appearance of crack in a rotor shaft causes slope discontinuity in the simply supported shaft response. The resulting slope discontinuity is very difficult to visualize. Derivative of the shaft elastic line amplifies any discontinuity in the shaft elastic line due to presence of crack. In the present work, the finite difference method (equation 1) is used for the localization of the cracks present in the shaft. The presence of local discontinuity is revealed as a spike at the crack location. Figure 4 shows the results obtained by applying the finite difference method, which is represented by formula in equation

1, on the cracked shaft response at different SNR values. The crack depth ratio *(α/D)* is taken 0.33 in this case, where '*α*' is the crack depth. Here, the dominant peak shows the position of the crack. The crack is detectable up to 85 dB of SNR.

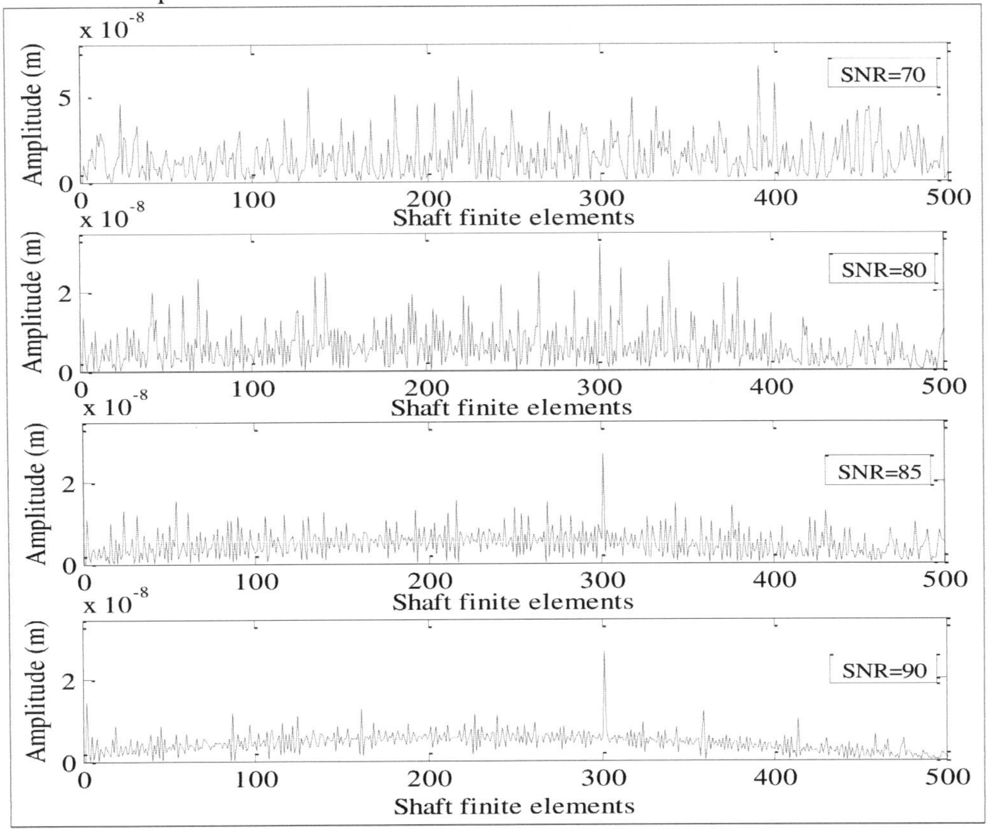

Figure 4. Crack identification for relative crack depth ratio (*α/D*=0.33)

When the crack depth ratio (*α/D)* is increased to 0.50 then the crack is detectable at lower SNR. In figure 5, the dominant peak corresponding to the crack location is observed at different SNR values.

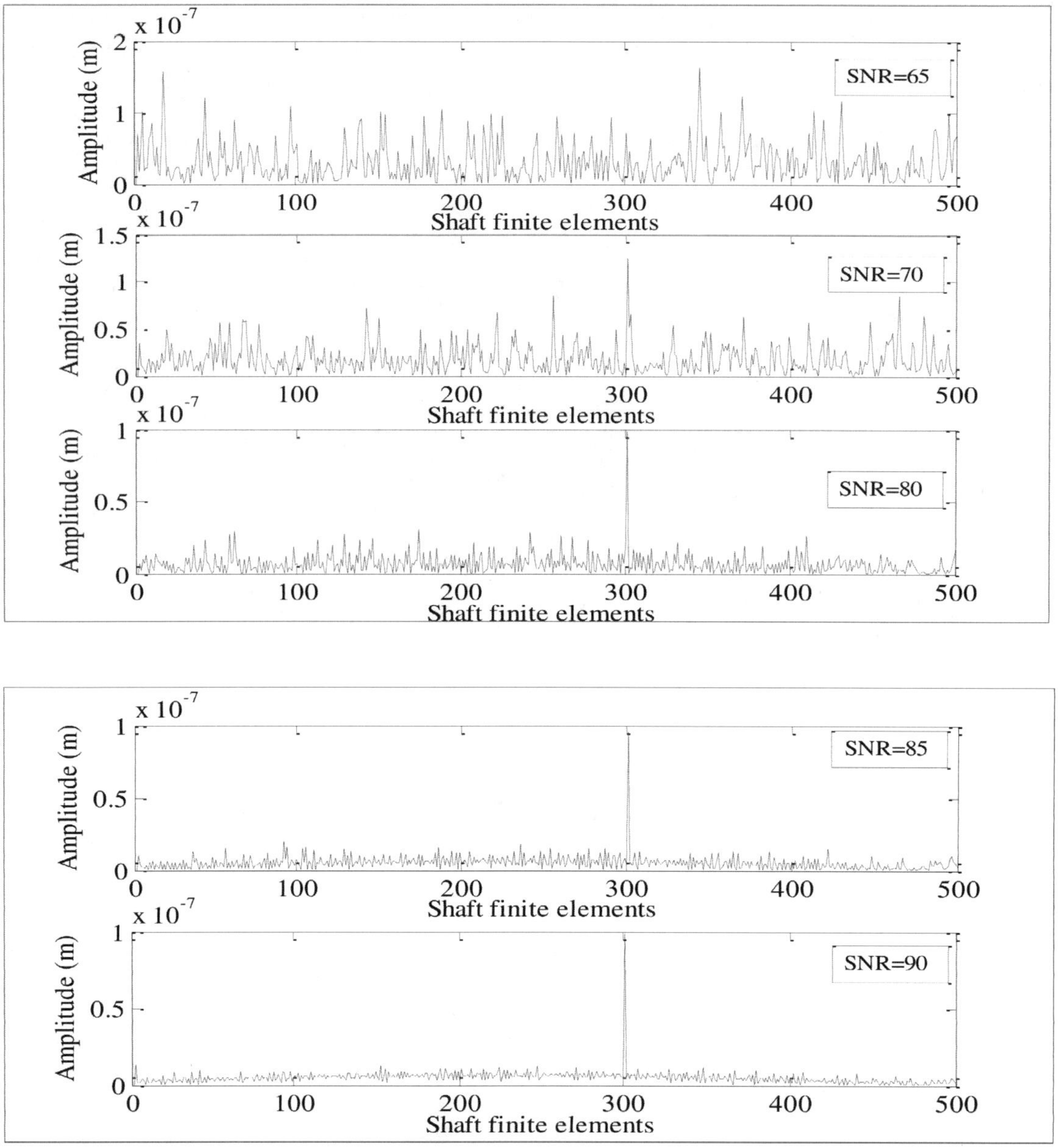

Figure 5. Crack identification for relative crack depth ratio (α/D=0.50)

Hence the present method is employed successfully to detect the crack position in the rotor shaft. This method can also be employed to detect the cracks in the beams and plates. The present algorithm needs high resolution measurement of shaft deflection. High resolution measurement of shaft deflection can be obtained by using scanning laser vibrometer based measurement and photographic methods.

5. Conclusion

A finite difference based approach is proposed to detect slope discontinuity in the shaft elastic line. Finite element modelling of a rotor shaft system has been performed for the study of vibration features of the rotor shaft with a single transverse open crack. The vibration response of the rotor shaft is analysed in the spatial domain. White Gaussian noise is added to the simulated response to imitate actual experimentation. Slope discontinuity is detected as a sharp peak in the second derivative of the shaft deflection. The finite difference method based evaluation successfully identified the crack position for varying crack depth ratios (*a/D*) at different signal to noise ratios (SNR) conditions.

Acknowledgment
The present work is financially supported by the Department of Science and Technology (DST), India under sanction order no. "YSS/2015/00155"

References

[1] Sekhar A S, Balaji P P 1999 Journal of Sound and Vibration **208** 457-474
[2] Sekhar A S, Prabhu B S 1994 Journal of Sound and Vibration **173** 415–421
[3] Dimarogonas A D 1996 Engineering Fracture Mechanics **55** 831-857
[4] Wauer J 1990 Applied Mechanics **43**
[5] Schultz A B, Warwick D N 1971 Journal of Composite Materials **5** 394-404
[6] Cawley P, Adams R D 1979 Journal of Strain Analysis **14** 49-57
[7] Singh S K, Tiwari R 2010 Mechanism and Machine Theory **45** 1813-1827
[8] Singh S K, Tiwari R 2012 Fatigue & Fracture of Engineering Materials & Structures **36** 85-91
[9] Chance J C, Tomlinson G R, Worden K 1994 In the Proceedings of the 12 International Modal Analysis Conference 778-785
[10] Raju G, Ramesh L 2016 International Journal for Innovative Research in Science & Technology **3** 282-289

IOP Conf. Series: Materials Science and Engineering **624** (2019) 012013 doi:10.1088/1757-899X/624/1/012013

Powder metallurgy process towards functional gradation of Al-Al_2O_3 metal ceramic mixture samples

R Nirmalkumar[1], Ritwik Sohgaura[1], Ravikiran Kadoli[1] and Sharnappa Joladarashi[1]

[1] Department of Mechanical engineering, National institute of technology Karnataka Surathkal, Mangaluru,India

Email ID: nirmalkumar2882@gmail.com

Abstract. Functionally graded materials (FGM) belong to the class of advanced materials whose properties vary along one or more coordinate. In the present study, an attempt has been made to manufacture FGM beams using a powder metallurgy process. The Al and Al_2O_3 in powder form are mixed to different volume fraction according to the rule of mixture. These samples are ball milled to form a proper blended mixture and to reduce the particle size for 1 hour at 250 rpm with tungsten carbide balls. A cold compacted die made of AISI D2 steel is used for preparing the sample. Different composition of Al and Al_2O_3 with Al_2O_3 varying from 0 to 10% is laid in the die using ethanol as the binder. This sample is compressed using hydraulic press with a load varying from 50-70 ton. The compressed sample is removed using a compression moulding machine, then it is sintered to a temperature of 650°C using a muffle furnace to form a smooth graded Al-Al_2O_3 FGM beam. Various defects such as surface cracks and bending were experienced during the preparation of samples; these defects were overcome by varying the parameters such as loading and sintering temperature.

1. Introduction

FGM is basically an advanced class of composite in which the volume fractions of two or more materials are varied continuously. Two types of gradation can be conceptualized: a continuous gradation and a stepwise gradation. Human bone is a best example for continuous gradation which is formed by variation in porosity and composition. Spark plug and gears, in which gradient is formed by changing its composition from a ceramic to metal comes under step wise graded material [2].

Different manufacturing process such as vapor deposition technique, plasma spray coating, powder metallurgy, centrifugal method and solid freeform fabrication method has been developed to produce the FGM [1]. Selecting the suitable manufacturing process depends on the volume of FGM production. Powder metallurgy is commonly used in mass production primarily being economical compared to other production process.

This paper gives an overview of manufacturing FGM beam samples using powder metallurgy process, steps involved such as ball milling the powder mixture, design and preparation of a die, powder layup into the mold, compaction of the layered sample, sintering and cooling. A stepwise gradation method is adopted and is shown in the figure 1.

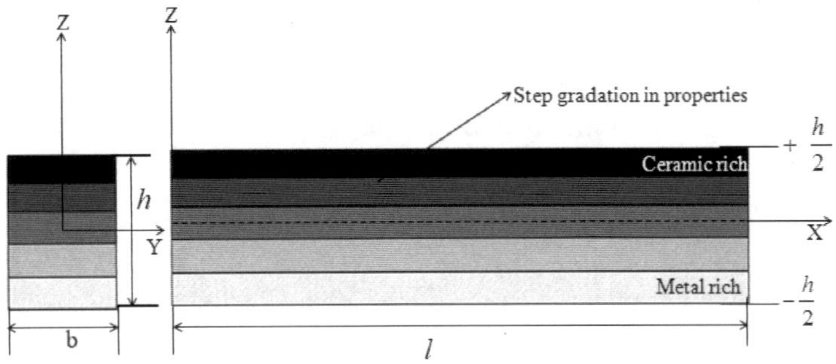

Figure 1. Step wise gradation along thickness of the beam.

2. Processing of Functionally graded beam

Schematic representation of the manufacturing process is shown in figure 2.

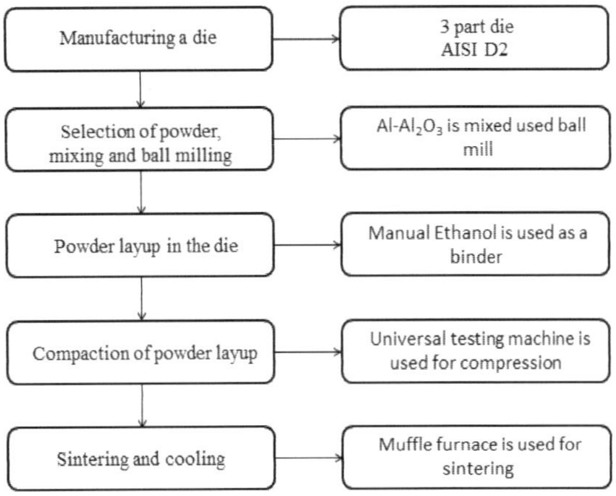

Figure 2. Flow of the powder metallurgical manufacturing process of FGM beam.

2.1. Features of powder compaction die

AISI D2 material is used for machining the die for cold compaction of powder samples. The die is made of three parts namely top die, middle die and bottom die as shown in figure 3.

Figure 3. Die to prepare FGM beam.

IOP Conf. Series: Materials Science and Engineering **624** (2019) 012013 doi:10.1088/1757-899X/624/1/012013

Top die can also be defined as a punch block of 210mm×60mm×21mm in dimension. It has a projection of 150mm×20mm×25mm and also has three holes through its thickness. The mid die is of 210mm×60mm×25mm and it has a cavity along its length where different powder compositions are laid. The dimension of cavity is 150mm×20mm×20mm. The cavity is provided with 2° draft angle at inner walls for easy removal of green compact. The bottom most part of the die is of 210mm×60mm×15mm in dimension.

2.2. Powder mixing

Commercially available aluminium and aluminium oxide powders are used for preparing FG beam samples. These powders are mixed according to required volume fraction proportion with regards to different layers and are blended at 250 rpm for 1 hour in high energy ball mill for proper mixing of aluminium and alumina oxide. The powder is placed into cylindrical jar consisting of 50 tungsten carbide balls having 10 mm size. The tungsten ball to powder weight ratio is 10:1. Particle sizes of ball milled powders according to their different volume fraction are measured with the help of scanning electron microscope.

2.2.1. Determination of powder mass

Powder mass measurements are taken according to the thickness of beam. In the present study samples with thickness 2.25 mm and 2.5 mm are fabricated. Thickness of each layer can be found out by dividing thickness of complete beam with number of layer. The volume of each layer is calculated according to the dimension of cavity and thickness of each layer. The density of each powder layer can be calculated with the help of law of mixture which is given by

$$\rho_{eff}(z) = \rho_m V_m(z) + \rho_c V_c(z) \tag{1}$$

Where

$$V_m + V_c = 1 \tag{2}$$

ρ_{eff} =effective density, ρ_m =metal density (Al), ρ_c =ceramic density (Al_2O_3), V_m =metal volume, V_c =ceramic volume

The mass of the Al-Al_2O_3 for preparation of the FGM beam depends on the dimensions of the beam i.e. length (l), width (b) and thickness (t). In the present study 2 different thicknesses of the beams are prepared. The dimensions and the mass of the beam prepared are shown in table 1. The percentage weight composition of Al_2O_3 is given in table 2.

2.3. Die filling

Die filling is done manually with the help of commercially available sieve. The stability of powder flow and apparent density is critical for better consistency in fabrication of samples. Various distinctive conditions such as humidity and temperature plus storage duration can all influence the performance of a powder mixture. It has also been found that environmental conditions during the compaction process can influence the performance of powder mixtures [3]. Hence for this purpose binders are used for better green compact strength in powder metallurgy products. In this study ethanol is used as binder medium for mixing the powder (Al- Al_2O_3).

In the die filling process firstly, die walls are cleaned by zinc stearate with acetone [4]. Zinc stearate is used to avoid the sticking of powder mixtures to die wall. Then different powder mixture stocks are mixed with ethanol binder and deposited in the die. The deposition starts with 10% Al_2O_3 and 90% Al layer and the last layer is of 100% Al composition. After laying one layer of certain powder mixture the powder layer is pressed with the help of top die so that layer is flattened to receive the next layer of mixture.

Table 1. Parameters involved in FGM beam preparation.

	Mass of Al-Al_2O_3 powder for different composition in gm.				
Parameters(mm)	100% Al	98.75% Al-1.25%Al_2O_3	97.5% Al-2.5%Al_2O_3	95% Al-5%Al_2O_3	90% Al-10%Al_2O_3
l=153,b=21,t=2.25	3.7582	3.7765	3.7948	3.8315	3.9047
l=153,b=21,t=2.5	4.514	4.5405	4.5665	4.6189	4.7234

Table 2. Mass of Al_2O_3 in FGM beam preparation.

	Mass of Al_2O_3 powder for different composition in gm.				
Parameters(mm)	100% Al	98.75 Al-1.25%Al_2O_3	97.5% Al-2.5%Al_2O_3	95% Al-5%Al_2O_3	90% Al-10%Al_2O_3
l=153,b=21,t=2.5	0	0.2571	0.3373	0.4982	0.8196

2.3.1. Cold compaction of FGM samples

After completion of the filling of die cavity with powder mixture, it is taken to a compaction machine. For this purpose a universal testing machine of 100 ton capacity is used. 60 tons load is applied gradually for 5 minutes, after which sample is removed by compression molding machine. . Samples removed by this process are termed as green compacts they possess sufficient strength to maintain the shape of beam.

2.4. Sintering process

The green compact samples which are obtained after compaction cannot withstand mechanical load. The green compacts are therefore kept in a furnace under controlled condition to obtain samples with better strength. Studies on the effect of various parameters such as particle size, sintering temperature and duration of sintering on physical properties of Al-Al_2O_3 based alloys fabricated with the help of powder metallurgy process have been reported [5]. They found highest relative density of 99.95% around 600°C. Hence sintering temperature was set up in between 600°C and 650°C for sintering of FGM beam samples. Studies on effect of step-wise sintering temperature increment on the properties of Al-SiC based alloys were reported by [6].

Based on these literatures the maximum sintering temperature was set as 650°C. During sintering the furnace temperature was gradually increased at a rate of 10°C per minute. Once the maximum temperature was attained, the samples were retained in the furnace for 4 hours. Further, samples were allowed to cool inside the furnace itself to make sure there were no cracks developed due to sudden cooling or air quenching. It was seen while operation that furnace nearly takes 1 to minute of time to reach 10°C of temperature.

3. Analysis of FGM beams

Various parameters such as load, binder quantity, re-compaction load and temperature were varied to manufacture a good FGM beam samples. The optimum load, temperature, binder required is shown in the table 3.

Table 3. Parameters involved in FGM beam preparation.

Parameters	Equipment used	Units
Compression load	Universal testing machine	60 ton
Binder	Ethanol	4-5 drops
Sintering temperature	Muffle furnace	650°C

3.1. Defects in the sample

All the samples prepared were of 5 layers. The composition of each layer is as given in the table 1. During initial stages of sample preparation lot of defective samples were obtained. The main defects of the FGM samples were top face peeling and cracks in the sample. The reasons for crack are due to the loading effect. Initially the load applied on the sample was around 40-50 tons which was found to be not sufficient for the proper binding of the powder particles [6]. So this defect was overcome by increasing the compression load. Face peeling is also because of the quantity of binder used. So this defect was overcome by re-compacting the sample after removal around 10-15 tons. 15 FGM samples were prepared and the brief analysis is given below:

Figure 4 FGM sample with face peeling.

Figure 5 FGM sample with powder erosion and crack.

Figure 4 and 5 gives the FGM samples with defects such as face peeling, irregular edges, surface cracks before sintering and powder erosion after sintering due more usage of ethanol respectively.

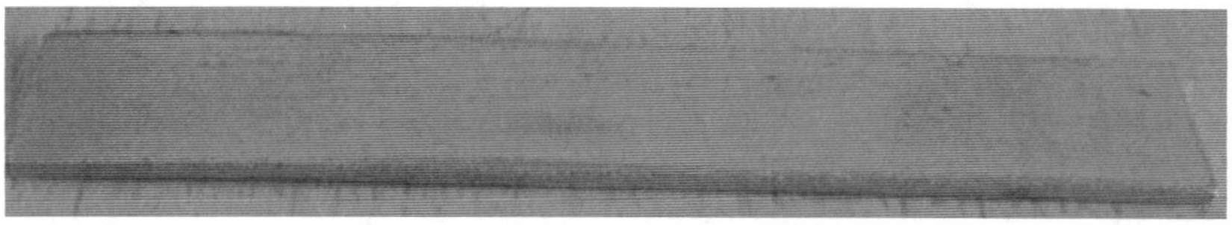

Figure 6 Defect free FGM sample.

Figure 6 is the sample which is obtained after maintaining all the optimum parameters. Minor defects such as warping during sintering were overcome by keeping the sample between two ceramic tiles to reduce the thermal stresses which led to buckling [8, 9 and 10].

The reason for restricting the maximum percentage of Al_2O_3 to 10% is because as the ceramic content is increased it was observed there was no proper binding between the powder particles and there was surface peeling on the ceramic side [8, +9 and 10].

From the SEM analysis of the FGM sample as shown in Figure 7 it can be seen that Al_2O_3 dispersion is sparse in the matrix of Al. The large white spots are the Al_2O_3 whose percentage is increasing layer by layer.

Aluminum-steel FGM were fabricated using powder metallurgy process [4] with 100% Al and 100% steel, these samples were having fewer defects. In current work the FGM is fabricated using metal ceramic mixture (Al-Al_2O_3), more the content of the ceramic more will be the defects. Thus the volume fraction of Al_2O_3 is limited to 10%. The various processing techniques and the problems faced during the fabrication process were discussed by [11 and 12]. It was observed that the preparation of FGM manufacturing is a costly process and mass production is not effective. Whereas the powder metallurgy process is the cost effective and the mass production is possible.

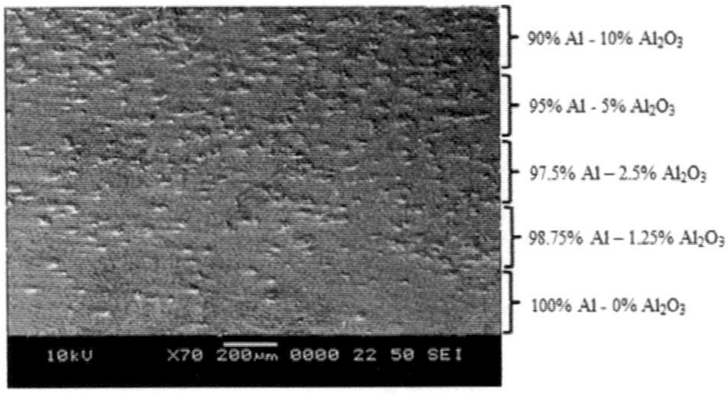

Figure 7 SEM image of the FGM sample.

4. Conclusions

It can be concluded that due to large difference in coefficient of thermal expansion of aluminium and alumina, the maximum volume fraction of alumina in aluminium is limited to 10% in order to avoid warping during sintering. The microstructure study of FGM powder mixtures shows the uniform distribution of Al_2O_3 in Al. The micrographs of FGM beam samples indicate the layer-wise gradation of aluminum and alumina powders.

References

[1] Mahamood R, Akinlabi E. T, Shukla M, and Pityana S. 2012. *Functionally graded materials:An overview World Congr. Eng.*, London U.K. Vol III, 2–6.
[2] Miyamato Y, Kaysser W A, Rabin B H, Kawasaki A, Ford R G 1999 *Functionally graded materials: Design, processing and applications* Ford R G Newyork Springer science+Business media LLC 1-339
[3] Thomas Y, St-Laurent S, Pelletier S, and Grlinas C. 2009. *Effect of atmospheric humidity and temperature on the flowability of lubricated powder mettalurgy mixes* PowderMet MPIF, Las vegas, Nevada, USA.
[4] Nemat-Alla M. M, Ata M. H, Bayoumi M. R and Khair-Eldeen W. 2011. *Mater. Sci. Appl.*, 02(12), 1708–1718.
[5] Coble R L and Kingery W D 1956. *J. Am. Ceram. Soc.*, 39(11), 377-385.
[6] Pieconka T. and Kaior J. (2014). *Int. J. Mater. Metall. Eng.*, 8(4), 285–288.
[7] Rahimian M, Ehsani N, Parvin N, and Baharvandi H M. 2009. *J. Mater. Process. Technol.*, 209(14), 5387–5393.
[8] Pravin M V., 2017, *Thermo-elastic response of thin functionally graded beams under various heat loads – theoretical studies and experimental validation*, Doctoral Thesis (PhD Thesis) National Institute of Technology Karnataka Surathkal India.
[9] Maruthi M D, 2016. *Powder mettalurgy technique to produce Al-Al_2O_3 FGM beams and their free vibration studies,* M.Tech dissertation National Institute of Technology Karnataka Surathkal India.
[10] Jagadish P, 2017, *Studies on static deflection and free vibration of functionally graded cantilever beams processed by powder mettlaurgy route*, M.Tech dissertation National Institute of Technology Karnataka Surathkal India.
[11] Kawasaki A and Watanabe R Y 1995 *Ceram.Inter.*, 23, 73-83.
[12] Kieback B, Neubrand A and Riedel H 2003 *Mater. Sci. Eng.*,362. 81-105

IOP Conf. Series: Materials Science and Engineering **624** (2019) 012014 doi:10.1088/1757-899X/624/1/012014

Influence of Link Lengths & Input Angles on the Foot Locus Trajectory of Klann Mechanism

N Prashanth[1*], R M Manoj[2], B Nikhil[2]
[1]Research Scholar, Department of Mechanical Engineering,
[2]Under Graduate Scholars, Department of Mechanical Engineering, Nitte Meenakshi Institute of Technology, Bengaluru, INDIA.

E-mail: prashanth.n@nmit.ac.in

Abstract. Locomotive drive mechanisms for robots have been an active research area in the field mobile robotics. Several drive mechanisms such as wheeled, legged, hybrid & reconfigurable mechanisms have been extensively studied. The legged locomotive drive mechanism has several advantages to its credit & it is often bio-inspired. The disadvantage of legged locomotive drive mechanism in general is it is often considered as complex in design due to its multi degrees of freedom which requires complex control scheme. This has led roboticists to explore minimal legged locomotive drive mechanisms consisting of single degrees of freedom. The present work focuses on synthesis and analysis of one such planar mechanism such as klann mechanism for a mobile robot application. In this work, we have synthesized the various linkages of klann mechanism to find the optimal foot locus trajectory. The synthesis and analysis of mechanism is performed using SAM (Synthesis and analysis of mechanism) software. We have described influence of various parameters such linkage length and input crank angle on the foot locus. We have compared the foot locus trajectory for three different foot locus by performing the kinematic analysis. Various parameters such as position, velocity and acceleration of linkages are compared & illustrated in detail. The information provided in this work will help the roboticists to select appropriate link length for klann mechanism to obtain various foot locus designs. This information may be further used to design efficient legged locomotive drive mechanisms for various terrain conditions.

1. Introduction

The legged locomotive drive mechanisms has been extensively studied in the field of mobile robotics, legged mobile robots(LMR) are advantageous compared to wheeled mobile robots(WMR) has it can be used in all terrain conditions [1]. The legged robots are broadly classified based on the degrees of freedom (DoF) it possesses. There are multi-degrees of freedom legged robot and single degrees of freedom legged robot. The multi degrees of freedom legged robots require more actuators for the motion and its complexity is more in terms of mechanical design and control scheme. Hence single degrees of freedom legged robots play an important role [2]. There are many single degrees of freedom legged robots based on several planar mechanisms, the most popular amongst them are Klann mechanism and Theo Jansen mechanism. The stability of legged locomotion determines the terrain in which the robot can move and efficiency of the robot locomotion. [3]. In the present work the focus is on synthesis and analysis of foot locus trajectory using Klann mechanism. Klann mechanism is the single degrees of freedom mechanism which is widely used as locomotive drive mechanisms for legged robots [4]. The linkage consists of a frame, a crank, two grounded rockers, and two couplers all connected by pivot joints [5]. As shown in figure 1. The stability of Robot is determined by the foot locus traced by the leg of the robot. The foot locus is the path traced by the leg of the robot for one cycle [6]. The optimisation and the Performance analysis of the mechanisms is performed in software called SAM (Synthesis and Analysis of Mechanisms). SAM was developed by Artas Engineering software's. Mechanisms can either be generated via the design wizards or they can be assembled from basic components including beams, sliders, gears, belts, springs, dampers and friction elements [7].

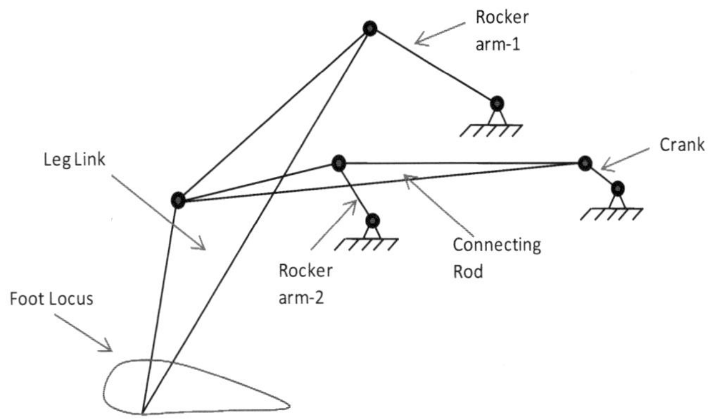

Figure.1 Klann mechanism designed in SAM software

2. Optimization Technique

Step Height & Stride Length are the two important parameters of the foot locus which are considered for optimization. Step height determines the lift of the leg along vertical axis while the stride length determines the horizontal distance covered by each leg for one complete revolution of the crank [8]. From the performance point of view, higher the step height better will be the all-terrain capability, while longer stride length will help in faster displacement of the robot. In the present work, since klann mechanism is non-reconfigurable and has fixed link length, the step height & stride length will also be fixed. The step height & stride length of mechanism can be varied by changing the input crank angle (θ_{in}) and link lengths(l_n) during synthesis stage with a limiting factor without compromising on the validity of the mechanism. Therefore, the objective function for the optimization of mechanism is maximizing the co-ordinates of step height (x_{sh}, y_{sh})& stride length (x_{sl}, y_{sl}).

$$Max(x_{sh}, y_{sh}, x_{sl}, y_{sl}) = f(\theta_{in}, l_{1,} l_{2,} l_3 \ldots\ldots l_{n,}) \tag{1}$$

The optimization process in SAM involves two step processes: 1. Exploration of the design space 2. Optimization of a specific solution. To begin with the entire parameter space of the mechanism is explored using a combination of pure Monte Carlo technique & evolutionary algorithm, which is derived from genetic algorithm. Further local optimization is performed using simplex technique. The target co-ordinates of the foot loci are specified by the designer along with the boundary conditions of the mechanism (constraints only). The combination of a global exploration strategy and a local optimization strategy with the designer in the loop for selecting the mechanism that is further optimized is believed to give the best trade-off between speed and coverage of the design space.

The SAM provides basic GUI required optimizing planar mechanisms. Following steps were followed for complete optimization of both mechanisms [9].

Step 1: Defining the Objective Type- The objective type which we chose was path specified in terms of nodes of the foot loci which covered both the peak node for stride height & stride length.

Step 2: Definition of the element/node property to be optimized

Step 3: Target Definition-The co-ordinates of the nodes for the required foot locus was defined using ASCII based file format. i.e.

$$\begin{matrix} X\_1 & Y\_1 \\ X\_2 & Y\_2 \\ \vdots & \vdots \\ X\_n & Y\_n \end{matrix} \tag{2}$$

Step 4: Specifying the Objective Performance Number: Maximum was chosen as the objective performance (other options include: absolute maximum, average, RMS, standard deviation)

Step 5: Optimization Target was specified to be maximum

Step 6: Further the parameter space of the mechanism was chosen. Parameter space represents the properties of the nodes which can be varied during the optimization.

3. Optimization Results

The Klann mechanism was optimized for different types of foot locus depending on the varying parameters such as stride length and step height, the variation in the parameters results in different foot loci with varying dimensions of the klann mechanism[10]. In the present work, we compared the foot locus trajectory of standard foot locus with the maximum & optimized foot locus trajectory. Iteration 1 results in foot locus with maximum stride length & step height. Iteration 2 represents the standard foot locus, while iteration 3 results in optimized foot locus. The dimensions of links length for maximum, standard and optimised foot locus is shown is shown Table1. Figure 2 shows the corresponding foot locus obtained for Iteration 1, 2 & 3 respectively.

Table.1 Link Length & Foot Locus Parameters of Klann Mechanism.

Links	ITERATION 1 Link Lengths (mm) For MAXIMUM Stride Length & Step Height	ITERATION 2- Link Lengths (mm) For STANDARD Stride Length & Step Height	ITERATION 3 Link Lengths (mm) For OPTIMIZED Stride Length & Step Height
1	150.29	123.304	150.290
2	99.49	87.794	99.49
3	240.958	209.557	240.940
4	55.26	48.814	55.268
5	56.373	37.568	56.373
6	96.895	99.5243	102.89
7	151.415	135.533	159.39
8	49.202	44.838	49.202
9	37.850	21.704	32.842
Result	Step Height = 53.18mm Stride Length = 168.57mm	Step Height = 26.529mm Stride Length = 63.394mm	Step Height = 45.094mm Stride Length = 137.137mm

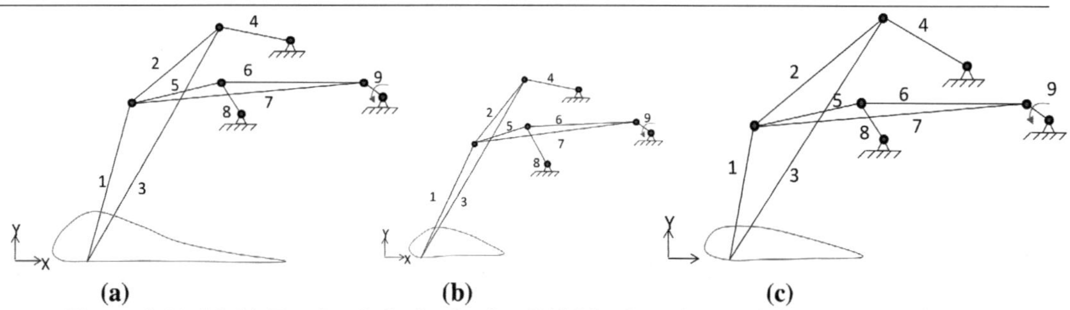

Figure.2 (a) (b) (c) Foot Loci obtained using SAM for Iterations 1, 2 & 3 respectively

4. Performance Analysis

The Performance analysis of the Klann mechanism is obtained by marking ideal nodes on the foot locus shown in Figure.3 and obtaining the velocity and acceleration values on that nodes.

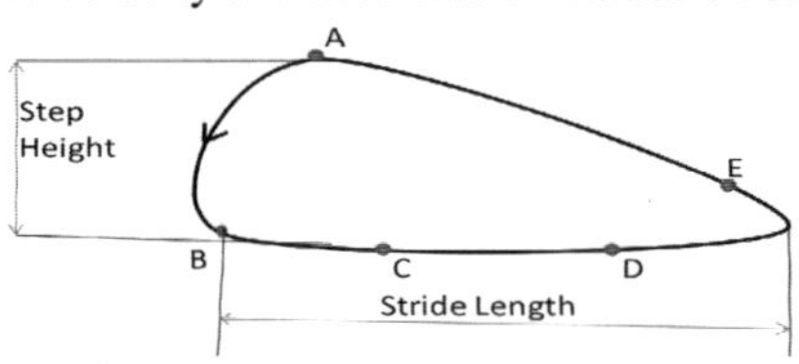

Figure.3 Major Points affecting the foot locus

Table.2 Nodes Identified for the comparison of Klann& Theo Jansen Foot Locus.

Nodes	Influence on Mechanism
A	Step Height-Determines Ability to overcome Obstacles
B	Touchdown-Determines Point of contact with ground
C & D	Determines the Stability during the strike length
E	Post Lift-off- Retardation from point D & approaches towards point A

The results of variation in the velocity and acceleration versus the crank angle are shown in the table 3.

Table.3 Performance Parameters of Klann mechanism for maximum stride length and step height

Nodes	ITERATION 1-MAXIMUM			ITERATION 2-STANDARD			ITERATION 3-OPTIMIZED		
	Crank Angle (rad)	Velocity (rad/sec)	Acceleration (rad/sec^2)	Crank Angle (rad)	Velocity (rad/sec)	Acceleration (rad/sec^2)	Crank Angle (rad)	Velocity (rad/sec)	Acceleration (rad/sec^2)
A	1.141	0.403	23.547	1.679	0.726	6.614	1.177	0.604	13.779
B	0.127	-0.711	11.522	0.131	-0.417	2.887	0.2064	-0.322	2.792
C	5.236	-1.578	3.224	5.415	-0.666	1.028	5.598	-1.032	3.460
D	4.174	-2.325	-0.315	4.723	-0.715	-0.117	4.239	-1.204	-2.003
E	2.852	1.946	-24.528	2.486	1.101	-2.327	2.822	1.119	-14.096

5. Kinematic Analysis

The kinematic analysis of the klann mechanism was performed using SAM and CATIA software. SAM was used to obtain velocity & acceleration of leg tracing the foot locus, while CATIA was used to model the optimized klann mechanism and check for any interference between the links. The interference check was performed using DMU (Digital Mock-up) workbench. Further, for the optimized klann mechanism, a foot locus was obtained in CATIA software to validate the foot locus obtained in SAM. It was observed that the foot locus obtained in CATIA was same as that obtained in SAM which is shown in the Figure.4

(a) (b)

Figure.4 (a) CATIA model of klann mechanism, **(b)** Foot Locus traced by Klann mechanism in CATIA DMU kinematics respectively

6. Modelling and Fabrication

The optimised foot locus was modelled in CATIA software after the interference check and then the mechanisms were fabricated and assembled to the chassis. The Chassis was modelled using Sheet Metal Workbench. The CATIA model of the assembled robot and fabricated robot is shown in the Figure.5

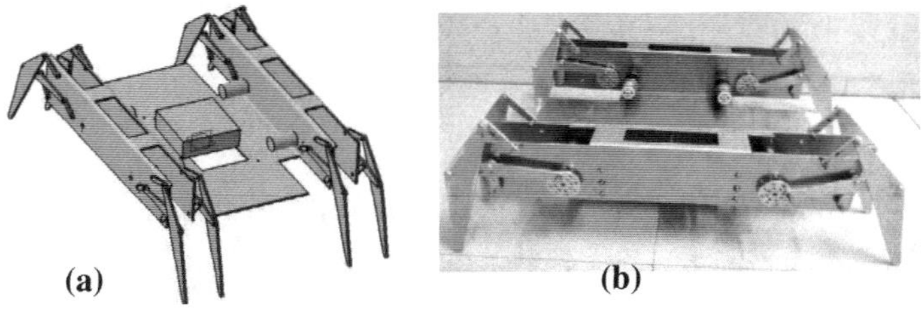

(a) (b)

Figure.5 (a) CATIA assembly model, **(b)** fabricated assembly respectively

IOP Conf. Series: Materials Science and Engineering **624** (2019) 012014 doi:10.1088/1757-899X/624/1/012014

7. Results and Discussions

The performance parameter such as angular velocity and angular acceleration of the Klann mechanism leg (link which makes contact with the ground) for three different foot locus i.e. maximum, standard and optimum stride length and step height is shown in the subsequent plots. Figure 6 shows the variation of angular velocity of the leg for iteration 1. The angular velocity is increasing as leg approaches from node A to D; here negative sign indicates change in direction of velocity vector. Figure 7 shows the variation of angular acceleration which increases while lifting the leg and it gradually decreases on the nodes B, C and D. Figure 8 shows the slope of angular velocity for iteration 2, angular velocity between nodes B, C and D is less compared to that of iteration 1. In Figure 9 the plots of node vs crank angle and acceleration is shown which infers that the magnitude of angular acceleration is less compared to that of iteration 1.

Iteration-1 MAXIMUM stride length and step height

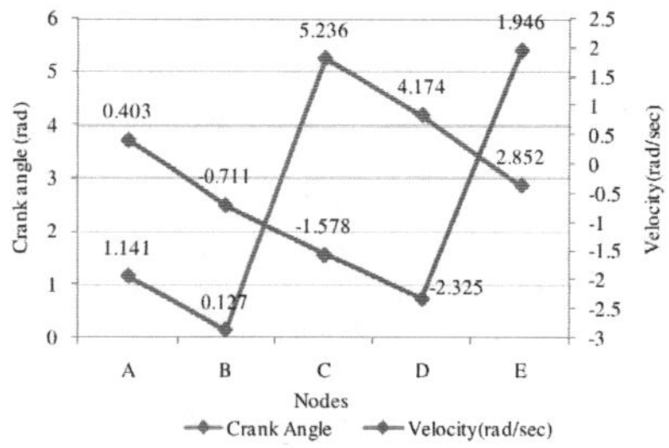

Figure 6. Node v/s Crank Angle & Velocity

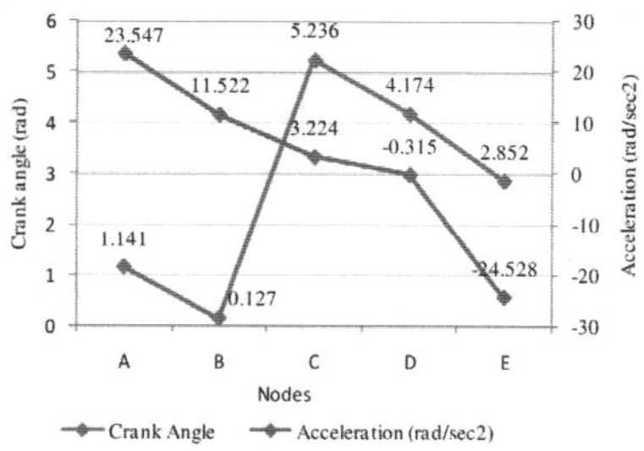

Figure 7. Node v/s Crank Angle & Acceleration

Iteration-2 STANDARD stride length and step height

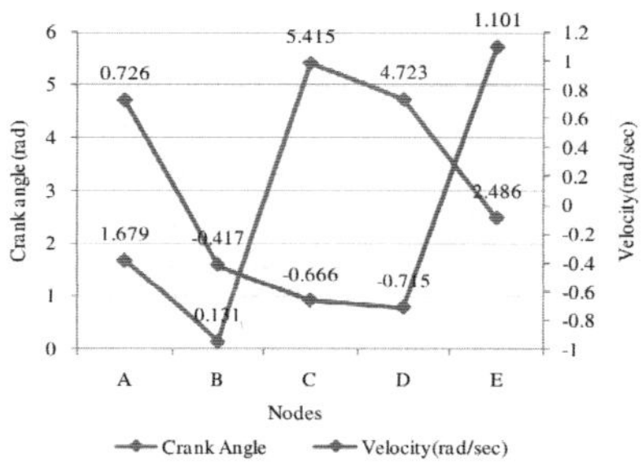

Figure 8. Node v/s Crank Angle & Velocity

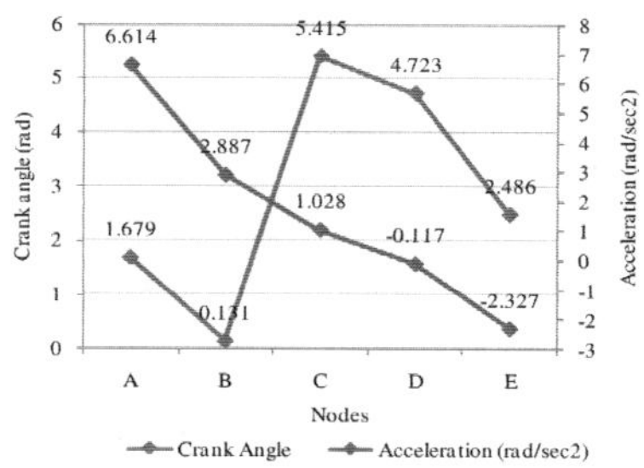

Figure 9. Node v/s Crank Angle & Acceleration

Iteration-3 OPTIMAL stride length and step height

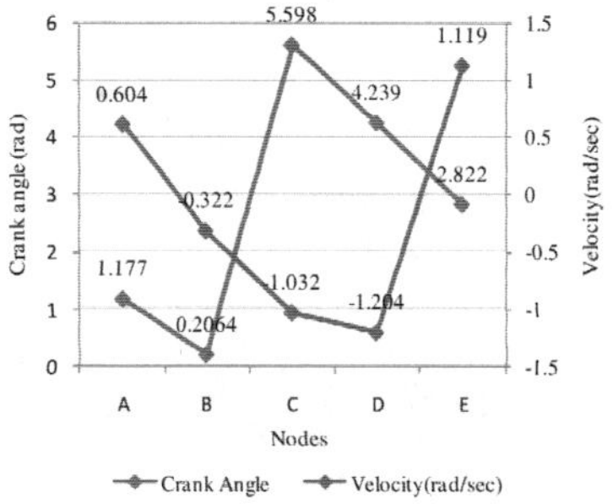

Figure 10. Node v/s Crank Angle & Velocity

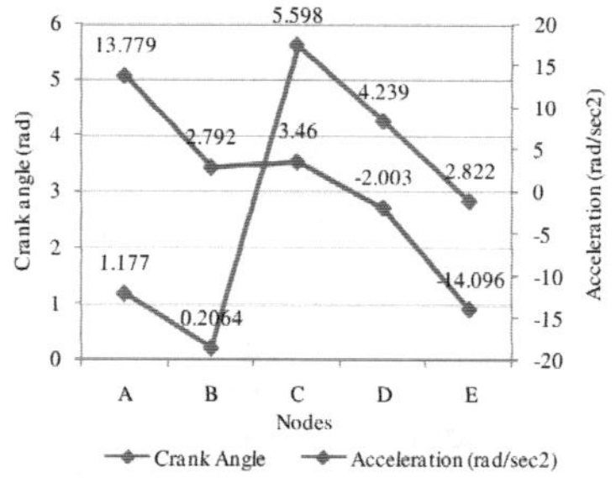

Figure 11. Node v/s Crank Angle & Acceleration

In iteration 3, the plots of node vs crank angle and velocity is shown in Figure 10 which infers that magnitude of angular velocity doesn't vary much between nodes C, D and E. In Figure 11, the plots of node vs crank angle and acceleration is shown which infers that the acceleration doesn't vary much between nodes B, C and D & acceleration between nodes A & E is almost same in magnitude, due to this phenomenon, there will be smooth transition in the acceleration from the upper phase of the foot locus to lower phase, as a result, the foot locus obtained under iteration 3 is considered as optimal. Further it has to be noted that the velocity & acceleration of links has significant influence on the interaction of robot leg with the ground. Since the frictional forces are greatly influenced by the acceleration of a body, leg making contact with the ground at very high acceleration should be avoided as it leads to slippage while walking.

8. Conclusions

The optimization of the Klann mechanism is performed for three different iterations of three different foot locus trajectory. The different foot locus helps in various applications where the foot locus with the higher step height is used in avoiding obstacles and foot locus with higher stride length helps in stability over the continuous locomotion of the robot. The results in this paper help in identifying the dimensions of the Klann mechanism which can be used for different robot Locomotive application. Further, the fabrication of the legged robot which utilizes the Klann mechanism has been completed and control system will be integrated with the robot & its performance analysis on different terrain will be tested.

References

[1] Jaichandar, Mohan Elara, Edgar Martínez-García, and Le Tan-Phuc Kulandaidaasan Sheba, ""Trajectory Generation and Stability Analysis for Reconfigurable Klann Mechanism Based Walking Robot.", *Robotics 5*, p. 13, 2016.

[2] J. A., Estremera, J., & De Santos, P. G Galvez, "A new legged-robot configuration for research in force distribution.," *Mechatronics*, pp. 8-9, 2013.

[3] A., Choi, P. P., & Whitaker, D Ghassaei, "The Design and optimization of a crank-based leg mechanism," Pomona College, 2011.

[4] J. C., Prieto, M., Armada, M., & De Santos, P. G. Grieco, "A six-legged climbing robot for high payloads," in *In Proceedings of the 1998 IEEE International Conference on Control Applications (Cat. No. 98CH36104)*, 1998, September.

[5] Jaichandar Kulandaidaasan, Mohan Rajesh Elara, Edgar Martínez-García, and Le Tan-Phuc Sheba, "Synthesizing reconfigurable foot traces using a Klann mechanism," *Robotica*, p. 11, 2017.

[6] Joseph E Shigley, "The Mechanics of Walking Vehicles," 1960.

[7] Artas Engineering Software. SAM Mechanism design. [Online]. https://www.artas.nl/en/sam/features/general

[8] W. Y., A. M. Liu, and D. Howard Jiang, ""Optimization of legged robot locomotion by control of foot-force distribution." , in *Transactions of the Institute of Measurement and Control 26.4* , 2004.

[9] M., Ratliff, N., Stolle, M., Chestnutt, J., Bagnell, J. A., Atkeson, C. G., & Kuffner, J. Zucker, "Optimization and learning for rough terrain legged locomotion.," *The International Journal of Robotics Research,* vol. 175-191, p. 30, 2011.

[10] Manuel Fernando, and JA Tenreiro Machado. Silva, ""A literature review on the optimization of legged robots.", *Journal of Vibration and Control 18.12*, vol. 1753-1767, 2012.

[11] Joseph E Shigley, "The Mechanics of Walking Vehicles," *Army Tank-Automotive Center Warren Mi*, p. 70, 1960.

[12] R. D., Nelson, G. M., Bachmann, R. J., Kingsley, D. A., Offi, J., & Ritzmann, R. E. Quinn, "Insect designs for improved robot mobility. Climbing and Walking Robots," *From Biology to Industrial Applications, 59.*, (2001).

Influence of vehicle parameters on handling characteristics and its control using torque vectoring

Mayuresh Mutha[1], Chinmay Karanjkar[1], Prasad Kadekar[1], Mandar Nikumb[1] and Sakthivel Palanivelu[2]

[1]Department of Automotive Engineering, VIT University, Tamil Nadu, India.
[2]Associate Professor, Department of Automotive Engineering, VIT University, India.
E-mail: muthamayuresh.ajay2018@vitstudent.ac.in

Abstract— Handling characteristics of any vehicle refer to its directional control and stability for given steering input. The vehicle parameters such as mass, mass moment of inertia and location of C.G. play a vital role and decide handling response of vehicle under different scenarios. Safe drive is ensured by vehicle's responses such as yaw rate, roll rate and vehicle lateral acceleration. This paper presents steady state and transient state simulation of XUV 500 model in accordance with ISO 4138 and ISO 3888 test standards using CarSim-Simulink Co-simulation. Influence of changes in above mentioned parameters on Vehicle response during both tests is investigated. Further, control of power train output using torque vectoring is implemented in both tests and its effect on vehicle response is evaluated. The results show that understeer tendency of the vehicle is reduced in steady state test (Constant radius test: ISO 4138). Critical vehicle responses of yaw rate, roll rate, lateral acceleration and path tracking showed improvement during the DLC manoeuvre (ISO 3888) hence, improving stability at higher lateral accelerations.

1. Introduction

The concept of vehicle handling has always been more subjective than objective, based on how a trained test driver describes his drive. There are however, objective metrics developed to understand a very particular quality of a vehicle which give an idea of how it will behave under specific conditions. Subjective handling tests are, moreover, comparison based and largely depend on driver. Sideslip angle data measured on vehicles clearly indicated that vehicles were safe in cornering but psychophysics showed that driver can get a perception of vehicle not being safe[1]. Thus, subjective tests can sometimes misguide the evaluation. Variation in tyre slip leads to perception of stability or instability. Thus, Objective metrics of vehicle also need to be evaluated.

1.1. Steady state circular driving behaviour-open loop test

ISO 4138 prescribes standard tests to determine Understeer characteristics of vehicle. Various tests prescribed in this standard are listed in the table below [2]. The gradient is measured as a slope of the plot of the Measured vs. Varied parameter.

Table 1. List of tests as per ISO 4138

Test	Radius	Steer angle	Speed
Constant radius test	Constant	Measured	Varied
Constant steer wheel angle test	Measured	Constant	Varied
Constant speed discrete turn radii test	Varied	Measured	Constant
Constant speed discrete steering wheel angle test	Measured	Varied	Constant

A real-time factor that was observed to affect the performance of vehicles in steady state cornering tests was torque distribution to the wheels. This directly related to how the power train was set up and which wheel received greater torque during turns. This bias of torque created a virtual assist/hinder to load transfer during these tests.

IOP Conf. Series: Materials Science and Engineering **624** (2019) 012015 doi:10.1088/1757-899X/624/1/012015

1.2. Double lane change test

ISO 3888 gives general rules to follow for a double lane change manoeuvre test [4]. DLC test is used to determine vehicle transient handling characteristics. It simulates emergency turning or cornering manoeuvre. Vehicle will only avoid accident if it is controllable during sudden manoeuvre, otherwise it can go out of control and can result into roll over.

1.3. Torque Vectoring to improve Vehicle Handling

To prevent the spinning and drifting of vehicle especially during sharp manoeuvres, there is a need to monitor the yaw stability of the vehicle. The yaw stability control system is either a differential braking based or torque distribution based [5]. But, the vehicle systems based on brakes, on sharp manoeuvres, deteriorate the longitudinal performance [6]. Hence, here we are dealing with a vehicle incorporated with torque distribution. Different control strategies can be used for torque distribution depending upon the control algorithm used. Due to independently controlling the drive torque to each wheel, vehicle's stability and handling characteristics are significantly enhanced. The main focus of vehicle's dynamic control system is to improve control of vehicle to prevent unintended vehicle behavior and thus to ensure safety of both driver and vehicle.

In this paper, the handling and stability performance of the vehicle is investigated by means of CarSim and Simulink co-simulation. A Simulink model is designed to vary the left and right torque distribution. Characteristic to be controlled is the understeer behavior of the vehicle [7]. The results indicate that the torque management strategy is generally very effective in reducing the understeer characteristics of vehicle and improve handling characteristics at higher vehicle speeds. In this Simulink model the resulting system shows a significant improvement over conventional driveline configurations (without torque vectoring control) under aggressive cornering.

1.3.1. Torque distribution strategy. Based on the handling and stability characteristics, a left and right torque distribution controller is modelled to optimize the vehicle maneouverability for various conditions proposed earlier like variation of position of the CG, mass, etc. Georg Rill investigated and analyzed different torque distribution ratios for front left and right wheels [8]. It was proved that greater torque to the outer wheel reduced the understeer gradient. Based on this, a controller is designed to give a fixed torque distribution of 90:10 (outer: inner) while cornering and steady state test is performed.

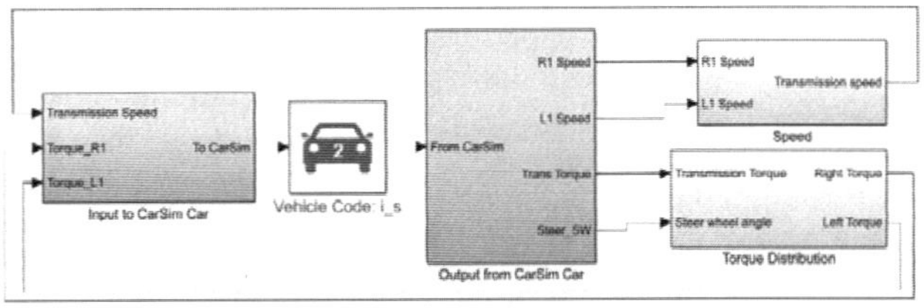

Figure 1. Simulink model of Torque vectoring strategy used.

1.3.2. Yaw rate Control strategy. Vehicle stability is achieved in this paper by controlling the vehicle yaw rate with the help of torque controller model designed in Simulink [9] - [12]. The yaw rate error (e) is formulated as follows,

$$e = r - r_d \tag{1}$$

Where, r = Actual Yaw rate input from CarSim and r_d = Desired Yaw rate

The desired Yaw rate is calculated by using,

$$\left.\begin{aligned} r_d &= r_d && |r_d| < \left(\frac{\mu \times g}{V_x}\right) \\ &= \left(\frac{\mu \times g}{V_x}\right) \times \operatorname{sgn}(r_d) && |r_d| \geq \left(\frac{\mu \times g}{V_x}\right) \quad \text{Where } r_d = \left(\frac{V_x}{\left(l \times \left(1 + \left(K \times V_x^2\right)\right)\right)}\right) \times \delta \end{aligned}\right\} \tag{2}$$

V_x: Longitudinal Velocity $\quad l$ = Wheel Base $\quad \delta$ = Front Wheel Steer Angle

The Stability factor (K) takes the form, $K = \frac{m}{l^2} \times \left(\left(\frac{l_r}{C_f} \right) - \left(\frac{l_f}{C_r} \right) \right)$ (3)

The desired yaw moment is then calculated using the following formula,

$$M_{zz} = \left(I_{zz} \dot{r}_d - I_{zz} \dot{r} \right) \tag{4}$$

From the above moment, corrective torque T_z is found as follows,

$$T_z = M_{zz} \times \left(\frac{R}{d} \right) \tag{5}$$

Where, R = Wheel radius d = Track width

This is implemented in CarSim Simulink as shown in figure 2 below and DLC test is performed.

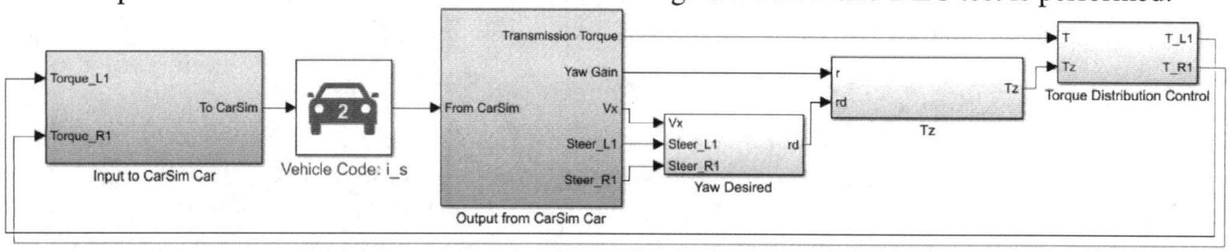

Figure 2. Simulink model of Yaw Control Strategy used.

2. Vehicle Model

CarSim software was used to simulate ISO Constant radius test and DLC test in order to study handling of vehicle. Mahindra XUV 500 was modelled to perform the tests.

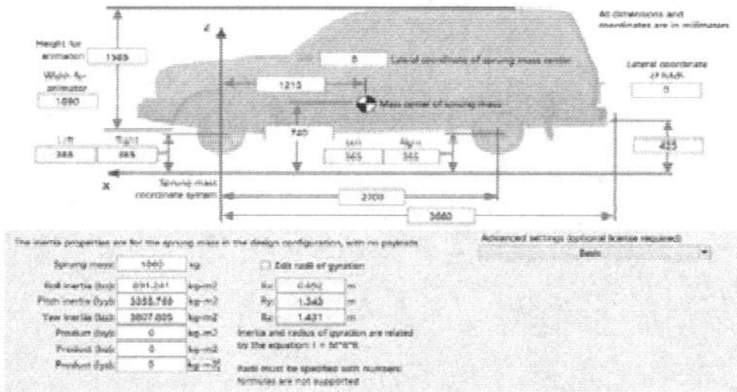

Figure 3. Vehicle Dimension, Sprung Mass and Inertia input screen in CarSim.

3. Results

3.1. Steady state circular driving behaviour-open loop test method- ISO 4138 Constant radius test

From figure 4, very less variation is observed in steer angle with variation in mass at lower values of lateral acceleration. It is clear that the understeer characteristics of vehicle show almost no change with variation in sprung mass. From figure 5, little variation is observed in required steer angle with variation in C.G. height at lower speed. Whereas at lateral acceleration above 0.65 g, there is some deviation in required steer angle to manoeuvre the same turn. Vehicle shows reduction in understeer with decrease in C.G. height and increase in understeer characteristics with increase in C.G. height.

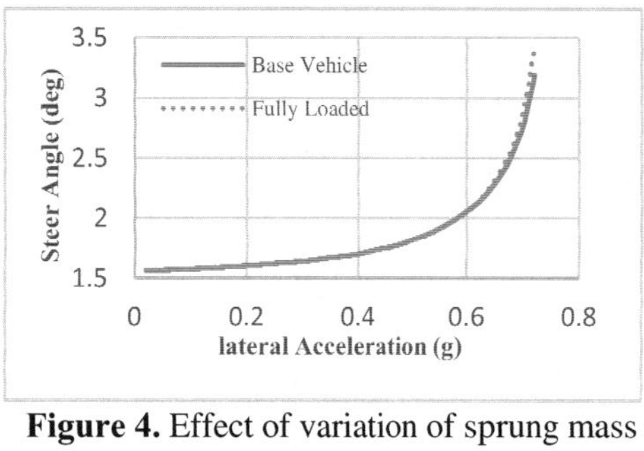

Figure 4. Effect of variation of sprung mass on Steer angle.

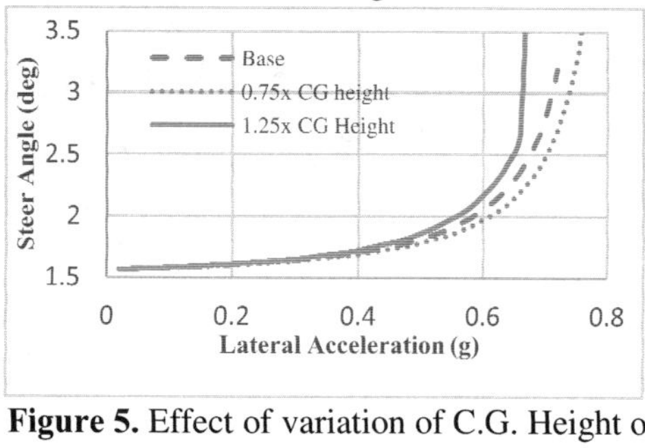

Figure 5. Effect of variation of C.G. Height on Wheel Steer angle.

From figure 6, vehicle's behaviour shows overall reduction in understeer with shifting longitudinal position of C.G. behind and increase in understeer characteristics with shifting longitudinal position of C.G. ahead.

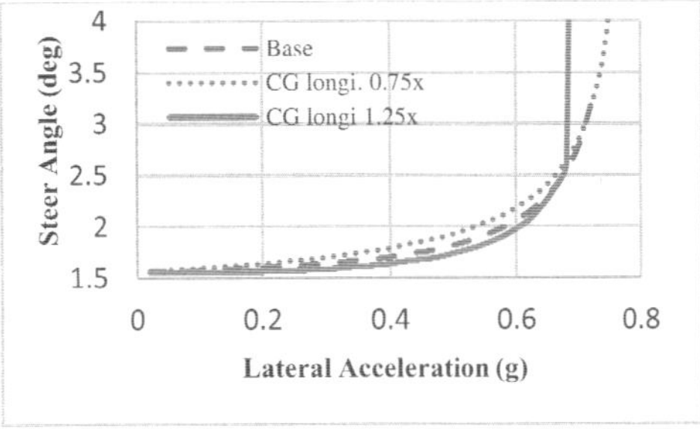

Figure 6. Effect of variation in longitudinal position of C.G. on Steer angle.

3.2. Torque Vectoring to improve Vehicle Handling

From figures 7 and 8, it is evident that even in extreme variations in location of C.G. in both vertical and longitudinal directions (which increases understeer behaviour), understeer tendency of the vehicle significantly reduces after torque vectoring. The lateral acceleration limit (after which the vehicle response overshoots as seen in the plots) increases from about 0.66g to about 0.73g in case of 25% increase in C.G. height and from 0.68g to 0.76g for 25% shift of C.G. towards front axle. Similar trend is observed in case of full load condition as seen in figure 8.

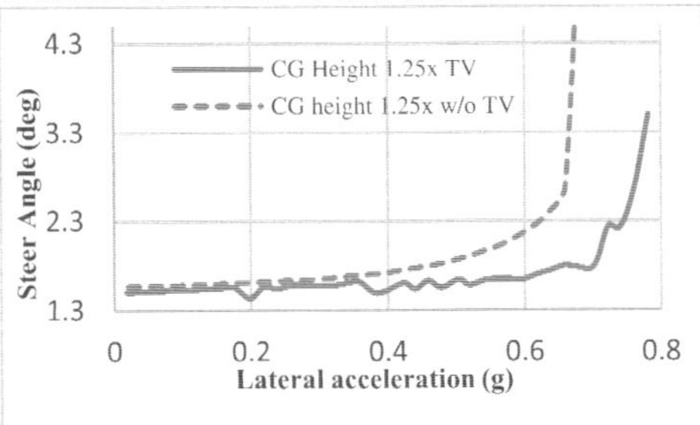

Figure 7. Variation of C.G. Height on Wheel Steer angle with & without Torque Vectoring.

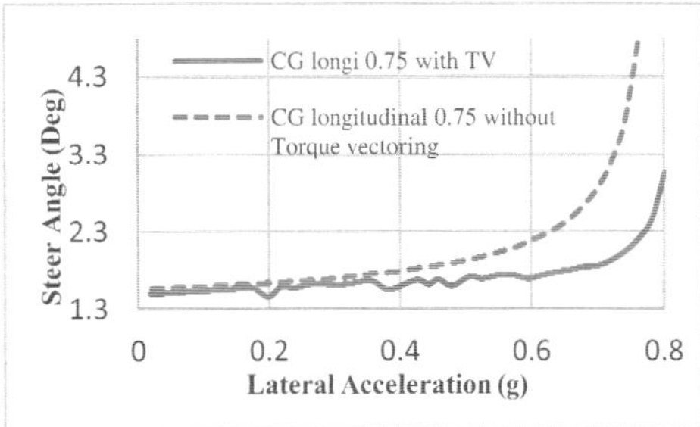

Figure 8. Variation in longi. position of C.G. on Steer angle with & without Torque Vectoring.

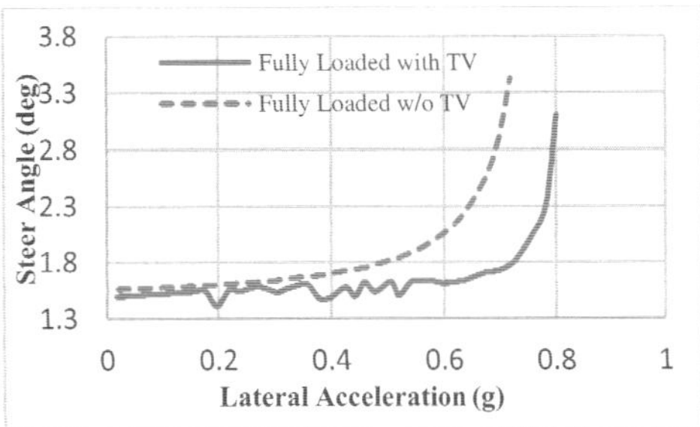

Figure 9. Effect of variation of sprung mass on Steer angle with and without Torque Vectoring.

3.3. Double lane change test- ISO 3888

It can be observed from figure 11 that roll rate of the sprung masses in the vehicle is extremely sensitive to changes in vertical C.G. location with a 50% increase in height of C.G. giving a RMS change of about 85% at 50kmph and 90kmph both. Such a severe change in roll rate will affect a real driver's response and results in an actual test may therefore vary.

All the characteristics (Roll rate, Yaw rate and steering input) show a significant increase when the vehicle speed used to simulate the test is increased. The vehicle follows the prescribed path closely at

50kmph and a greater deviation is seen when the vehicle speed is increased to 90kmph. Therefore, it can be concluded that the compliance to the prescribed path is a major function of vehicle speed

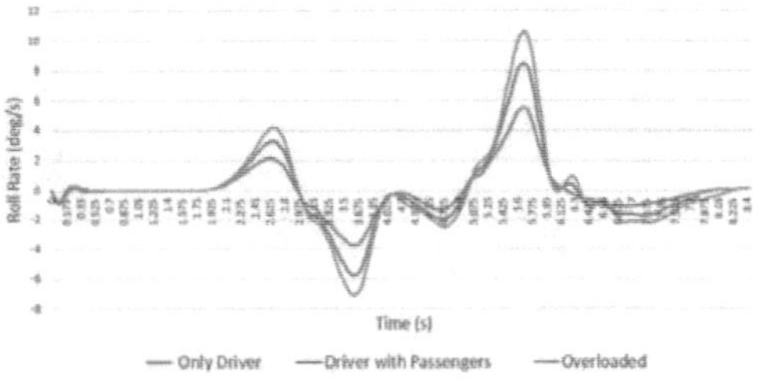

Figure 10. Roll rate vs. time for Driver only, Fully loaded and Overloaded Condition.

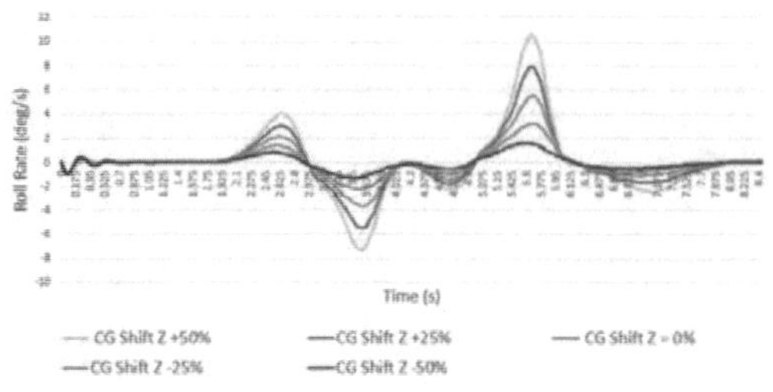

Figure 11. Roll rate vs. time for Variation in C.G. Height.

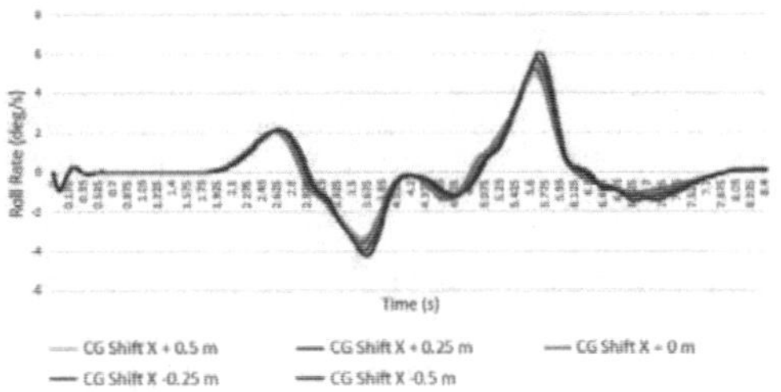

Figure 12. Roll rate vs. time for Longitudinal Variation in C.G.

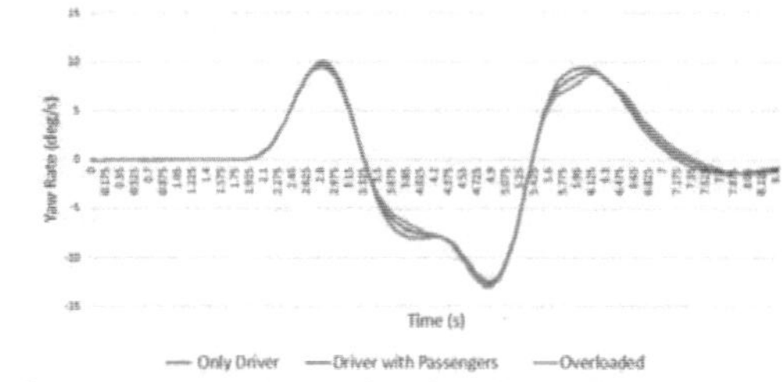

Figure 13. Yaw rate vs. time for Driver only, Fully loaded and Overloaded Condition.

3.4. Yaw rate Control strategy

Vehicle yaw rate, Lateral Tracking, roll angle, Lateral acceleration for the vehicle with and without torque vectoring are compared. Figure 14 represents the lateral tracking of the vehicle during performing DLC test from which it is observed that the vehicle with torque distribution strategy implemented on it resembles to the target path more closely as compared with the vehicle without implementing the torque distribution. Figure 15 is a measure of vehicle's stability in terms of its yaw rate experienced while performing the test. Vehicle with control strategy happens to follow the desired track as closely as possible due to corrective torque applied unlike the vehicle without any control strategy. Roll angle of the sprung masses as shown in the Figure 16 for the vehicle with and without control strategy states about the tendency of the vehicle to roll while maneouvering. Less roll angle in case of the vehicle with control strategy ensures the vehicle stability. Comparison of the lateral acceleration given in the Figure 17 shows that controlled vehicle experiences less lateral acceleration which is desirable during maneouvering as compared to the uncontrolled vehicle i.e. without any control strategy.

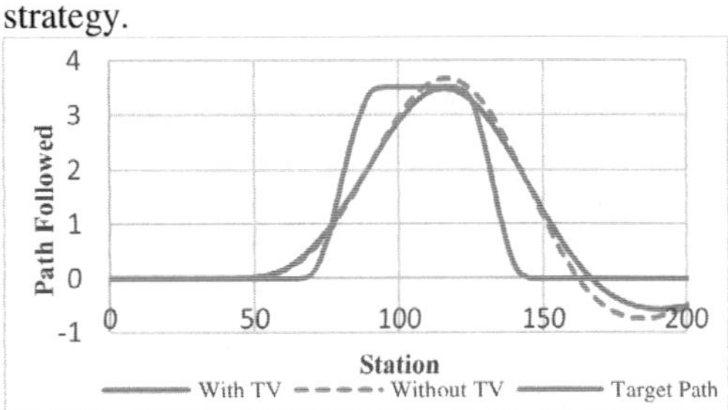

Figure 14. Lateral tracking for vehicle with and without Torque Vectoring

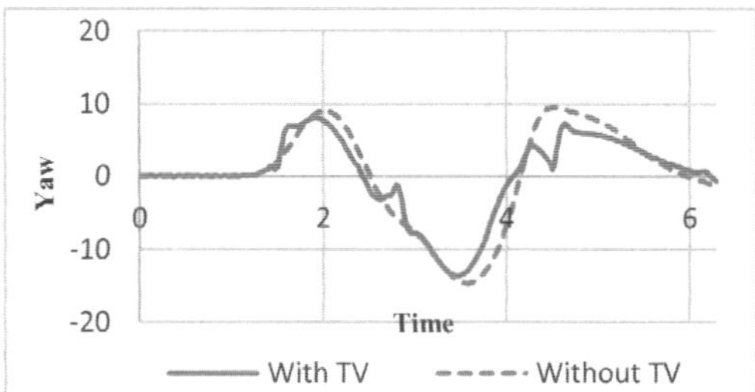

Figure 15. Yaw Rate vs. time for vehicle with and without Torque Vectoring

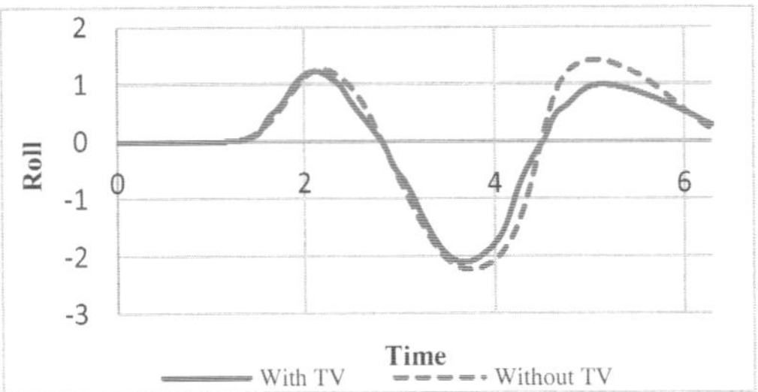

Figure 16. Roll angle vs. time for vehicle with and without Torque Vectoring

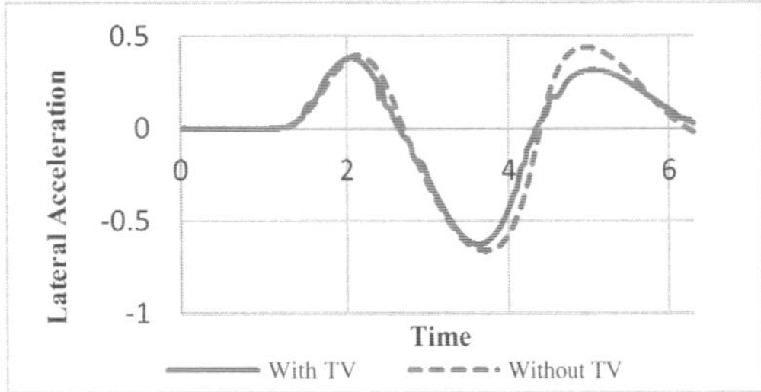

Figure 17. Lateral Acceleration vs. time for vehicle with and without Torque Vectoring

4. Conclusion

Torque vectoring is observed to be effective in steady state as well as transient scenarios. With implementing torque vectoring control strategy it was observed from the respective graphs that in steady state test, understeer gradient of vehicle was reduced from 0.5114 deg/g to 0.3288 deg/g for fully loaded condition, 0.7516 deg/g to 0.5168 deg/g for C.G. longitudinal variation and from 0.6211 deg/g to 0.4259 deg/g for C.G. vertical variation case. While in transient test, RMS values of yaw rate, lateral acceleration and roll angle of the vehicle indicated reduction of 18%, 14% and 15% respectively. This system can be designed for any vehicle and optimized to improve handling performance and safety characteristics.

Acknowledgments

Authors would like to acknowledge the facilities provided by Automotive Research Centre VIT to carry out this research work.

References

[1] Ganzarolli F and Leal AT 2004 Influence of Understeer Gradient Variation during Cornering in the Vehicle Stability Perception *SAE Technical Paper* 2014-36-0209
[2] ISO 4138: 2012. Passenger cars–steady-state circular driving behaviour–open-loop test methods
[3] Liang CG 2011 A Constant Radius Constant Speed Simulation Methodology-Yaw Rate Control *SAE Technical Paper* 2011-01-0738
[4] ISO 3888 Passenger cars- Test track for a severe lane change manoeuvre
[5] Latif R, Azim RA, Baqai AA and Shafi I 2014 Vehicle Modeling and Performance Evaluation Using Active Torque Distribution *SAE Technical Paper* 2014-01-0103
[6] Hancock MJ, Williams RA, Gordon TJ and Best MC 2005 A comparison of braking and differential control of road vehicle yaw-sideslip dynamics *Proceedings of the Institution of Mechanical Engineers, Part D: Journal of Automobile Engineering* pp 309-327
[7] Karogal I and Ayalew B 2009 Independent torque distribution strategies for vehicle stability control *SAE Technical Paper* 2009-01-0456
[8] Rill G 2017 Reducing the cornering resistance by torque vectoring *Procedia engineering* pp 3284-3289
[9] Huang F, Fu C, Tang X and Hu M 2017 Study on stability and handling characteristics and control of four-wheel-drive electric vehicles *Advances in Mechanical Engineering*. 9(12):1687814017737725
[10] Drakunov SV, Ashrafi B and Rosiglioni A 2000 Yaw control algorithm via sliding mode control. *InProceedings of the 2000 American Control Conference* Vol 1 ACC pp 580-583 (IEEE Cat. No. 00CH36334)
[11] Ding S, Liu L and Zheng WX 2017 Sliding mode direct yaw-moment control design for in-wheel electric vehicles *IEEE Transactions on Industrial Electronics* 64(8) pp 6752-6762
[12] Canale M, Fagiano L, Ferrara A and Vecchio C 2008 Vehicle yaw control via second-order sliding-mode technique *IEEE Transactions on Industrial Electronics* 55(11) pp 3908-3916

IOP Conf. Series: Materials Science and Engineering **624** (2019) 012016 doi:10.1088/1757-899X/624/1/012016

A Novel Methodology to Optimize Design Parameters of an Asymmetric Gear

Umar Farook N, Parthasarathy R, Murshid P K, Anikesh Kumar Rathod, Naveen Raghava Krishnan S B, Anoop Kumar Verma, K P Vineesh, Sreenath A M
Department of Mechanical Engineering, National Institute of Technology Calicut, India

E-mail: thoppil2693472@gmail.com

Abstract. Gears play an important role in mechanical power transmission. It is critical in several design scenarios to optimize its weight without compromising the power transmission requirements. It has been noted that the majority of the gears transmit power only in one direction during most of their lifetime. However, conventional gear design procedures do not account for it and assume equal power transmission in both directions. This results in equal strength in both direction and becomes over design in one direction. It may be possible to reduce the weight of the gear if the power transmission is predominantly in one direction. In the present work, an asymmetric pair of gears were designed for this purpose. The gears were checked and found satisfactory in various aspects such as interference, backlash etc. Design of experiments principle is applied to reduce the number of combinations of the asymmetric gears. The selected gears are analyzed using the finite element method and the torque capacity of asymmetric gears are estimated. Asymmetric gears reduce the bending stress up to 25% in one direction when compared to symmetric gears of comparable size and weight. The selection of suitable design parameter for a pair of asymmetric gears for given power transmission is also the scope of the present work.

1. Introduction

The conventional gear designs were carried out with a symmetric gear profile and equal strength in both directions. In some cases, such as automobile gearbox, the transmission duration and the torque levels are vastly different in two directions. Designing gear teeth for the maximum value of torque and duration may grossly over design the gears. This is especially true in automobile applications, in which a low contact ratio (LCR) gears are used.

The classical solution to this problem is to use the asymmetric gears, with different geometry at two sides of the gear teeth. This gear design philosophy was not popular, due to the complexities in design, lack of standards in design and manufacturing, and other related factors. Due to improvements in the powder metallurgy and injection moulding processes, it is becoming practical to manufacture gears through these process in considerably large scale itself. These improvements open up the possibility to use the asymmetric gears in large-scale applications.

The simplest asymmetric gear structure employs the involute profile of different pressure angle at two sides of the gears. It is expected that the side with the higher-pressure angle will be able to take a higher load than the other side. The gear pairs may be checked for various kinematic parameters such as interference, the minimum number of teeth etc.

One of the major challenges in implementing the asymmetric gears is the lack of standards in the design and manufacturing of the gears. All the design factors such as Lewis form factor etc. will be varied significantly when compared to the standard counterparts. The scope of this paper is to develop such design data and standards for the design of asymmetric gears.

Adequacy in bending load carrying capacity of a gear unit is a major problem. The critical section of a gear tooth experiences bending stresses that cause most of the failures [1, 2]. It is observed that over 90% of all power gear trains have only one main direction for torque transmission [3]. Therefore gear experiences only one directional loading. As the loading is unidirectional, it is not expected for the drive side to be symmetric with the coast side which permits the designers to design the gears with asymmetric teeth [4]. These gears provide flexibility to the designers due to the non-standard geometry of the gear tooth. If designed judiciously, these gears can make important contributions to the improvement of designs in aerospace industry, automobile industry, marine drives and wind turbine industry, where weight to torque ratio plays a critical role [5, 6]. Due to their geometry, these gears allow for selection of different pressure angles on the drive side and coast side, which is very crucial in obtaining key properties, such as high load carrying capacity and low weight. The high-pressure angle on drive side enables to reduce the bending stresses as well as contact stresses whereas low-pressure angle on drive side helps to reduce the bending stresses and keep the contact stresses on the same level as for symmetric teeth.

As far as the research on asymmetric gears, till date, there hasn't been a standard that defines all relations between parameters such as torque capacity, pressure angles, Lewis form factor, bending stress and weight. The current research is an attempt to formulate the same.

2. Methodology

2.1. Kinematic Considerations

The gear geometry was modelled as two involute profiles with different pressure angles. The equations of the involutes were used to determine the profile of teeth. The equation for the profile of the gear teeth is given as

Involute profile equation:

$$x_2 = x_1 \cos\phi + y_1 \sin\phi + r(\sin\phi - \phi\cos\phi) \quad (1)$$

$$y_2 = -x_1 \sin\phi + y_1 \cos\phi + r(\cos\phi + \phi\sin\phi) \quad (2)$$

Where,

x_1, x_2 = x-coordinate of involute profile in parametric equation

y_1, y_2 = y-coordinate of involute profile in parametric equation

r = radius of the base circle

ϕ = involute pressure angle

A MATLAB code was made to generate asymmetric gear profiles for given gear parameters. The gear was modeled in Solid Works based on the profile generated by MATLAB.

2.2. Development of Design process

Design of the asymmetric gears is carried out in following steps.

a. Selection of asymmetry factor ($\cos\phi1 / \cos\phi2$)

b. Check for kinematics feasibility such as interference, undercutting etc.

c. Selection of Lewis form factor and other design values.

d. Checks for fatigue, dynamic loadings etc.

FEM analysis was conducted for various combinations of the asymmetric gears and the Lewis form factor for the various asymmetry is calculated from estimated stress values from FEM results.

2.3 Numerical experimentation

The coast side pressure angle is taken as 200 since this is the widely used pressure angle for symmetric gear. Drive side pressure angle is varied from 20 to 480 and the number of combinations of drive side pressure angle is selected to cover the maximum range of values in drive side pressure angle. The module is taken as 4 mm, higher value than widely used, so that the width of the gear can be reduced while keeping the addendum constant. Number of teeth of gears are limited to 40 to minimize the number of finite element simulations. Details of selected values are given in Table 1.

Table 1. Input parameters for analysis

Module	**Gear ratio**	**No. of teeth**	**Drive side pressure angle**	**Coast side pressure angle**
4 mm	1	Varied from 10 to 40	Varied from 20 to 48 degree	20 degree

The asymmetric involute profile was generated by using MATLAB. This data is converted as solid model using the SolidWorks software. The file is then imported to ANSYS for analyzing for the stress strain response. A sample of the meshed part is given in Figure 1. -

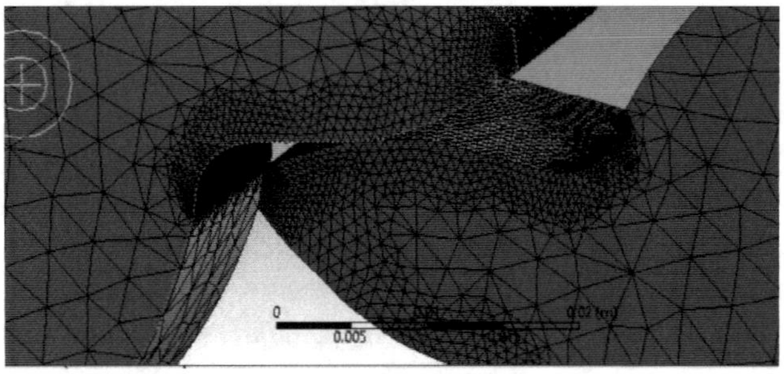

Figure 1. Meshing of gear for analysis

Mesh refinement was done near the tooth profile near the tooth surface instead of on the whole gear as shown in Figure 1. Torque (1 kN-m) is applied at the highest point of single tooth contact. Fixed support is given to the shaft of the driven gear and friction-less support is given to shaft of driver gear. Rim region is not fixed to include effects of rim deformations. The material properties used for finite element analysis are given in Table 2.

Table 2. Material properties used for finite element analysis

Material	**Density**	**Compressive Yield Strength**	**Tensile Yield Strength**	**Poisson's Ratio**	**Number of elements**	**Mesh Type**
Structural steel	7850 kg m^{-3}	250 MPa	250 MPa	0.3	72101	Tetrahedrons

3. Results and Discussion

3.1 Lewis form factor for asymmetric gear

Table 3 shows the Lewis form factor for various values of asymmetry factor and number of teeth. Asymmetry is the ratio of cosine of drive side pressure angle to cosine of coast side pressure angle. Values of asymmetry has been varied from 1 to 1.5. Total of 36 combinations have been checked and their Lewis form factor is evaluated from finite element result. For symmetric gear, asymmetry is one, Lewis form factor estimated are in good agreement with the standard design data book. Estimation of Lewis form factor is estimated only up to 40 teeth.

Table 3. Lewis Factor values for each asymmetry factor, C and number of teeth, N

N C	1.000	1.100	1.200	1.300	1.400	1.500
10	0.201	0.232	0.237	0.252	0.263	0.274
12	0.245	0.272	0.282	0.294	0.306	0.319
15	0.289	0.314	0.326	0.339	0.352	0.366
20	0.320	0.354	0.371	0.385	0.400	0.416
30	0.358	0.399	0.413	0.429	0.445	0.463
40	0.389	0.403	0.417	0.432	0.448	0.466

3.2 Weight and Bending stress variations with respect to asymmetry

Figure 2 shows the weight increase of the gear with increase of drive side pressure angle, increase of asymmetry factor. Coast side pressure angle is kept at 200 and the drive side pressure angle varies from 20 to 480.

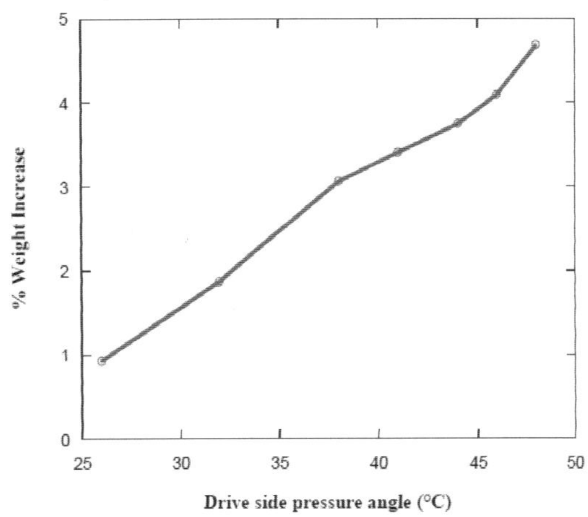

Figure 2. % Weight increase of asymmetric gear with respect to drive side pressure angle.

Percentage increase of weight with respect to weight of symmetric gear of pressure angle 200 is compared. It can be seen that % increase in weight is almost linearly increasing with drive side pressure angle. Even with increase of pressure angle of drive side to 400 from 200, weight is increased only by about 3.5%. Figure 3 shows the reduction of % bending stress with respect to increase of drive side pressure angle. Stress developed in symmetric gear with 200 pressure angle is taken as the basis of stress calculation. Even for an increase of drive side pressure angle to 260, which causes an decrease of bending stress by about 15%, but the weight of gear is increased only by about 1% compared to symmetric gear. For a pressure angle of 480, bending stress is reduced by 25.5 % while the weight of gear is increased by 4% only.

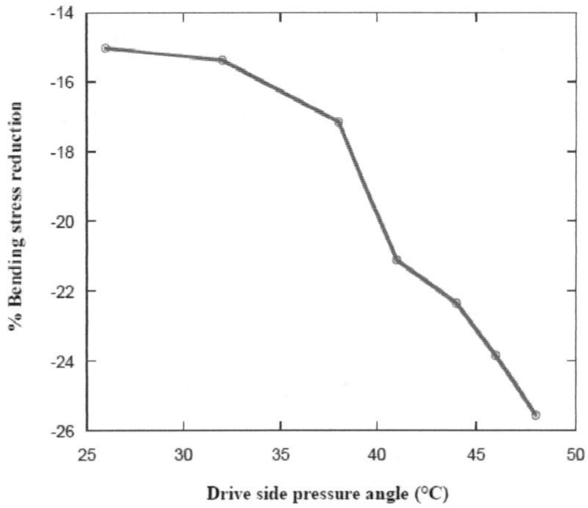

Figure 3. % bending stress reduction with respect to drive side pressure angle.

4. Conclusions

Geometry of the asymmetric gear was modelled using involute profile with different pressure angle on coast and drive side. Lewis form factor is estimated for different asymmetry and number of teeth for facilitating the design process. Maximum bending stress at the root during power transmission can be reduced by using asymmetric gear.

5. References

[1] Bolotovsky IA, Vasil'eva OF, Kotelnikov VP (1984) Involute gears with asymmetric teeth. Vestnik Mashinostroeniya 4:15–17

[2] Definition of involute spur gear profiles generated by gear-type shaper cutters , Mechanics Based Design of Structures and Machines, 38: 481–492, 2010

[3] Kumar, V. S., Muni, D. V., & Muthuveerappan, G. (2008). Optimization of asymmetric spur gear drives to improve the bending load capacity. Mechanism and Machine Theory, 43(7), 829-858.

[4] Evaluation of load capacity of gears with asymmetric tooth profile , Tomova Masuyama and Naoki Miyazaki Masuvama and Miyazaki, International Journal of Mechanical and Materials Engineering (2016) 11:11

[5] Kapelevich AL (2000) Geometry and design of involute spur gears with asymmetric teeth. Mech Mach Theory 35:117–130

[6] Vulgakov EB, Rivkin GV (1976) Design of gears with asymmetric tooth profile. Mashinovedenie 5:35–39 Vulgakov EB, Rivkin GV (1976) Design of gears with asymmetric tooth profile. Mashinovedenie 5:35–39

IOP Conf. Series: Materials Science and Engineering **624** (2019) 012017 doi:10.1088/1757-899X/624/1/012017

Gear loss factor using the load distribution model for varying contact ratio in spur gear drive for improved bending strength

Ravivarman R[*1], Palaniradja K[2], Prabhu Sekar R[3]

[1,2] Department of Mechanical Engineering, Pondicherry Engineering College, India
[3] Mechanical Engineering Department, Motilal Nehru National Institute of Technology, Allahabad, India
E-mail: varman92@gmail.com

Abstract. Gear loss factor is one of the imperative parameter in estimating the power loss in a meshing gear pair. This paper demonstrates the prediction of gear loss factor parameter in spur gear drive using the load distribution technique. The aptness of two distinct contact ratios is studied and compared in this paper: Normal Contact Ratio (NCR) and High Contact Ratio (HCR). The described model is studied for gearing strength which is improved through profile modification technique. The ultimate focus is to enhance the gearing strength of the pinion in a spur gear drive by discovering the optimal modification point in the tooth profile with at most balanced stresses in the gear pair considered for the study. Using contact analysis, the stresses developed over the contracting surfaces are predicted through finite element method and it is balanced between the gear and pinion. A gear loss factor is predicted along the path of contact for a complete meshing cycle for this balanced drive. It is uncovered that the methodology utilized encourages significant decrease in the gear loss factor of the HCR pair which is confined with the literature models. The values predicted using the load distribution model shows that the high contact ratio gear drive with increased addendum height has the gear loss factor minimum compared with a normal contact ratio gear drive.

1. Introduction

Gear drives are the most widely familiar transmission systems. Abundant researchers have focused on different highlights of the gears and gear trains. The main motive for the designers in spur gear transmission is to develop an efficient drive with improved life period. Design of gear drives with minimum power loss and enhanced life is a thrust for designers in gear design for developing more reliable and efficient mechanisms. In recent times more focus has been given to the HCR gear pair due to the increased load share with high strength and reduced weight. Several models and studies are available on the HCR spur gear pair [1, 2] in calculating the losses. The introduction of loss factor term in predicting the power loss gets its significance from the early stages of research in transmission drives which is majorly caused by the load acting in the gearing system. Buckingham [3] and Ohlendorf [4] was the first to introduce the loss factor (H_v) parameter in predicting the power loss but the effect of the load acting on the tooth during meshing and the tooth profile changes were not considered in their model. The contact load plays a vital role in the prediction of the gear loss factor acting along the path of contact. Many researchers have discussed about the prediction of load share in the gear drive. Load share ratio is estimated along the contact path by Pedrero et al. [5] in his study. Li [6, 7] used contact analysis to predict the load distribution through Finite Element Methods (FEM). Thus the expansion of numerical simulation allows the designers to predict the load distribution accurately as if in the physical model using the Finite Element Methods. Here the load and stresses are estimated using FEM and the predicted stresses are balanced through profile modification technique (Figure 1) which has been elaborated in the previous studies made by the authors [8-11]. This paper presents a model for the calculation of gear loss factor by including the load distribution. This load distribution is evaluated for an improved bending strength pinion by providing variation in the tooth profile. With this equation, the gear loss factor, and accordingly the gear loss factor at any point of the contact path has been evaluated, both for normal contact ratio and high contact ratio spur gears.

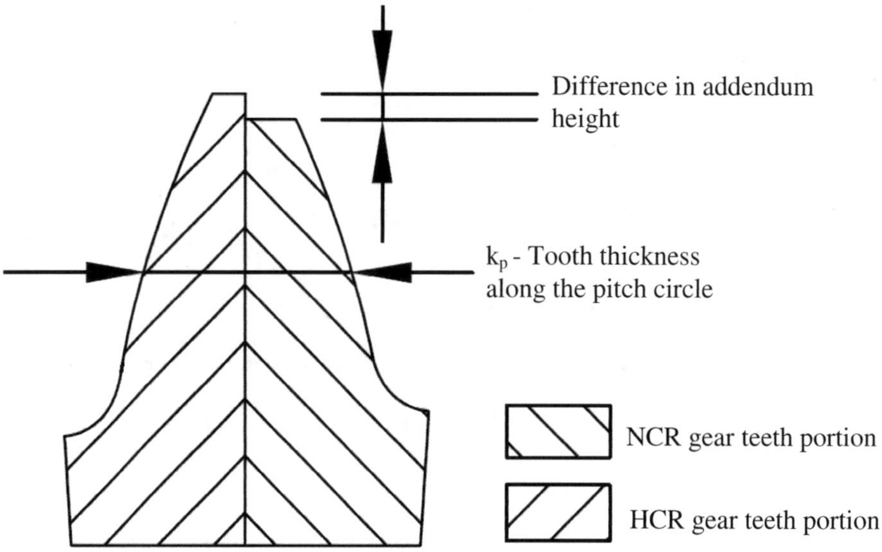

Figure 1. Variation in teeth height for varying contact ratio.

2. Load Share Ratio and Gear Loss Factor for Balanced Gear Drive

In spur gear drives, the teeth come in contact at engagement points during meshing by sharing the load for an instant of time. This sharing of tooth load take place in different pattern based on the type of gear pair and contact ratio. Normally in spur gear drive the tooth load sharing is divided based on the contact ratio (ε). For a normal contact ratio gear pair (ε<2) the tooth load is shared by single pair and double pair regions which occurs for particular time period during meshing. Whereas for a high contact ratio gear pair (ε>2) the tooth load is shared by double and triple pair regions. Hence it become important to predict the way in which the load sharing takes place in a gearing pair for the varying contact ratio. The load share ratio can be evaluated from the equation given below and the contact forces are assessed based on the contact analysis technique.

$$\text{Load Share Ratio, } LSR = \frac{F_i}{F_n} \tag{1}$$

Where,

F_i- Contact force at any instantaneous point
F_n- Cumulative contact force

Gear loss factor (H_v) is an important parameter in predicting the efficiency of the drive. In this study, the gear loss factor is estimated by including the load share ratio into account which has been neglected in the previous methods of predicting the loss factor. The gear loss factor (H_v) of a spur gear pair during meshing can be expressed from the load share ratio (*LSR*) as:

$$(H_v)_i = \frac{1}{\varepsilon p_b}(LSR)_i\left(\frac{(v_s)_i}{(v_b)_i}\right) \tag{2}$$

Where, p_b is the base pitch, v_s and v_b is the sliding and base velocity.

The equation (2) is a generalized form of the Wimmer [12] approach in predicting the loss factor along the contact path for one complete meshing cycle.

3. Description of the finite element model

The finite element model used for this study is shown in the figure 2 which includes the gear and pinion with contact initiated. Here an APDL program is developed and used in ANSYS for computation. Generation of the tooth profile is initiated by the trochoidal portion followed by the involute portion. A Bi-dimensional FE study has been carried with element PLANE 82 to form the model. An equivalent torque is applied in the pinion hub which will induce a relative motion among the drive through tooth deflection. In terms of constraints, pinion can rotate around the axis whereas motion of the gear is prevented. CONTA172 and TARGE169 are the considered elements at the contact zones. The total number of element count is around 158,000 for NCR and 316,000 for HCR transmission drive studied. In which the critical areas at the contact zones are fine meshed with convergence study. Material considered for the study was steel with a modulus of rigidity of 210 GPa and Poisson's ratio of 0.3. The parameters taken for the study is as follows: Module (m) =1, Number of teeth in pinion (z_p) =50, Gear ratio (i) =1.5, Pressure angle at pitch circle (α_o) =20^0, Addendum height (h_a) =1 (For NCR) and 1.2 (For HCR).

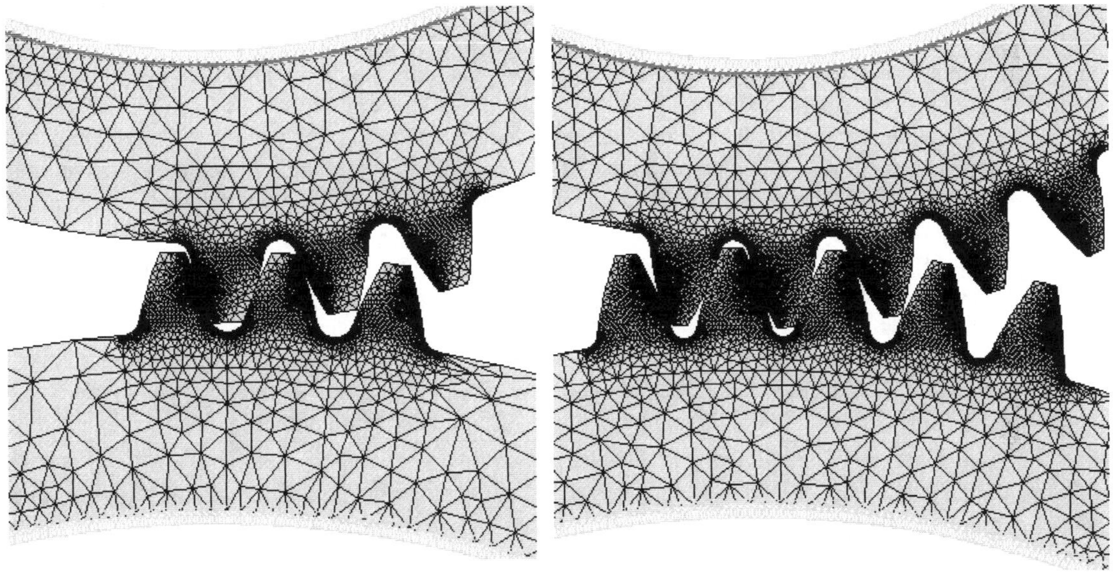

(a) NCR spur gear pair. (b) HCR spur gear pair.

Figure 2. FE model study for analysis.

4. Gear loss factor- a comparison with literature models

The gear loss factor is evaluated for the normal and high contact ratio spur gear pair using equation (2) by including the load share ratio as mentioned. Figure 3 shows the comparison of the gear loss factor (H_v) with the literature models. As discussed in the introduction section Buckingham [3] and Ohlendorf [4] evaluated the gear loss factor using analytical equations. But these models failed to include the load sharing effect which is a vital parameter in estimating the gear loss factor (H_v). From the figure 3, it can be assessed that the loss factor is achieved minimum for the present work for both the NCR and HCR gear pair.

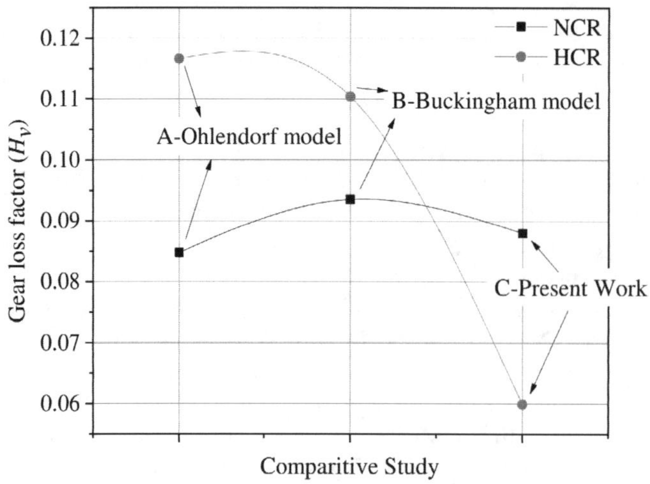

Figure 3. Gear loss factor (H_v) vs. various literature models.

5. Results and discussions

For the considered parameters, m=1, z_p=50, i=1.5, α_o=20^0 the balanced root stress is attained by varying the tooth thickness along the pitch circle of the gear pair. The tooth thickness factor achieved for NCR and HCR with balanced root stress is listed in table 1. It can be noticed from the figure 4(a) that the tooth thickness factor is 0.518 for a balanced NCR gear pair that is primarily due to increased tooth thickness at the critical region which takes more load in it during balancing of stresses. Whereas for an HCR gear pair the stress is balanced at a thickness factor of 0.4854, here the effect of increased radius of curvature at the fillet region gets signified when compared to the effect of tooth thickness. From the table 1, it can be noticed that the contact stress achieved is minimum for an HCR gear pair when compared with NCR gear pair drive. Figure 4(b) shows the load distribution along the contact path for the NCR and HCR gear drive. It is observed from the trend that in NCR gear pair, the load share is distributed at the A'B' and C'D' region as double pair region whereas the region B'C' takes the full load as single pair region. For the case of HCR gear pair [10], it can be noted that there exist a triple-double-triple-double-triple pair regions for one mesh cycle. The obtained value of load share is maximum of 60% in HCR when compared with NCR of 100% which is a significant enhancement in load carrying ability of the drive.

Table 1. Comparison of NCR and HCR computed data.

Type	**Contact Ratio (ε)**	**Tooth thickness factor (k_p)**	**Gear loss factor (H_v)**	**Maximum Balanced root stress $(\sigma_b)_{maxp} = (\sigma_b)_{maxg}$ MPa**	**Maximum contact stress $(\sigma_s)_{max}$ MPa**
NCR	1.785936	0.518	0.088	21.231	256.431
HCR	2.1074545	0.4854	0.0599	15.269	199.724

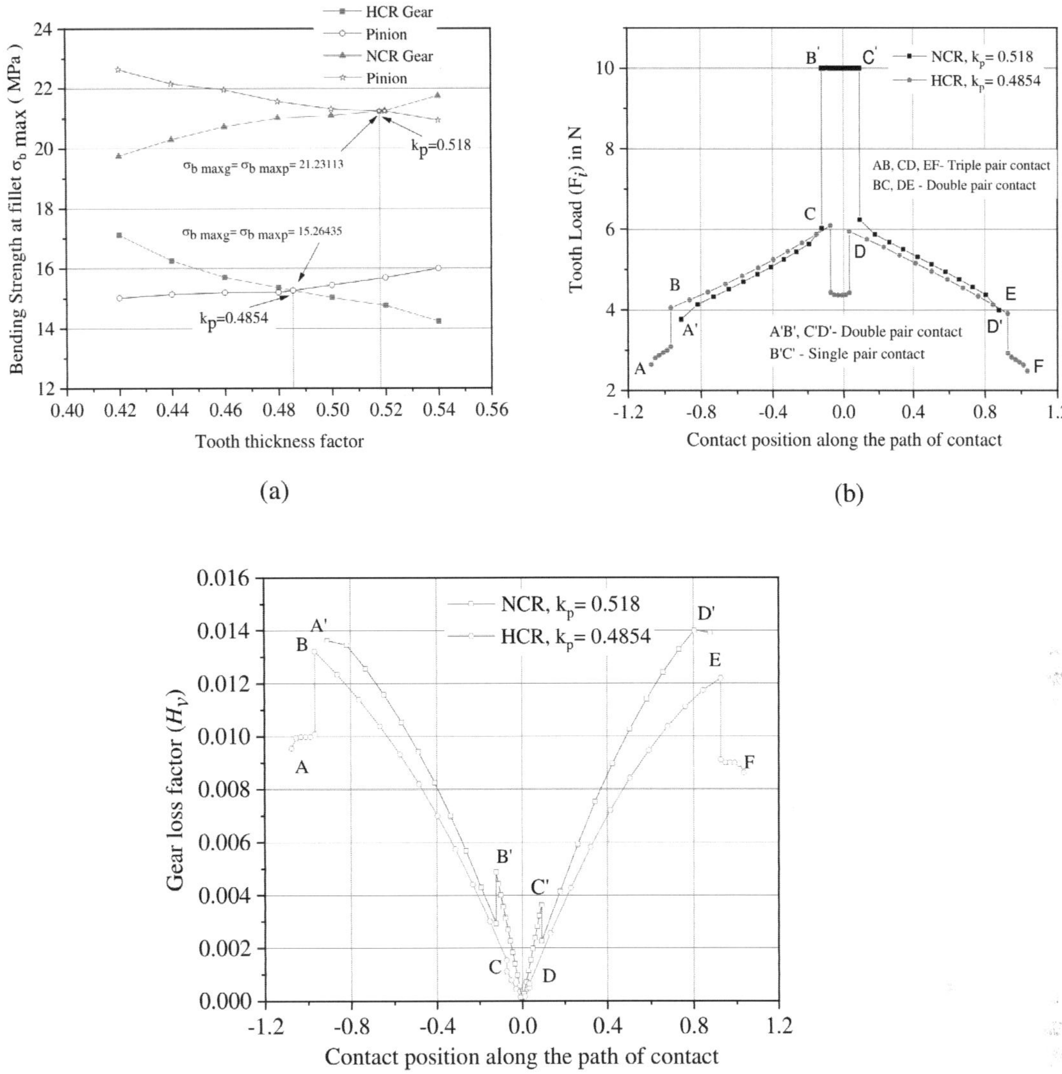

(c)

Figure 4. Non-standard NCR and HCR spur gear drive (m=1, z_p=50, i=1.5, α_o=20^0, h_a=1m (NCR) and 1.2m (HCR)).

The gear loss factor (H_v) is estimated along the contact path for one complete mesh cycle and compared between the NCR and HCR gear pair as shown in Figure 4(c). The H_v obtained for the HCR gear pair is less compared to the NCR gear pair throughout the mesh cycle due to reduced tooth load. The contact ratio increases significantly when the drive gets converted from NCR to HCR gear drive. It is also noticed that the maximum contact stress $(\sigma_s)_{max}$ decreases due to the decline in tooth load for the change in drive. The gear loss factor increases in the regions of increased load share at B'C' for NCR and at B and E for HCR spur gear pair which evidently proves that the gear loss factor depends on the load distribution.

IOP Conf. Series: Materials Science and Engineering **624** (2019) 012017 doi:10.1088/1757-899X/624/1/012017

6. Conclusions

The load share based estimation of gear loss factor has been proposed for improved root strength drive by varying tooth thickness factor. The contact load capacity is enhanced for the high contact ratio drive compared to normal contact ratio drive. The gear loss factor obtained using the proposed technique is validated with a comparative study carried among the earlier literature models. For the HCR spur gear pair, a noticeable reduction in loss factor (almost 50%) was achieved when compared with the Buckingham and Ohlendorf model.

7. References

[1] Yildirim N, Gasparini G and Sartori S 2008 An improvement on helicopter transmission performance through use of high contact ratio spur gears with suitable profile modification design *P. I. Mech. Eng. Part G: J Aerospace Engineering* **222(8)** 1193-1210.

[2] Rameshkumar M, Sivakumar P, Sundaresh S and Gopinath K 2010 Load Sharing Analysis of High-Contact-Ratio Spur Gears in Military Tracked Vehicle Applications *Gear Technology* **1(3)** 43-50.

[3] Buckingham E 1963 Efficiencies of gears. Analytical mechanics of gears *New York, NY, USA: Dover Publications* 395–425.

[4] Ohlendorf H, Verlustleistung und Erwarmung von Stirnradern 1958 (Ph.D. thesis) Dissertation TU Munchen, Munchen.

[5] José I. Pedrero, Miguel Pleguezuelos, Mariano Artés, Juan A. Antona 2010 Load distribution model along the line of contact for involute external gears *Mechanism and Machine Theory* **45 (5)** 780-794.

[6] Shuting Li 2007 Effects of machining errors, assembly errors and tooth modifications on load-carrying capacity, load-sharing rate and transmission error of a pair of spur gear, *Mechanism and Machine Theory* **42** 698–726.

[7] Shuting Li 2015 Effects of misalignment error, tooth modifications and transmitted torque on tooth engagements of a pair of spur gears *Mechanism and Machine Theory* **83** 125-136.

[8] Sekar RP and Muthuveerappan G. A 2015 Balanced Maximum Root stresses on Normal Contact Ratio Spur Gears to Improve the Load Carrying Capacity through Non-Standard Gears *Mechanics based design of structure and machines*, *Taylor and Francis* **43** 150-163.

[9] Sekar RP and Sathishkumar R. 2017 Enhancement of Wear Resistance on Normal Contact Ratio Spur Gear Pairs through Non-Standard Gears *Wear* **380-381** 228-239.

[10] Ravivarman R, Palaniradja K and Sekar RP. 2018 Evolution of balanced root stress and tribological properties in high contact ratio spur gear drive *Mechanism and Machine Theory* **126** 491-513.

[11] Ravivarman R, Palaniradja K and Sekar RP. 2018 Influence of gear ratio on wear depth of nonstandard HCR spur gear drive with balanced fillet stress, *Materials Today- Proceedings* **5** 17350–17359.

[12] Wimmer AJ. Lastverluste von Stirnradverzahnungen 2006 PhD thesis, Fakultät für Maschinenwesen der Technischen Universität München, München.

Acknowledgments

This research was supported by INSPIRE Fellowship scheme, Department of Science and Technology (DST) Government of India, for the corresponding author by providing financial assistance.

IOP Conf. Series: Materials Science and Engineering **624** (2019) 012018 doi:10.1088/1757-899X/624/1/012018

Design Optimization and Loaded Tooth Contact Analysis of Multispeed Gearbox for Low Noise Behaviour

Mahendran M[1], Murali V[2], Baskar N[3], Srinath V[4]

[1,2,4] -NPD Dept., Shanthi Gears Ltd, Coimbatore, India
[3] - R&D., Shanthi Gears Ltd, Coimbatore, India

E-mail: mahendranm@shanthigears.murugappa.com

Abstract. This project aims in the development of Multispeed Gearbox to exhibit low noise behavior during its operation under various load and speed conditions, by optimizing the design parameters with tooth micro-geometry profiles -using Loaded Tooth Contact Analysis (LTCA). Low-noise design is the most important criteria defined by the German customer, for their new 9 -speed gearbox fitted in the high precision tool room CNC Lathes. In loaded condition, gearbox develops noise and vibrations due to the tooth deformations, shaft deflection, internally generated gear tooth loads from non-conjugate meshing action of the gear mesh. Hence selecting the best values during micro-geometry optimization for a multi-speed gearbox becomes complex and needs a unique combination of FEM and analytical tools. LTCA is a very useful optimization tool to study the gear tooth strength evaluations and gear meshing contact characteristics under light and significant loads.

1. Introduction

Gears are critical components in every rotating machine and the sizing of gears within the given tight boundary for maximum torque carrying capacity will be a challenging job to designers. Multispeed gearboxes will have different load paths and also fluctuating torque loads at various operating speeds. Hence, the study on gear design parameters by analysing transmission errors with appropriate dynamic models are essential to control vibrations and noise [1].

The design has to take care of the reduction of the excitations from its source and also its effect, with corresponding load spectrums at different speeds. The gear optimization for the better behaviour of vibration and noise has been analyzed for

- Gear Teeth Combination z1/z2.
- Gear Macro Geometry Parameters.
- Gear micro-geometry Parameters
- Gear Quality /Accuracy of Manufacturing

2. Methodology

Develop, optimize and validate the design parameters of the multi-speed gearbox for low noise behaviour, in a wide range of operating load spectrum with the simulation of tooth surface behaviour under Hertzian deformation by

- Analyze the macro geometry of the Multi-speed gearbox to suit customer required power and transmission ratios in the provided space constraints in Layout [5].

- Compute the Loaded Tooth Contact Analysis (LTCA) by simulating and evaluating with various micro-geometry parameters (profile-lead modifications) and approximated by a cantilever beam model with FEA software
- Develop a computational approach for analyzing the load spectrum and review the quality of the meshing contact by computing non-uniform distribution of load and stress levels
    - along the lines of contact (Transverse load distribution factor)
    - across the gearing face width (Face load distribution factor)
    - elastic deformations and misalignment of the axes of rotation of the pitch cylinders of the mating gear (Mesh Alignment Factor)
- Design iterations and micro-geometry optimization to have Low noise design by ensuring
    - low-level Peak-to-peak transmission error (PPTE)
    - minimized contact shocks due to shaft deflections
- Evaluating results and choosing the best solution among various iterations and develop strategies to validate analytical results in test bench/load tests.

3. Gearbox Design Architecture

The multi-speed gearbox has to be designed to operate smoothly in all 9 speeds while fitted in high precision tool room CNC Lathes, in Aerospace components manufacturing plant.

During the design, gears and gearbox components were sized for their macro- geometry with the help of the latest software tools. Shifter configuration also ensured for linkages and proper engagements.

Hunting teeth combination is designed to avoid the frequency of same pinion teeth engaging with same teeth in Gear. The optimal size (module) gears were selected for better root strength and pitting strength. Shaft strength analysis (DIN 743) and bearing life calculations (ISO 281) ensures the adequate safety of gearbox components fitted in the housing.

Maintaining high-level accuracy in gears during profile grinding (DIN 3961-Q5) and Precise machining of housing bores with high-level concentricity ensures the smooth meshing of teeth at no load condition.

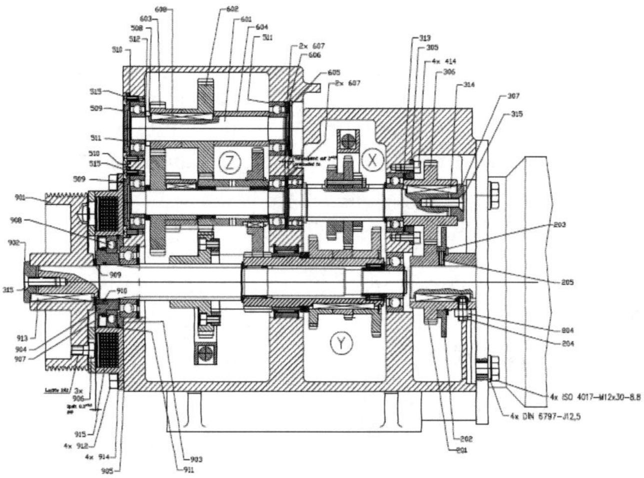

Figure 1. Gearbox Configuration.

4. Load spectrum

Load spectrum, an important tool for conducting accelerated reliability tests and fatigue life prediction of the mechanical structure has been evaluated to analyze the real load signals that reflect dynamic characteristics of the inertial force and the impact loads [2].

IOP Conf. Series: Materials Science and Engineering **624** (2019) 012018 doi:10.1088/1757-899X/624/1/012018

The testing load spectrum provides the first-hand data for fatigue life prediction, studying fatigue damage and accelerated reliability test of the transmission system.

Motor 1.9 kW @ input speed 1500 rpm

Table.1. Speed and lever position for 9 speeds

SPEED POSITION	A1	A2	A3	B1	B2	B3	C1	C2	C3
OUTPUT SPEED (RPM)	1189	1880	748	302	477	190	75	119	47

Multi-speed gearboxes will be subjected to various loads and speeds in operating nature. In order to estimate the proper fatigue damage, it is important to consider real service loads of the gearbox than considering the single value for torque derived from motor power.

The service fatigue life of the gears is depending on both the variable amplitude load conditions and the assumptions during fatigue design. The total design life and its estimated time and spindle speed for total life 40,000 hours are given in table 2. Load spectrum calculation has been computed for required safety factors for pitting, tooth breakage and scuffing as per ISO-6336 standard [3] using KISSsoft [4].

Table.2. Load Spectrum –Spindle Speed Vs Torque vs Time

	Time at level per bin (hour) against spindle torque (Nm)								
Torque (Nm)	**9.7**	**15.3**	**24.3**	**38**	**60.1**	**95.5**	**152.5**	**241.9**	**386.1**
Spindle(rpm)									
47	34	27	648	412	816	3196	2108	2712	1128
75	37	38	912	120	912	360	192	432	368
119	42	36	864	120	1008	696	744	408	792
190	42	13	312	744	1008	672	552	816	584
302	6	20	480	336	144	768	480	672	128
477	4	12	288	744	96	552	480	768	124
748	42	36	864	576	1008	576	144	96	304
1189	5	5	120	120	120	408	648	744	856
1880	17	17	408	264	408	504	336	576	824

5. Design Optimization with LTCA

The LTCA optimization program has been developed with objective function to ensure the even load distribution under given load by iterating micro geometrical parameters. The variables are profile corrections and lead corrections like tip relief, end relief, lead crowning, helix modification. The constraints are shaft deflections, gap analysis by comparing lead variation. The algorithm is a one-dimensional contact (mesh gap) analysis with two dimensional deflection & torsional analysis as defined in ISO 6336-1, Annex E

The contact ratio under load shall be the same as the theoretical contact ratio, for the gears with perfectly modified tooth geometry which includes flank line modifications, crowning to compensate tolerances and profile modifications. The tooth pair gap with contact stiffness $C\gamma\beta$ is illustrated with a spring model. When the tooth flanks are brought together, both flanks have a common contact width $bcal$.

Face load distribution factor KHβ calculation was done based on ISO-6336 with various assumptions on load distribution in the plane of action, at operating pitch cylinder. Based on FVA reports by Weber/Banaschek [6], tooth pair spring stiffness has been computed with an Analytical approach, considering

- Deformation δ, along with the path of contact
- Three deformation components
  - Bending
  - Tilting
  - Hertzian flattening

This approach also complies with Hooke's law: F=$C{\cdot}\delta$, where F is restoring force, C is stiffness tensor and δ is deflection parameter.

IOP Conf. Series: Materials Science and Engineering **624** (2019) 012018 doi:10.1088/1757-899X/624/1/012018

6. Spectrum Analysis of Transmission Error

Transmission Error (TE) indicates the variation on the contact stiffness. The amount of TE is determined by the way how the contact point is displaced on the path of contact in µm.

The amplitude of the transmission error is a decisive criterion in determining how quietly a gear unit runs. (PPTE=Peak to Peak Transmission Error)

Transmission Error is considered to be the primary cause of whining noise in Gear Boxes. TE curve is important, and a steeper slope in the resultant TE will produce higher accelerations and vibrations.

The gearbox TE is optimized with various load levels and found significant improvement with LTCA analysis.

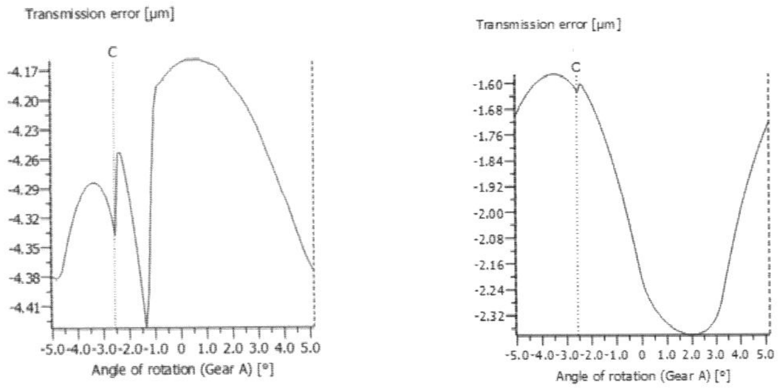

TE, before LTCA modifications **TE, after LTCA modifications**

Figure 6. Transmission Error Analysis

7. Contact Pattern Analysis

The analysis made for the input shaft deflections has shown the significance of bending and torsional moments in the system. The amount of contact varies with different load and speed conditions of the application. The contact pattern is analyzed for various load level and the spread also visually studied in a graphical pattern.

Since it is bi-directional, both the flanks of the gear teeth are compensated to have full involute contact at the desired range of operating load. The best contact pattern will be in mid of tooth portion and no stress concentration at edges.

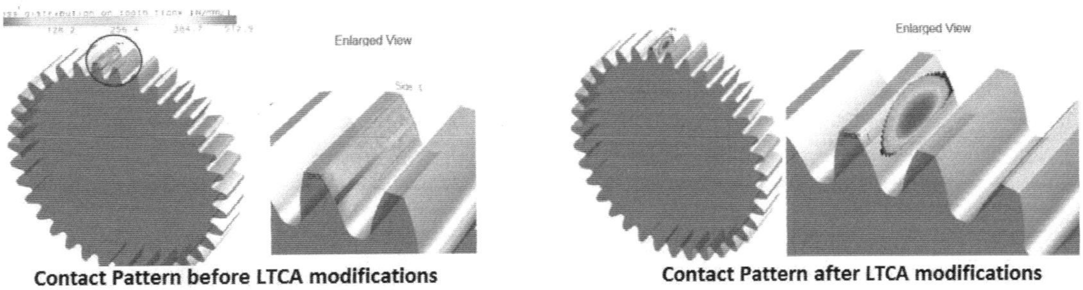

Contact Pattern before LTCA modifications **Contact Pattern after LTCA modifications**

Figure 7. Simulated Contact pattern with LTCA design parameters

8. Effect of Tooth modifications on Vibration and Noise

The optimal modification of the tooth profile and lead has been obtained by iterating in a wide range of values in KISSsoft software. The vibration measurements are taken on housing, near to the bearing positions. The data is analyzed to understand the effects of LTCA modifications, and the vibrational amplitudes that related to the input stage gear mesh frequency (product of the number of teeth and the shaft rotational speed) has been improved towards better characteristics with profile and lead modifications.

Harmonic analysis is used to determine the response of the structure under a steady-state sinusoidal (harmonic) loading at a given frequency. An order refers to a frequency that is a certain multiple of a reference rotational speed. In fig. 8, the amplitude in the order of harmonics is compared for vibration at the initial condition and after LTCA modification, which shows better characteristics with optimal LTCA parameters.

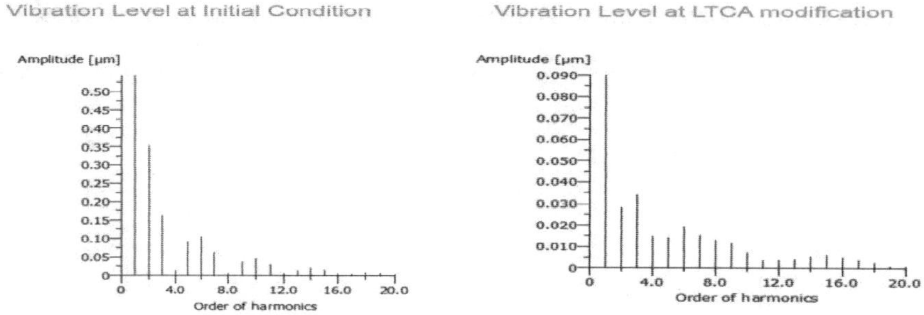

Figure 8. Comparison of Vibrational level at initial and after LTCA modification

9. Result

The legacy gearbox, when used in the new application, developed noise level and tonality above 85dB is analyzed for improvements with possible modifications in tooth data. The new set of gears with micro-geometry modifications are manufactured, assembled and tested in the loaded condition. The noise level is controlled within the customer acceptance norms (<80dB), thanks to LTCA with optimum micro-geometry parameters. Test results are validated with analytical results and proven that the LTCA approach is reliable to design low noise gearboxes in the first prototype itself.

Table 3. Measured Sound Level at various speeds

DIRECTION = COUNTER CLOCKWISE		GEAR SPEED POSITION								
		A1	A2	A3	B1	B2	B3	C1	C2	C3
INPUT RPM=1400	OUTPUT SPEED (RPM)	1189	1880	748	302	477	190	75	119	47
	SOUND LEVEL (dB)	72	71	68	68	68	69	67	68	67
INPUT RPM=2800	OUTPUT SPEED (RPM)	2350	3752	1490	602	952	379	150	237	93
	SOUND LEVEL (dB)	75	75	73	73	74	73	72	74	73

10. Conclusion

The studies, research and software iterative analysis on the transmission error, gear mesh stiffness variation has revealed that minimization of the transmission error will minimize noise significantly.

The dynamic simulations, Multi-Body Dynamics (MBD) and Topography modifications with CMM interaction over gear grinding process will be more useful to precisely understand the dynamic behaviour of the gears towards noise and vibrations characteristics under the influence of time varying loads.

This project also has ample scope for future continual research in the same arena by validating the micro-geometry design parameters thru Highly Accelerated Load Tests (HALT) in a Back to Back load test arrangement with a Supervisory Control and Data Acquisition (SCADA) system.

References

[1] Niemann,"*Maschinenelemente*," Band II, Springer Verlag, 1985.
[2] Hao Chen, Yu Sun*, Kai Wu , "*Load spectrum testing and analysis for transmission system of closed high-speed press*", Nanjing University of Science and Technology, ICTP 2014.
[3] ISO 6336, Part 1,"*Calculation of load capacity of spur and helical gears*" ISO Geneva, 2006.
[4] KISSsoft; *Calculation software for machine design*, www.KISSsoft.AG
[5] Litvin, F. L., "*Gear Geometry and Applied Theory*," Prentice Hall, 1994.
[6] Weber, E., Banaschek, K., "*Change in shape and Profile modification in straight and helical*

gears", Series Drive Engineering 11, 1953.

INVESTIGATION ON INFLUENCE OF CIRCUMFERENTIAL VELOCITY AND LUBRICATION OIL GRADES ON SCUFFING AND THERMAL EFFICIENCY OF GEARBOX

Vaidhyanathan D[1], Pushparaj A[2], Mahendran K[3], Ragunathan V[4]

[1,2,3,4] – New Product Development., Shanthi Gears Ltd, Coimbatore, India
E-mail: vaidhyanathand@shanthigears.murugappa.com

Abstract. Lubrication oil grades plays major role in reducing the friction to prevent scuffing in gears. Gears transmit power through rolling/sliding contacts between teeth, which must operate reliably for designed life with reduced friction so that there will be minimum wear. Lubrication of tooth contacts, both as a means of minimizing scuffing and as a means of reducing friction due to sliding, is therefore of crucial importance. The circumferential velocity of high speed gearboxes and oil grades plays major role in thermal efficiency of gearboxes. In gearboxes with oil sumps, churning losses contribute for significant portion of the total power loss. Present empirical approaches for examining the churning losses are often limited to certain constraints and operating conditions. Furthermore, they do not provide any information about the oil distribution in the gearbox. In order to overcome these shortcomings, we investigated a steel mill gearbox failure with analytical tools and simulations with the help of renowned software packages and predicted the thermal behavior of the gearbox as function of oil grade and circumferential velocity, also validated later at application.

1. Introduction

The gear fatigue failures like pitting, tooth breakage will take time to happen, whereas a single overloading of gear can lead to scuffing failure immediately. Scuffing is defined as localized failure caused by solid - phase welding that occurs between surfaces in relative motion. Scuffing most commonly occurs at one of the two extreme end regions of the contact path or near the points of single tooth contact where sliding is maximum.

Circumferential velocity between the mating gears determines the contact time between gear teeth. The type of lubrication (splash, fully oil immersed and forced lubrication) depends on the range of operating speed and oil grade selection also vital to have better thermal rating.

Thermal rating of the gearbox (torque /power) is considered as the torque level at which the system stablishes the oil temperature at 95 degrees, during the full load test. It depends on several factors like oil parameters, casing surface area exposed to external environment, air velocity, altitude, free space available in the gearbox surrounding, etc.

The scuffing is caused by the high frictional heat generated by high hertizan stress and high sliding velocity. This increased temperature with minimal film thickness at the areas of tooth contact surface along with the higher loads and thin lubrication film leads to welding of surfaces. The following are the factors that influences scuffing.

IOP Conf. Series: Materials Science and Engineering **624** (2019) 012019 doi:10.1088/1757-899X/624/1/012019

1. Sliding velocity
2. Lubrication film thickness
3. Load sharing factor
4. Temperature
5. Effective tip relief

1.1. Sliding Velocity

The sliding velocity of gear tooth is defined as a difference between the rolling velocities of gear and pinion in mesh. Pure rolling is present only at the pitch point where the pitch circle intersects the line of action. The amount of sliding increases when the gear tooth contact point moves away from the pitch point. The sliding velocity of a gear mesh can be determined using the following equation:

$$v_S = v_{Rp} - v_{RG}$$

$$v_R = v_T * \sin\alpha$$

Where: v_S – Sliding velocity(m/s)

v_{Rp} – Rolling velocity at pinion

v_{RG} – Rolling velocity at gear

v_T – Tangential velocity

α – Pressure angle

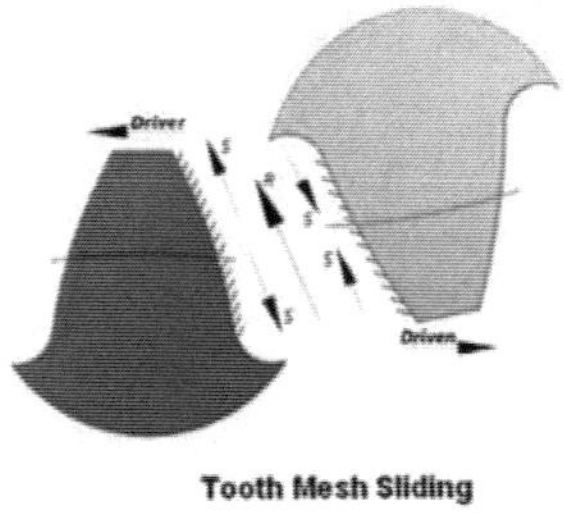

Figure 1. Tooth Mesh Sliding

In reference to figure 1, the driver connects the mating gear initially near root, and contact rolls up towards tip. In this motion the sliding is happening always from pitch line. In contrast, the initial contact happens near to tip for the driven gear. In driven gear the contact rolls down towards root and sliding is happening towards pitch line. Crack grows opposite direction of sliding and converges at pitch line.

For optimum specific sliding, the value suggested in KISSsoft [1] shows the profile shift for a cylindrical gear pair that has been balanced for a specific sliding between the pinion and the wheel.

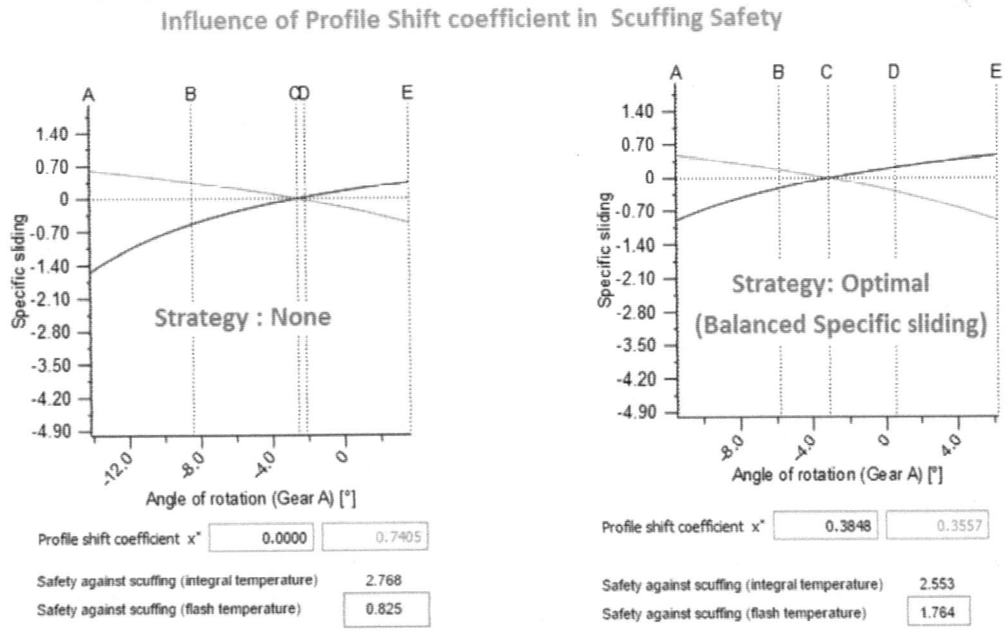

Figure 2. Specific Sliding influence on Flash Temperature

IOP Conf. Series: Materials Science and Engineering **624** (2019) 012019 doi:10.1088/1757-899X/624/1/012019

1.2. Lubrication film thickness

When the gear mesh occurs, the lubricant between the mesh surface will experience the hertizan pressure which in turn increases the viscosity of the oil. The lubricant with increase in the viscosity behaves as a solid phase, thus increasing the temperature.

When the lubricant thickness reduces asperity contact increases. As shown in the figure 3, there will be no scuffing, when a thick lubricant film separates the gear teeth by minimizing the surface asperities. As lubricant film thickness decreases, asperity contact increases and scuffing becomes more probable.

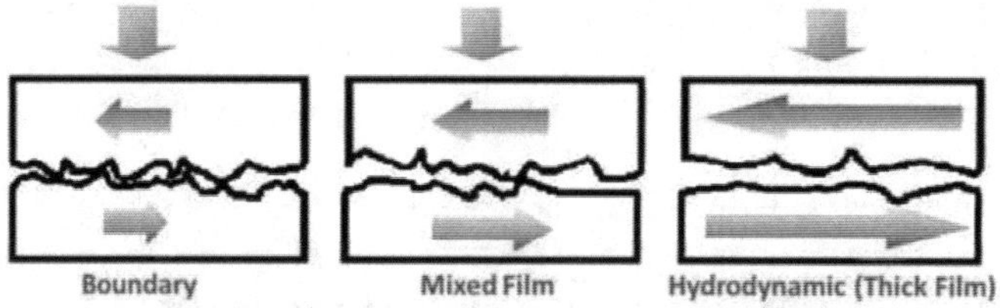

Figure 3. Effect of lubricant on scuffing

The poor surface roughness of the gear tooth causes the lubrication film thickness to increase the bulk temperature, thus results in scuffing. The adequate care has been take to have smooth profiles with profile grinding Ra<0.8, and tooth gap with the recommended specifications of DIN 3967.

1.3. Load Sharing Factor

The scuffing usually occurs when a single tooth contact happens to transfer the load at specified speed. The figure.4, shows the gear tooth load sharing factor from Lowest Point of Single Tooth Contact (LPSTC) to Highest Point of Single Tooth Contact(HPSTC). The load sharing factor is highly influenced by Tip relief and profile modifications for high load capacity and smooth contact. The Loaded tooth contact analysis has been computed for optimal combination of profile modification values.

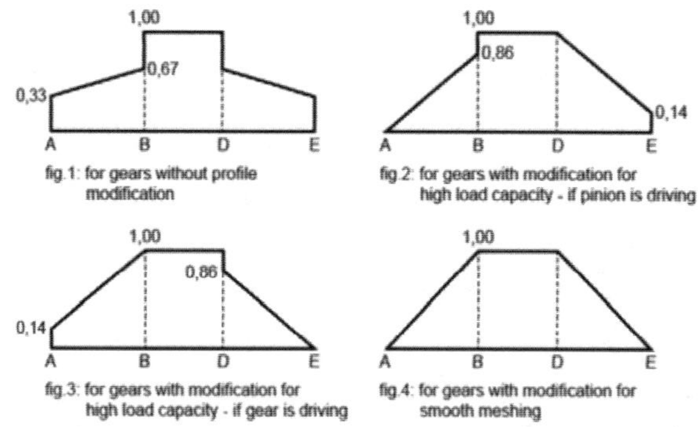

Figure 4. Load sharing factor Xgam

1.4. Temperature

1.4.1. Flash temperature: The flash temperature is the calculated increase in gear tooth surface temperature at a given point along the line of action resulting from the combined effects of gear tooth geometry, load, friction, velocity and material properties during operation.

1.4.2. Integral temperature: The integral temperature method has been proposed as an alternative to the flash temperature method by which the influence of the gear geometry imposes a critical energy level based on the integrated temperature distribution along the path of contact and adopting a steady gear tooth temperature. The safety factors for scuffing is calculated according to ISO TR 13989 [2,3] and values are compared.

1.5. Effective Tip Relief

As the scuffing occurs the most two end region of the gear pair, providing a tip relief at the gear tooth reduces the risk of occurrence of scuffing. The tip relief reduces the entry impact & exit impact of the driven gear and driving gear respectively. Usually, tip relief is applied to both the gear and pinion as specified in the ISO 15144.

2. Methodology

The gearbox failed under scuffing at the application has been analyzed for the Macro Geometry [4] parameters and found absolutely OK for pitting and root strength [5] for the rated capacity. However, it has lower service factors on starting torque which is higher than rated capacity and reaches peak while application is preloaded. Further investigation with detailed calculations throws light on the gaps in micro-geometry parameters. Hence the study has been extended with three zones of different tip relief conditions and the respective scuffing safety is analyzed using KISSsoft to choose the best parameters for profile modification.

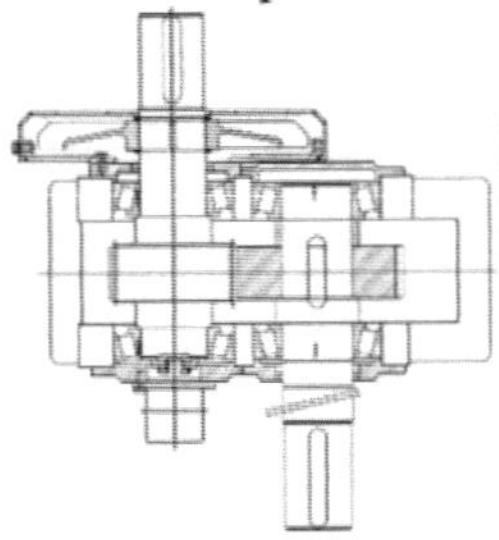

POWER	500 kW
G/B RATED CAPACITY	920 kW
INPUT SPEED	1000 RPM
NOMINAL RATIO	1.4:1
OUTPUT SPEED	714.28 RPM
TYPE OF LUBRICATION	SPLASH
OIL QTY (APPROX.)	22 lts.

Figure 5. Investigated Gearbox Layout

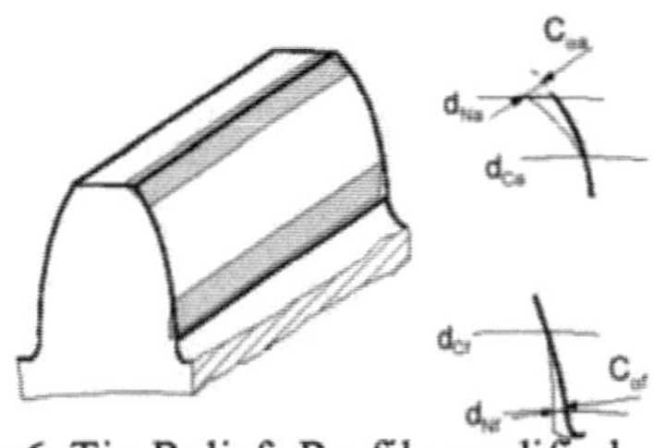

Figure 6. Tip Relief -Profile modified

2.1. Effect of Tip Relief and oil quantity on Scuffing

Table 1 represents the scuffing safety factors [6] against integral temperature and flash temperature. Case 1 data has been computed without tip relief in an oil bath system and Case 2 results are obtained with an optimum tip relief value in a gear system where as gears are completely immersed in lubrication oil.

Table 1. Strength against scuffing under different cases

Effect of Lubrication & Tip relief on Gear Scuffing Safety factors (AGMA 925-A03)				
Profile Modification: (Ca) in microns	Caa=		none	66
Lubrication Type			Oil bath Lubrication	Fully immersed in oil
Scuffing temperature	$[\theta_S]$	°C	296	296
Tooth Temperature	$[\theta_M]$	°C	146.2	89.5
Maximum Flash Temperature	$[\theta_{flmax}]$	°C	136.09	134.85
Maximum contact Temperature	$[\theta_{Bmax}]$	°C	282.3	224.36
Probability of Scuffing	$[P_{scuff}]$	%	38	5.33
Result : Risk of Scuffing (AGMA 925, Table 5)			**HIGH**	**LOW**
Safety Against Scuffing (Integral Temperature)			2.171	2.628
Safety Against Scuffing (Flash Temperature)			1.817	2.275

As the tooth temperature decreases in immersed condition with higher tip relief, the chances of scuffing are reduced, thus improving the scuffing safety.

2.2. *Effect of Lubrication on scuffing*

The gear pair is tested with two types of approved oils - Brand A & Brand B, to check which one has better properties against scuffing safeties using the KISSsoft calculations.

Table 2. Lubrication oil specifications

		Brand A	Brand B
Relative Density at 15.6°C		0.87	0.86
Flash Point	°C	226	240
Pour Point	°C	-36	-39
Kinematic Viscosity oil at 40°C [nu40]	mm²/s	320.8	334.4
Kinematic Viscosity oil at 100°C [nu100]	mm²/s	34.2	37.4
Lower limit service temperature	°C	-30	-30
Upper limit service temperature	°C	95	95
Lubricant base – synthetic (Polyalphaolefin – PAO)		PAO	PAO
Scuffing test – FZG TEST A/8.3/90;ISO 14635-1 (normal) (stage passed)		14	12
Micropitting test -C-GF/8.3/90 with Ra=0.50 theta = 90deg. (FZG)		10	10

Brand A has better in scuffing parameters due to its higher load stage in FZG scuffing test [7]. The table shows the technical specifications of the lubricant oil which are selected for the comparison.

Table 3. Lubrication oil specifications

Safety for	Brand A	Brand B	% diff
Scuffing (integral temp.)	2.628	1.903	27%
Scuffing (flash temp.)	2.275	1.732	24%

Since the circumferential velocity of the gearbox which is under investigation has not exceeded beyond the recommended limit of 15m/s, there is no influence of circumferential velocity in this gearbox.

2.3. *Thermal Efficiency*

The previous supply gearbox housing was analysed for thermal rating [8] to determine the power that can be continuously transmitted under the operating temperature without resulting any damage to the internals. The FEA[9] done with ANSYS software and found that the housing bottom portion is subjected to higher temperature than standard practice (85°C) values.

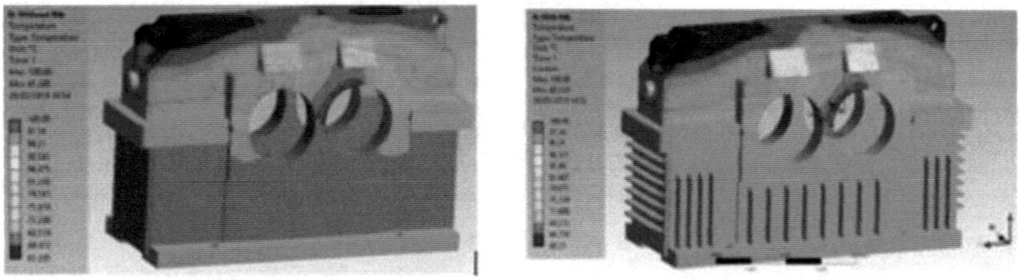

a) OLD CASING b) NEW CASING

Figure 7. FEA thermal analysis in ANSYS. a) Old casing vs b) New casing with Ribs

A new gearbox housing was made with additional ribs and fins and FEA shows significant reduction in temperature (73°C) within the same boundary dimensions. The thermal efficiency improved 14% with the new housing.

3. Result and Discussion

1. Providing Optimum Specific sliding of profile shift coefficient gives better scuffing safety on the gears thereby ensuring the reliability of the gears. (figure 2)
2. As the tooth temperature decreases in immersed condition with higher tip relief, the chances of scuffing are reduced, thus improving the scuffing safety. (table 1)
3. If the circumferential velocity increases above 15m/s care has to be taken to ensure that the gears are well lubricated by either external forced lubrication system or in such a way that the gears are not immersed in oil. The lubrication is done with the help of shaft driven pump. [1]
4. Housings made with more ribs and fins gives significant reduction in oil sump temperature. (figure 7)
5. Lubrication oil parameters and their influence is tabulated below

Oil Parameter	**Scuffing Safety factor for**	
	Integral	**Flash**
Density of oil - increased	increased	increased
Upper limit of kinematic viscosity - increased	increased	increased
Lower limit of kinematic viscosity - increased	increased	increased
Load stage scuffing -increased	increased	increased
Load stage micro pitting -increased	no change	no change

It is observed from the above table that when kinematic viscosity of the oil is increased, it has better scuffing safety. So oil grade ISO VG-320 has better scuffing safety than ISO VG-220 and so on.

6. The higher the stage passed in FZG scuffing test A/8.3/90, ISO 14635-1, will have better scuffing rating. (table 3)

4. Conclusion

With calculations we conclude that gearboxes with pitch line velocities closer or higher than 15 m/s are to be provided with forced lubrication system with suitable oil viscosity grades (low viscosity oil for light load) based on lubrication film thickness calculations. While gears are loaded higher torque with low pitch line velocities, we have to use high viscous oils to negotiate high loads and long contact times. It is also realised that the higher kinematic viscosity oil has better scuffing safety. While gearbox housing designed with higher number of ribs exposed to external atmosphere, the sump oil temperature gets reduced and thermal rating increased. This research has ample scope for validating all the results with FEA tools later compared with Highly Accelerated Load Tests(HALT).

References

[1] KISSsoft, "*Calculation software for machine design*", www.KISSsoft.AG
[2] ISO TR 13989, Part 1, "*Calculation of scuffing load capacity of cylindrical, bevel and hypoid gears - Flash temperature method*", ISO Geneva, 2000.
[3] ISO TR 13989, Part 2, "*Calculation of scuffing load capacity of cylindrical, bevel and hypoid gears - Integral temperature method*", ISO Geneva, 2000.
[4] Litvin, F. L., "*Gear Geometry and Applied Theory*", Prentice Hall, 1994.
[5] ISO 6336, Part 1,"*Calculation of load capacity of spur and helical gears*", ISO Geneva, 2006.
[6] DIN 3990, Part 4, "*Calculation of scuffing load capacity*".
[7] Bernd-Robert Hoehn, Peter Oster, Thomas Tobie, Klaus Michaelis, "*Test Methods For Gear Lubricants*".
[8] ISO/TR 14179, Part 2, "*Thermal load carrying capacity*".
[9] Björn Bauer, Ralf Hambrecht, Andreas Kube and Simon Becka, "*Thermal Analysis and Optimization of Gearboxes by Simulation*", Power Transmission Engineering, December 2018.

IOP Conf. Series: Materials Science and Engineering **624** (2019) 012020 doi:10.1088/1757-899X/624/1/012020

Newtonian approach towards mathematical modelling and tuning of a Continuously Variable Transmission

Prajwal B Bharadwaj[1] and Jeyaraj Pitchaimani[1]
[1]Department of Mechanical Engineering, National Institute of Technology Karnataka, Surathkal, Mangalore, India
Email: prajwalb9897@gmail.com

Abstract: Mathematical modelling of a Push - Belt Continuously Variable Transmission (CVT) has been done previously using energy-based methods. However, this paper presents a force based (Newtonian) approach as it eliminates the process of solving differential equations and hence, increases computational speed while eliminating the errors arising out of numerical solutions. The methodology for obtaining the engine rpm for optimum power is elucidated. An offset slider crank model has been used for conversion of rotary motion of flyweights to linear motion, in order to compress the pressure spring in the driving pulley. Mass of flyweights and stiffness values of pressure springs in driving and driven pulley are set as parameters which govern the constant engine rpm. The plot of engine rpm against vehicle speed is observed to be a saturating curve which gives the constant engine rpm obtained by CVT action.

1. Introduction

A CVT is a system of belt and pulleys which is used to achieve infinite gear ratios between the two transmission shafts. The CVT consists of two pulleys namely, the primary (driving) and the secondary (driven). Both the pulleys have one movable sheave and one stationary sheave. A belt (V shaped) goes over these pulleys. The movable sheave makes the belt to slide up or down in the pulley as it moves towards and away from the stationary sheave. The main reason of CVT usage is that, after a certain speed is achieved, the rpm of the engine remains constant irrespective of the increase in the speed of the vehicle.

In the clutch tuning handbook by Olav Aaen [1], CVT tuning is defined as setting a particular engine rpm as the working rpm, which is dependent upon the masses of the flyweights and the stiffness values of the pressure springs. Tuning is done in order to set the working rpm to be the one which gives the maximum power when looked up the power curve [2].

Sarah Crosby et.al [3], formulated a mathematical model for a complete vehicle equipped with CVT. A separate model for every component of the vehicle: the engine, clutch, CVT, car differentials, wheels and car body, is constructed. The differential equations for the angular velocities for every component in the system are numerically solved and the plots show how the system becomes steady after a short period of time with time.

Kim and Marshek [4], reported that, in a flat belt power transmission effective belt tension decreases with increasing belt velocity. Since the friction force between the belt and pulley is proportional to the normal pressure, the amount of torque load transmittable decreases with increasing belt velocity until the system is not able to transmit power. If the torque load is constant, the distributions of the normal pressure and tangential friction force change with belt velocity.

IOP Conf. Series: Materials Science and Engineering **624** (2019) 012020 doi:10.1088/1757-899X/624/1/012020

Oleivera and Nivaldo A Lemos [5] considered the simple case of a rope on a pulley with masses attached to the two ends of the rope. It says that ropes don't just convey the forces from pulley to pulley. Each element exerts a force on the part of the pulley it is in contact with. The net force and torque on the pulley is the integration of the infinitesimal forces over the angle of wrap.

Abhijeet Sanchawat et. al [6] proposed a model of the CVT assuming a pressure spring in the primary pulley and the torque sensing ramp in the secondary. The governing equations have been derived using Lagrangian method assuming a conservative system. Gear ratio, Vehicle acceleration and velocity is plotted against time. Velocity is observed to saturate as time proceeds while acceleration graph is decaying with time.

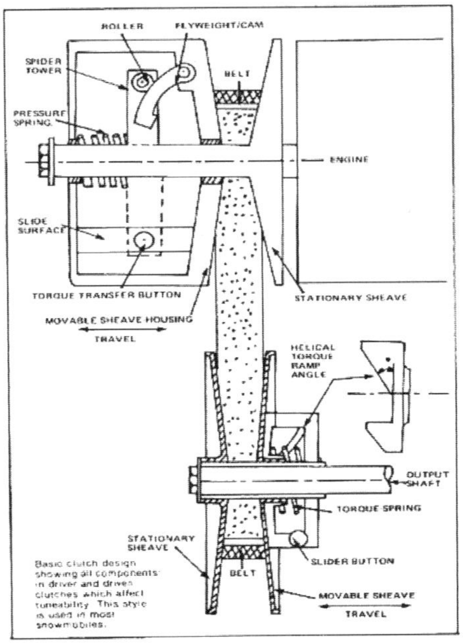

Figure 1. Components of a CVT [7]

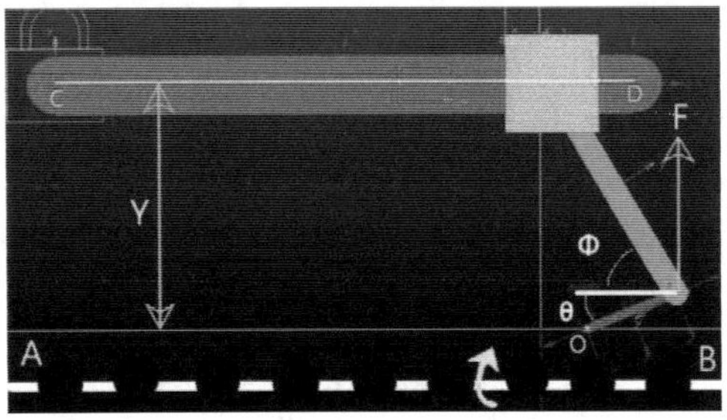

Figure 2. Offset slider crank mechanism used for spring compression

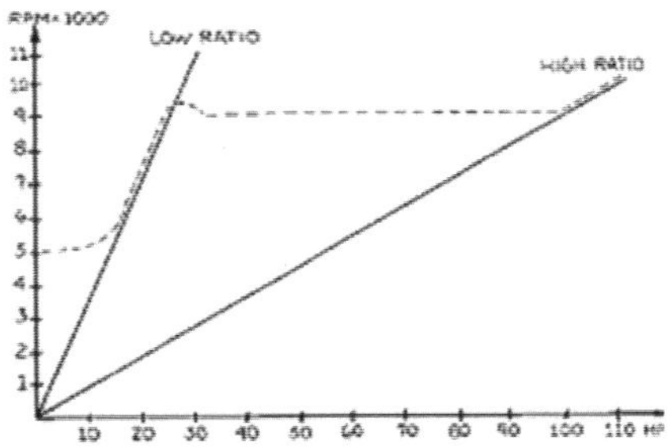

Figure 3. Ideal graph of engine rpm vs. vehicle speed for a CVT (in dotted line) [7]

This paper presents a mathematical model formulated using a force-based approach. It also incorporates the method of finding the effective normal force by integrating the infinitesimal normal forces over the angle of wrap of belt, as presented in [5]. In order to apply a force along the axis of the pressure spring, an offset slider crank mechanism has been used which converts the rotary motion of the flyweights to linear motion to compress the spring.

2. Mathematical model

The formulae to calculate belt drive variables are mentioned below.

Angles of wrap for primary and secondary pulley, α_p and α_s are given by,

$$\alpha_p = \pi + 2\beta$$
$$\alpha_s = \pi - 2\beta$$

Where angle β is,

$$\beta = \sin^{-1}\left(\frac{d_p - d_s}{2C}\right) \tag{1}$$

where d_p = primary pulley diameter, d_s = secondary pulley diameter, C = centre - centre distance between pulleys.

Length of open belt, L_0 is given by

$$L_0 = 2C + \frac{\pi(d_p + d_s)}{2} + \frac{(d_p - d_s)^2}{4C} \tag{2}$$

Tensions in the tight and slack side of pulleys respectively, $T_1$and T_2 are given by

$$\frac{(T_1 - m_b v^2)}{(T_2 - m_b v^2)} = e^{\frac{\mu\theta}{\sin\left(\frac{\alpha}{2}\right)}} \quad (3)$$

$$(T_1 - T_2)v = P \quad (4)$$

where μ is the friction co-efficient, θ is the angle of wrap, α is the angle of v-belt, m_b is the mass per unit length of v-belt, v is the velocity of belt and P is the engine power.
Since equations (3) and (4) are linear in T_1 and T_2. They are solved simultaneously to get values of T_1 and T_2.

Equations (1), (2), (3) and (4) have been referred from "Belt Drives" Module of IIT Kharagpur Lecture series [8].

Effective normal force on the secondary pulley, F_n^{string} is given in [5] as,

$$F_n^{\text{string}} = \int_0^{\pi+2\beta} nR\boldsymbol{r}d\theta = \int_0^{\pi+2\beta} T(\theta)(\cos\theta\,\boldsymbol{x} + \sin\theta\,\boldsymbol{y})d\theta \quad (5)$$

$T(\theta)$ = tension in the belt as a function of angle of wrap of secondary pulley
$\boldsymbol{r}$ = unit radial vector = $\cos\theta\,\boldsymbol{x} + \sin\theta\,\boldsymbol{y}$
n = normal force on an infinitesimal section of the belt
R = secondary radius
$d\theta$ = angle of wrap for the section of the belt

Primary spring compression distance, X is given by (refer figure 2)

$$R_{fly}sin(\theta) + Lsin(\phi) = Y \quad (6)$$

$$k_1\begin{pmatrix} Lcos(\phi)\sin(\phi)\sin(\theta) - Lcos^2(\phi)\cos(\theta) \\ -R_{fly}cos(\theta)\sin(\theta)\sin(\phi) + R_{fly}cos^2(\theta)\cos(\phi) \end{pmatrix} = Fcos(\phi)\sin(\theta) \quad (7)$$

$$F = mw^2 R_{\text{fly}} \quad (8)$$

$$X = Lcos(\phi) - R_{\text{fly}}\cos(\theta) \quad (9)$$

where k_1 is the primary spring stiffness, m is the effective mass of flyweights, R_{fly} is the crank radius, L is the length of connecting rod, F is the centrifugal force on the flyweight, Y is the vertical distance between stroke line CD and pivot point O of crank, w is the angular velocity of primary pulley and θ, ϕ are the angles as mentioned in figure 2.
The non-linear simultaneous equations (6) and (7) are solved using newton raphson method to get the values of θ and ϕ which can be used to calculate the spring compression using equation (9).

Belt rise in primary pulley, Δr_1 is given by

$$\Delta r_1 = \frac{x}{2\tan\left(\frac{\alpha}{2}\right)} \quad (10)$$

x = Primary spring compression distance calculated from equation (9), α = V-belt angle

Belt descent in secondary pulley, Δr_2 is given by

$$\Delta r_2 = \frac{F_n^{string}cos\left(\frac{\alpha}{2}\right)}{2k_2\tan\left(\frac{\alpha}{2}\right)} \quad (11)$$

F_n^{string} = Effective normal force on the secondary pulley calculated from equation (5)
k_2 = Secondary spring stiffness, α = V-belt angle

3. Methodology

Figure 2 shows the offset slider crank mechanism used to compress the primary spring. The whole mechanism rotates about the primary pulley axis AB. The centrifugal force (*F*) pushes the flyweight outwards which pushes the slider towards left, compressing the spring. The effective mass of the flyweights (*m*) is assumed to be concentrated at the end of the crank. The equations for primary spring compression have been derived by drawing free body diagrams and force balance for zero acceleration on each link [9]. Primary avg. speed vs. Secondary speed is plotted and results are analysed. Primary avg. speed is calculated by averaging the two engine rpms encountered in an iteration of the loop. The engine rpm is increased as shown in equation (12) and (13).

$$engine\ rpm_{new} = engine\ rpm_{old} + c \tag{12}$$

$$engine\ rpm_{new} = engine\ rpm_{old} + c\left(\frac{secondary\ radius_{old}}{primary\ radius_{old}}\right) \tag{13}$$

where *c* is a constant. In equation (12), the engine rpm is increased in equal steps in every iteration. In equation (13), *c* is taken to be the constant increase in secondary speed and hence is multiplied by the reciprocal of gear ratio while incrementing the engine rpm. Here, the increase in engine rpm is such that there is constant increase in secondary speed in every loop iteration. Error in belt length is calculated since the engine rpm increase is not continuous. Primary speed and engine rpm, secondary speed and vehicle rpm are used interchangeably in the following lines and each of them is expressed in terms of rpm.

3.1. *Parameters* [10]

1. k_1 (Stiffness of primary pulley spring) = 5 kN/m, 10 kN/m, 15 kN/m
2. k_2 (Stiffness of secondary pulley spring) = 20 kN/m, 25 kN/m, 30kN/m
3. *m* (Effective mass of flyweights) = 78g
4. α (Angle of V- belt) = 12°
5. *C* (centre-centre distance between pulleys) = 310 mm

3.2. *Assumptions*

1. The vehicle rpm remains constant during the gear ratio change.
2. The belt is assumed to be non-elastic and rigid.
3. Slippage is neglected.
4. Friction of the belt and pulley in radial direction is neglected [6].
5. There are no losses in power transmission.

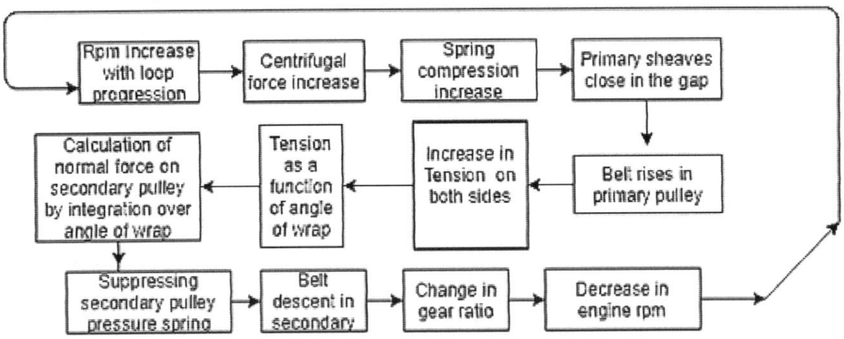

Figure 4. Methodology of modelling the processes

4. Results

Figure 5, 6, 7 and 8 show the plots where the engine rpm is increasing in accordance with equation (12). In figure 5, each of the plots are saturating upon reaching their respective constant working rpms. It is observed that the constant working rpm increases with increase in primary spring stiffness (k_2). In figure 6. too, it is observed that the constant working rpm increases with increase in secondary spring stiffness (k_1). The saturation observed in figure 5 and figure 6 is similar to figure 3.

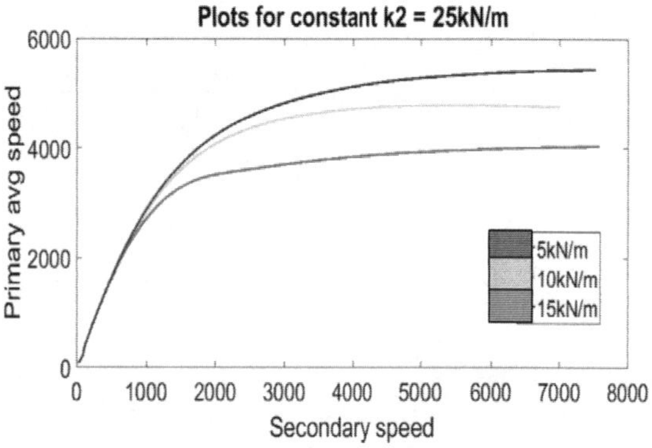

Figure 5. Primary avg. speed vs. Secondary speed plots for 3 values of primary stiffness

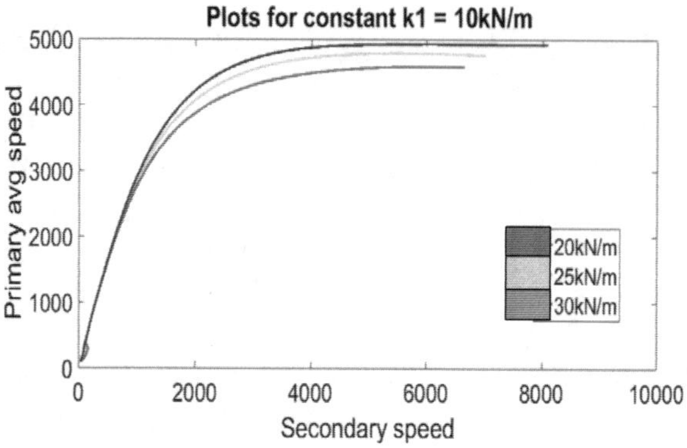

Figure 6. Primary avg. speed vs. Secondary speed plots for 3 values of secondary stiffness

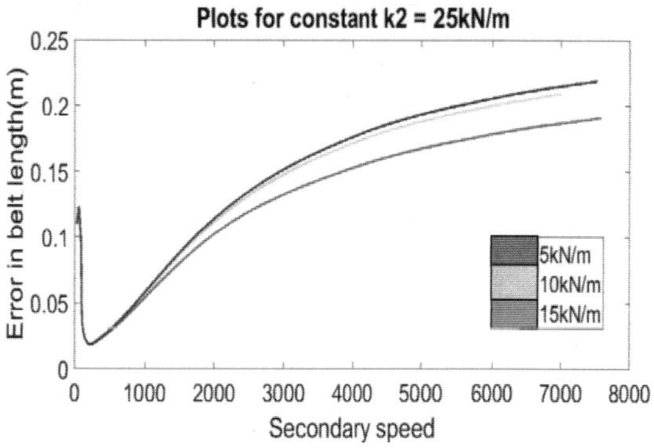

Figure 7. Error in belt length vs. Secondary speed plots for 3 values of primary stiffness

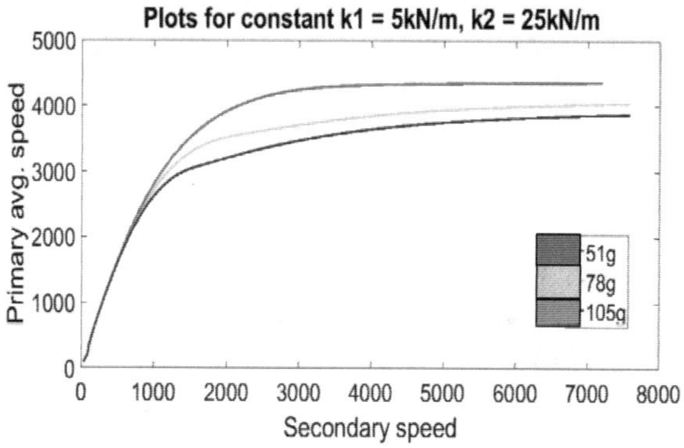

Figure 8. Primary avg. speed vs. Secondary speed plots for 3 values of effective flyweight masses

Figure 7 shows that the error belt length is increasing with secondary speed. The error has arisen since the increase in primary radius and decrease in secondary radius are such that belt length calculated in the new iteration is lesser than the actual belt length i.e., the secondary radius is decreasing at a faster rate than desired. In figure 8, it is observed that the constant working rpm increases with decrease in flyweight mass.

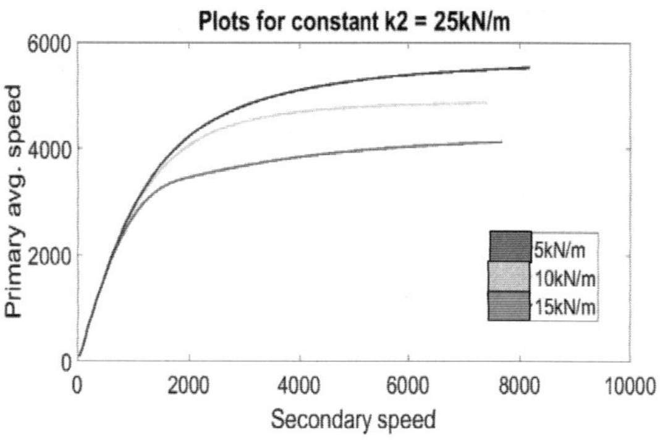

Figure 9. Primary avg. speed vs. Secondary speed plots for 3 values of primary stiffness

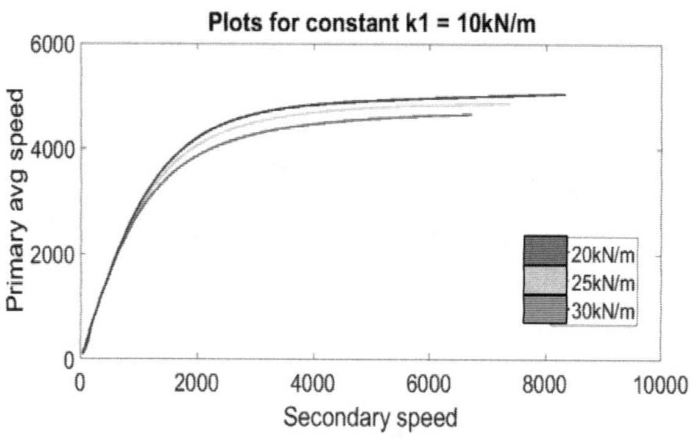

Figure 10. Primary avg. speed vs. Secondary speed plots for 3 values of secondary stiffness

Figure 9 and figure 10 show the plots where the engine rpm is increasing in accordance with equation (13). Upon comparison of these two figures with figure 5 and figure 6 respectively, it is observed that the curves are very similar to each other and that the method of using equation (12) leads to earlier saturation of the curve.

5. Conclusions

A force-based mathematical model for the working of a CVT has been developed in this paper. In order to utilize the centrifugal force to compress the spring in the primary pulley, an offset slider crank mechanism has been implemented for converting the rotary motion of flyweights to linear motion. Governing equations for belt rise and fall in the pulleys are formulated by using free body diagrams for zero acceleration case. Belt-pulley equations are used to calculate the tensions on the tight and slack side. Integration method is used to calculate the effective normal force of the belt on the secondary pulley. Since increase of engine rpm during the working of a real CVT happens continuously, in order to mimic the desired characteristics, two methods of increasing the engine rpm have been explored. One of the methods being constant increase of engine rpm in every iteration. In the second method, the engine rpm is increased such that the secondary speed increase will be constant in every iteration of the loop.

It is observed from the results that both the methods lead to saturation of the curve, however the first method leads to early saturation. The constant working rpm is found to be increasing with increase in either primary or secondary spring stiffness whereas it is decreasing with increase in the effective flyweight mass. This is in conjunction with the literature presented in [1]. Since the engine rpm is not increased continuously but in steps, there is bound to be a difference in the belt lengths calculated in two consecutive iterations. The increase of error in belt length with respect to secondary speed indicates that the secondary radius is decreasing at a faster rate than desired.

6. References

[1] Aaen, Olav. (n.d.). CVT Clutch Tuning Basics -Part 1: Background and Basics.
[2] Willis, C. R., Reinholtz, C. F., Goff, R. M., Ahmadian, M., Christopher, T., & Willis, R. (2006) A Kinematic Analysis and Design of a Continuously Variable Transmission
[3] Crosby, S. (2016). Mathematical Modeling of Continuously Variable Transmission(CVT) System (Vol. 5, pp. 53–62).
[4] Kim, H., & Marshek, K. M. (1987). The effect of belt velocity on flat belt drive behavior. In *Mechanism and Machine Theory* (Vol. 22, pp. 523–527). https://doi.org/10.1016/0094-114X(87)90047-4
[5] Oliveira, T. De, Fluminense, U. F., Lemos, N. A., & Fluminense, U. F. (2017). Force and torque of a string on a pulley. https://doi.org/10.1119/1.5016040
[6] Sanchawat, A., Agarwal, R., Chandrasekar, R., & Baskar, P. (2013). Simulation of a drivetrain of a vehicle comprising continuously variable transmission. In *International Journal of Engineering and Technology* (Vol. 5, pp. 3041–3050).
[7] Timothy R. DeGreenia. The Continuously Variable Transmission: A Simulated Tuning Approach
[8] Kharagpur IIT. (n.d.). Introduction to Belt drives, NPTEL Lecture series.
[9] Norton, R. L. (1999). an Introduction To the Synthesis and (p. 120 122).
[10] Mini-baja, S. A. E., Luck, G., & Free, T. (n.d.). Mini-Baja Sponsorship Program.

IOP Conf. Series: Materials Science and Engineering **624** (2019) 012021 doi:10.1088/1757-899X/624/1/012021

Study and control of factors influencing casting shrinkage using DOE and numerical simulation

Sarabjit Singh[1, 2*], Dr. Rajesh Khanna[1] and Dr. Neeraj Sharma[3]

[1] Department of Mechanical Engineering, DAV University, Jalandhar, India
[2]International Tractors Ltd., Hoshiarpur, India
[3]Department of Mechanical Engineering, Maharishi Markandeshwar (Deemed to be University) Mullana, Ambala, India

E-mail: ssasht@yahoo.co.in

Abstract. The purpose of this study is to predict the effect of various factors like alloying elements, Inoculation and Carbon Equivalent (CEV) on green sand casting shrinkage defect in gearbox housing, without changing the casting methoding system. Foundries are prone to have defects related to feeding and gating system, sand and process parameters. In some engineering applications, it is very difficult to alter methoding system due to specific casting geometry requirements. If not controlled during development stage, the rejections may seriously affect the productivity. This research work involves the optimization of various factors by DoE and numerical simulation to reduce the shrinkage defect. Unlike the prototype or sample lot trial, mass production of gearbox housing castings has been done in a foundry to achieve the research purpose. Effect of these factors on final casting hardness and tensile strength has also been studied and found to be on positive side, as these parameters are important for satisfactory performance under dynamic loading. Outcome was the reduction in shrinkage porosity from 10.9% to 3.2% with overall saving of 41k USD over the period of 4.5 months. Implication of this study involves standardization of optimum range of alloy elements and inoculation for different grades of grey cast iron.

1. Introduction

For the purpose of power and speed transmission, gear box housing is an integral part of any automobile. The gear box housing not only provides rigid mechanical supports to high speed gears, but also provide needed lubrication to gears and protection from external surrounding. The gear box housing, which is normally of grey cast iron and made from green sand casting, needs to be sufficiently strong to avoid gears and shafts deflection under dynamic loading conditions.

Globally green sand casting is used for automotive housings due to process being economical, intricate profiles may be cast and high productivity. The normal grades of grey cast iron are FG260 and FG300. The casting process brings some inherent defects with them also, like blow holes, sand inclusion and shrinkage porosity. During machining of these housings, parts with visual defects like blow holes, pin holes, inclusion may be rejected. But hidden defects like shrinkage porosity pose a serious threat to housings, as they may not only cause leakage problem but also breakage at highly loaded points of gearbox housing. Shrinkage porosity defect poses a serious threat to performance and reliability of the housing and its control is possible only through knowledge of source causing the porosity.

In practical terms, shrinkage is found to be deeply related with casting gating and feeding system, chemical composition, inoculation, pouring temperature and part geometry. Combination of any of these parameters causes the volumetric changes in casting during solidification process. Higher amount of alloying elements like phosphorus and molybdenum tend to increase the shrinkage porosity, whereas high pouring temperature also increases liquid shrinkage. Increasing carbon content may reduce shrinkage tendency and inoculation may help in increased level of nucleation.

2. Literature Review

In grey cast iron castings, the volumetric contraction is classified in three distinct forms, namely liquid shrinkage, liquid to solid shrinkage and solid shrinkage. High pouring temperature results in heat loss in liquid metal and results in liquid shrinkage, which is first shrinkage step. With decreasing temperature, molten metal viscosity decreases and feeding to isolated places becomes difficult. Due to dendrites formation, these voids get entrapped in dendrite arms. Solid shrinkage is related to pattern shrinkage allowance and may be controlled during pattern designing. Mould hardness (both horizontal and vertical) is also of importance to keep the mould stable due to volume change during solidification.

Metal composition has effect on freezing range and has role in porosity as well. Higher levels of inoculation reduces fading problem but may result in shrinkage. Although phosphorous helps in molten metal fluidity and feeding, yet higher levels results in formation of low melting point iron phosphide (steadite) and thereby enhancing shrinkage [1]. Phosphorus levels above 0.2% are found to be contributing towards shrinkage growth. Shrinkage increases with under-inoculation due to shortage of expanding graphite. It also increases with over-inoculation as large number of eutectic cells are formed early during solidification and causing lesser precipitated graphite [2].

Among alloying elements, silicon is most important alloy. It transforms carbon to graphite, resulting into softer iron, which results in lower strength but less shrinkage. High levels of sulphur causes short run defects and increase metal hardness. Manganese is added to counter harmful effects of sulphur by making manganese sulphide instead of iron sulphide. Alloys like titanium and zirconium are added to increase molten metal fluidity that eventually helps in reducing shrinkage as metal reaches far away voids. Molybdenum helps in graphite flakes refinement, reduction in carbides and matrix strengthening. Magnesium helps in improving flowability in ductile irons and thereby reduces shrinkage [3].

Cast iron inoculation changes the physicochemical state of the molten metal. Good control on the microstructure and properties of cast iron is achieved by inoculants through increase in nucleation sites for growth of graphite flakes in FG Iron and graphite nodules in SG Iron. The basic iron solidification mechanism tend to make chilled structure (carbide structures) in case of inadequate inoculation. This results in difficulty in part machining due to excessive tool wear or frequent tool breakage [4]

Casting solidification simulation software are widely used in foundries processing different materials like aluminium, copper, iron and steel. The processes adopted by different foundries include green sand casting, resin and shell-bonded sand casting to investment and gravity die-casting. With the help of metal and design data, both designer and foundry expert can help in effective utilization of simulation techniques for better quality castings [5]. Use of casting numerical simulation tools is providing a much needed support to foundry industry to reduce casting rejection by doing offline trials for parameters like gating and feeding system, chemical composition etc. In order to effectively predict casting defects, density and mass transportation calculations are linked with solidification simulation. Effect of various parameters like alloying elements, molten metal temperature and chemical composition needs to be considered to predict solidification behaviour for shrinkage control [6]. In simulation studies, formation and growth of porosity during solidification has been studied through multi-phase model, which is valid for both microporosity and macroporosity. The model predicted molten metal flow, melt pressure and location and percentage of porosity. A momentum equation was derived by combining Darcy's law (for fluid flow in mushy zone) and Stokes' law (for flow flowing pure liquid) [7]. Solidification simulation is extensively used for detection of hotspots and predict the feed paths.

On the basis of filling results and solidification results, one can evaluate good options available to reduce defects. These good options can be converted to optimum option through use of optimization techniques like Taguchi techniques. DoE is a widely used tool to identify the most important process parameters, and to identify the levels at which these parameters to be kept to optimize the response. In addition, DoE helps in predicting significance of input variables as well as variables working in combination with each other [8].

3. Shrinkage porosity defect in Gearbox housing

During machining of the gearbox housing, shrinkage porosity was reported at the critical bearing bores and surfaces that are critical load carrying areas of the castings. Presence of sub surface shrinkage porosity, if goes undetected during machining at these areas, may adversely affect the field performance of the component also.

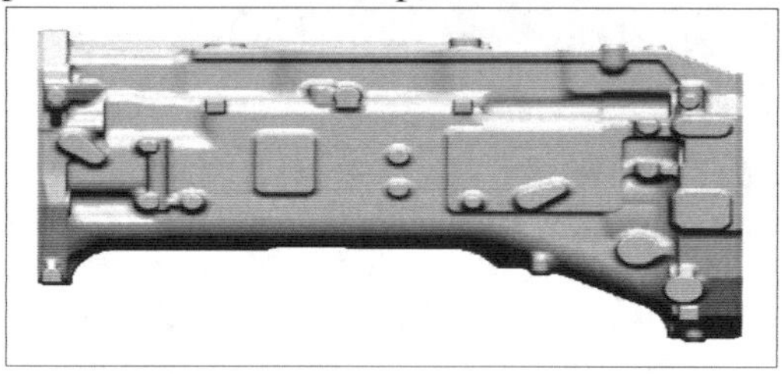

Component Details
Casting wt.:-105Kg
Yield: - 80%, Cavity: - 1 cav.
Pouring Time: - 15.0sec
Pouring Rate: - 8.8kg/sec
Wall Thickness:-6 mm
Liquid Metal Wt.:-131.25kg

Figure 1: Gearbox housing detail

Shrinkage porosity defect was investigated with a scanning electron microscope (SEM) and energy dispersive spectrometer (EDS) was used for chemical characterization of the cavities.

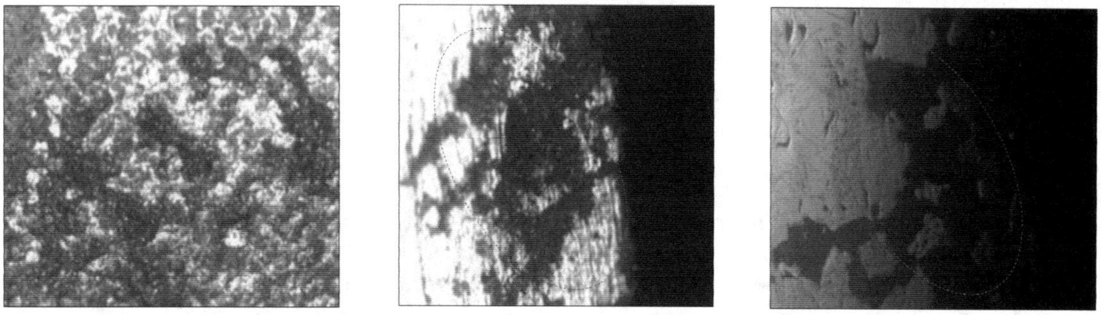

Figure 2: (a): Shrinkage at Mag. 20x Macroscopic View, **(b)**: Shrinkage at Mag. 32x Macroscopic View **(c)**: Shrinkage at Mag. 100x Macroscopic View

4. Research Methodology adopted for the work

The experiment for the research work is done at a foundry with automatic sand plant, high pressure moulding line and molten metal press-pour. Mass production tooling (patterns and core boxes) of gearbox housing are used for doing the casting.
Method adopted for the present work may be outlined in following way:

4.1. *Pareto study for shrinkage porosity defects*

Study done for quantity (% rejection) and concentration diagram for defects locations for castings with existing parameters.

IOP Conf. Series: Materials Science and Engineering **624** (2019) 012021 doi:10.1088/1757-899X/624/1/012021

4.2. *Identification of potential causes responsible for shrinkage porosity*

Following matrix was evaluated for validation purpose by taking in consideration all contributing parameters:

Table 1. Matrix for defect causes and validation tool

Domain	Cause	Validation Tool
Design	Feeding Modulus	Feed Mod > Neck Mod > Casting
Design	Pouring Temp	Temp. 1415- 1425 deg C
Design	Gating Location	Bottom gate, Flow is proper
Machine	Pouring Temp	Temp. 1430- 1435 deg C
Machine	Venting	All machine vented
Machine	Inoculation amount	(0.9 to 1 %) of Cast weight
Material	Inoculation amount	Composition (Ferro-silicon)
Material	Cr & Mo Optimize	Influence in shrinkage defects
Material	Other alloys	Graphitization retard influence
Material	Alloy composition	Added vs Customer Spec

4.3. Numerical simulation for old and new casting system parameters

Simulation study was conducted as per following table through change in alloying elements (Cr and Mo), Carbon Equivalent CE and inoculation (supersede grade)

Table 2. Change details for chemical composition and process parameters

Element	Previous	Proposed	Remarks
Carbon (%)	3.35 - 3.50	3.35 - 3.45	C less by 0.05 for maintaining Physical Prop
Silicon (%)	1.95 - 2.10	2.10 - 2.20	Silicon increased to counter Shrinkage
Manganese (%)	0.60 - 0.90	0.60 - 0.80	No Change
Chromium (%)	0.25 - 0.35	0.2 - 0.3	Chromium less by 0.05 to counter Shrinkage
Phosphorus (%)	0.050 - 0.090	0.04 - 0.05	Limits provided for P Content.
Sulphur (%)	0.070 - 0.10	0.070 - 0.10	No Change
Copper (%)	0.40 - 0.70	0.40 - 0.70	No Change
Tin (%)	0.040 - 0.070	0.040 - 0.070	No Change
Molybdenum (%)	0.20 - 0.30	0.20 - 0.30	Moly on lower side to counter Shrinkage
Titanium (%)	0.02 - 0.050	0.030 - 0.050	Titanium up to Counter Nitrogen Fissure
CEV (%)	4.02 - 4.22	4.06 - 4.20	Narrowed down CE Value
Inoculation (SS)	0.10%	0.15%	No Change
Pouring Temperature	1420 – 1435	1420 – 1435	No Change

4.4. Optimization of identified process parameters through DOE using Taguchi, with L9 orthogonal array

Different levels of alloying elements were taken for carrying 27 iterations.

Table 3. Variation taken in alloying elements and inoculants

Element	Designation	Level 1	Level 2	Level 3
Chromium (Cr)	Cr	Traces to 0.15	around 0.2	around 0.3
Molybdenum (Mo)	Mo	Traces	around 0.2	around 0.3
Inoculation (SS)	SS	0.1 %	0.15 %	0.2%

Table 4. L9 orthogonal array for trails

Expt No.	Level 1	Level 2	Level 3	Iteration A	Iteration B	Iteration C
1	Cr-1	Mo-1	SS-1	1A	1B	1C
2	Cr-1	Mo-1	SS-1	2A	2B	2C
3	Cr-1	Mo-1	SS-1	3A	3B	3C
4	Cr-2	Mo-2	SS-2	4A	4B	4C
5	Cr-2	Mo-2	SS-2	5A	5B	5C
6	Cr-2	Mo-2	SS-2	6A	6B	6C
7	Cr-3	Mo-3	SS-3	7A	7B	7C
8	Cr-3	Mo-3	SS-3	8A	8B	8C
9	Cr-3	Mo-3	SS-3	9A	9B	9C

5. Key Results

On the basis of changes done in optimized identified process parameters, results were outlined as:

5.1. Elimination of shrinkage defect depends on CEV, inoculants and alloying elements
5.2. Tensile decreases with increasing CEV, and hence range was limited
5.3. High CEV (carbon equivalent) also give rise to Open-Grain Structure
5.4. CEV and Inoculation also promotes Graphitization and hence expansion and shrinkage reduction
5.5. Chromium and Molybdenum limits prescribed to reduce the effect of carbide formation and Inoculation limits to achieve graphitization

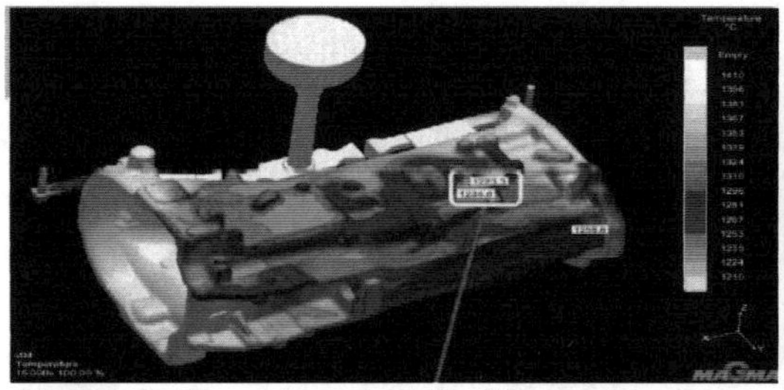

Figure 3. Temp less than liquidus temp upto 1223 deg

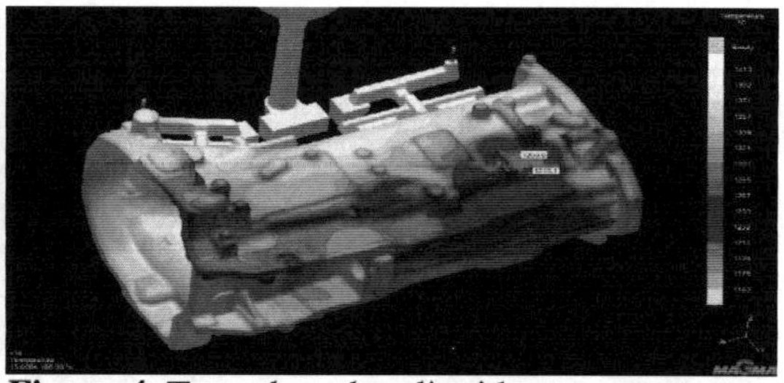

Figure 4. Temp less than liquidus temp upto 1210 deg

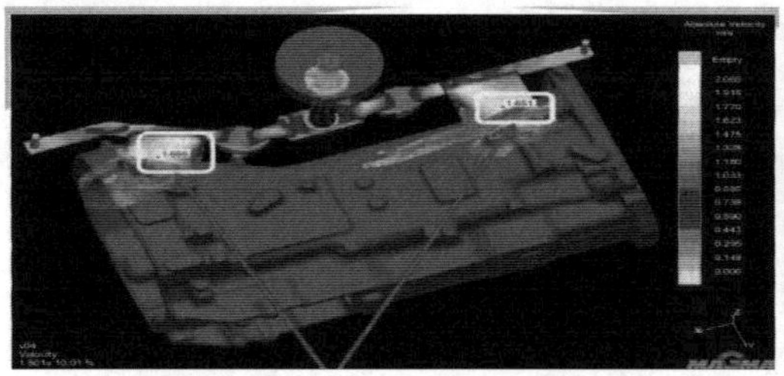

Figure 5. Velocity is 1.65 m/s

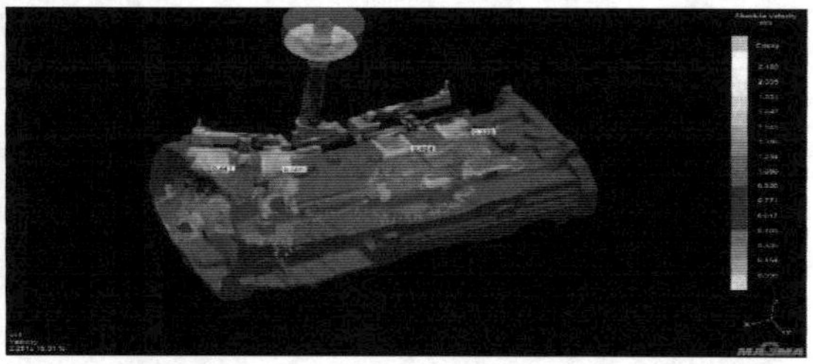

Figure 6. Velocity is from 0.33 m/s to 0.6 m/s

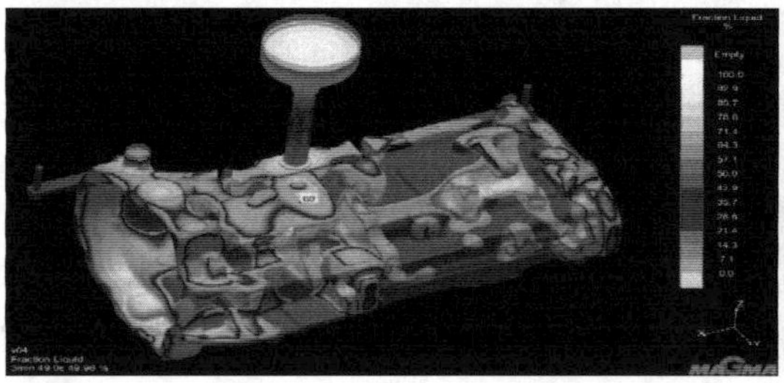

Figure 7. Fraction liquid result with old casting system

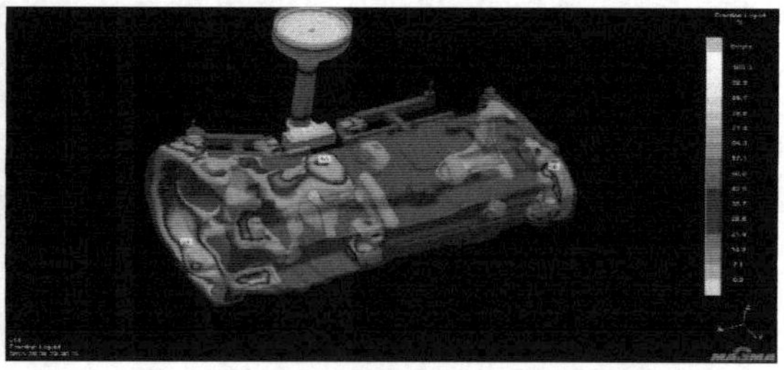

Figure 8.Fraction liquid result with new casting system

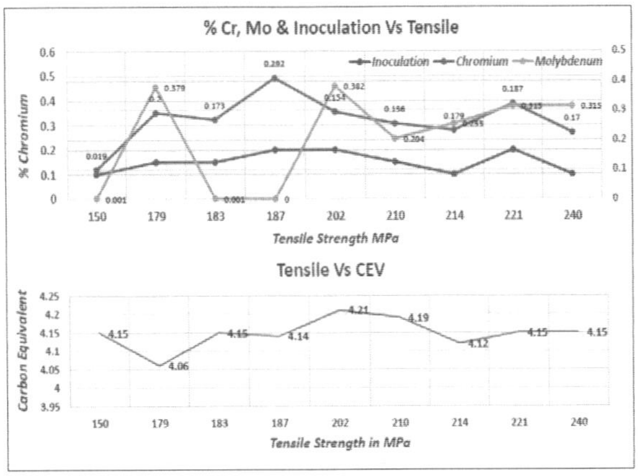

Figure 9. Effect of alloying elements, CEV on casting tensile strength

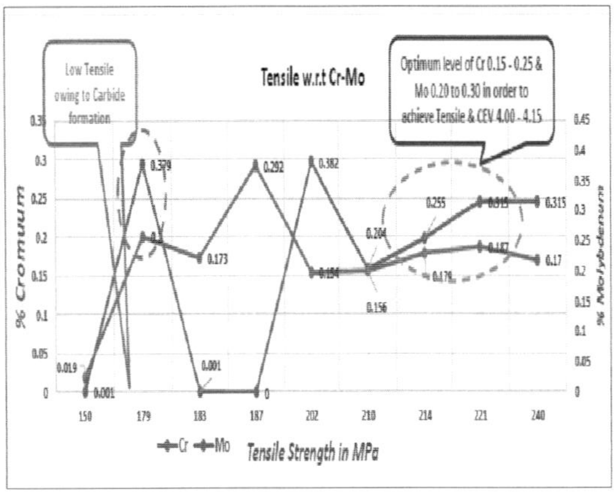

Figure 10. Optimum range for tensile strength with respect to alloying elements

6. Conclusions

Casting process parameters like alloying elements, inoculants and CEV help in controlling the shrinkage porosity defects in areas, where profile or dimensional changes are not possible due to specific part requirements. Optimization of these parameters helped in achieving the hardness and tensile strength as per functional requirements. Improvement resulted in the reduction in shrinkage porosity from 10.9% to 3.2% with overall saving of 41k USD over the period of 4.5 months.

Acknowledgments

The authors are thankful to Management of M/s International Tractors Limited (Punjab, India), for providing support in carrying casting pouring process and machining results of old and new casting system. We also gratefully acknowledge the support and facilities provided by M/s DCM Foundry (Punjab, India) to carry out all MAGMAS simulation trials and actual pouring of casting components.

References:

[1] Sadaf Vazehrad 2012 Shrinkage Porosity characterization in compacted cast iron components *Division of Casting of Metals, Royal Institute of Technology Stockholm* SE-100 44

[2] Lennart Elmquist 2010 On the formation of shrinkage porosity in grey iron castings, *Key Engineering Materials* ISSN: 1662-9795 **457** pp 416-421

[3] Muzzamil Ahamed 2014 Mechanical properties of sub-zero cast chilled iron useful for wear functional requirements of engineering components *Int. conf. on advances in manufacturing and material engineering Science Direct Procedia Materials Science* **5** pp 540-549

[4] Prof S.C.Borse and Y.E.Mangulkar 2014 Review on grey cast iron inoculation, *Int. Journal of Innovative Research in Science, Engineering and Technology* **3** Special Issue 4

[5] S. K. Paknikar 2014 Innovative simulation of castings –A technology to improve quality of castings as per global specifications with case studies *Indian Foundry Journal* **8** pp 28-32

[6] Dr. Ing. Jorg C. Sturm 2010 Cast iron - a predictable material *World Foundry Congress, Hangzhou, China*

[7] Kent D. Carlson and Zhiping Lin 2002 Modelling of porosity formation and feeding flow in steel casting, *Proc. of the 56th SFSA Technical and Operating Conference* Steel Founders' Society of America Paper No. 4.4

[8] Amol S.Karape and K.H. Inamdar 2016 Parameter optimization of sand casting process using Taguchi method *Asian Journal of Convergence in Technology* **2** Issue III, Issn No.:2350-1146

Evaluation of lubricant properties of vegetable oils as base oil for industrial lubricant

E Sneha [1], S Rani[1] and Muhammed Arif[2]
[1]Advanced Tribology Research Centre, Department of mechanical engineering, College of engineering, Trivandrum, India
[2]Department of chemistry, College of engineering, Trivandrum, India

E-mail id: snehaedla@gmail.com

Abstract. The possible insufficiency of oil resources in the future and the poor bio-degradability of mineral oil based lubricant resulted in the need of bio lubricant. Bio-based products are typically produced from vegetable oils. For the present study rice bran oil [RBO], jatropha oil [JO] and karanja oil [KO] are selected as the base oils for the industrial lubricant. Friction and wear characteristics are evaluated on four ball tester and the initial findings show that wear scar of RBO is lowest, this is because of the presence of natural anti-oxidants like gamma oryzanol. The coefficient of friction and wear of all the oils were evaluated as per ASTM D5183 and ASTMD 4172 respectively. Viscosity of the oils are tested on Brookfield rheometer and KO is having high viscosity when compared to other two oils but the change in viscosity with temperature is low for both RBO and JO. Chemical properties of the oils are conducted as per IS: 548 (Part 1) –1964. The iodine value for KO is less when compared to RBO and JO. Less iodine value is required to impart better oxidation stability. Thermogravimetric analysis showed that RBO is stable till 320 °C and KO, JO started degrading at 200 °C. The thermal properties of all the oils are measured and RBO has shown better properties at wide range of temperatures. The disadvantages of vegetable oils like oxidative stability need to be improved by adding additives and by chemical modifications.

1. Introduction

Growing environmental considerations and the regulations to reduce contamination and pollution will further propel the need for renewable and biodegradable lubricants. Most of the lubricants in the market are petroleum based, they became prominent because of their superior quality at low price [1]. The major concern by using mineral oil based lubricants is they are non-renewable and non-biodegradable. Because of these reasons an alternate environmentally friendly lubricants gained prominence. Most of the advantages and properties possessed by vegetable oils like biodegradability [2], good lubricating properties, high thermal stability [3], less toxicity, high viscosity and high flash point [4] are making them a potential alternate to be a good lubricant. The major disadvantages with the vegetable oils are poor low temperature behaviour and low oxidative stability which can be improved by modifying the oils chemically and mix with additives [5]. Oxidative stability of vegetable oils can be improved by chemically modifying the vegetable oils by transesterification or epoxidation [6].

Molecules of vegetable oils consists of glycerol part and fatty acid part. The structure of vegetable oils determines several properties to be a good lubricant. Stearic acid as additive in sunflower oil is useful in reducing the wear and coefficient of Friction [7]. Coconut oil under oxidative environment has shown better oxidative stability and high pour point when compared to the other vegetable oils considered in the study [8]. Smaller percentages of linoleic acid in soybean oil resulted in having low abrasion rate and coefficient of friction [9]. The natural anti-oxidant resulted in the low wear scar of rice bran oil compared with other vegetable oils [10]. Jatropha curcas L. oil (JCL) has the lowest followed by raw rapeseed oil (RME), Hydro treated Vegetable oil (HVO). JCL has shown low COF which is attributed to high amount of stearic acid [11]. Cutting fluids based on Palm Kernal Oil and Cotton Seed Oil showed better results when compared with commercial cutting fluids [12]. Avocado oil was shown low friction and wear when compared with other selected oils under research like canola (rapeseed), peanut, safflower, corn, olive, sesame, and soybean oils [13].

The monounsaturated acid that is found in the vegetable Oils called oleic acid is thermally stable than polyunsaturated fatty acids and shows good lubricating properties therefore it is highly desirable. Rice bran oil [RBO], karanja oil [KO] and Jatropha oils [JO] are having high percentages of oleic acid which is the major property to be a good lubricant. Karanja and Jatropha trees are grown in India, they are non-edible and drought-resistant plants with faster growth rate. Rice bran oil, which is grown much in countries like India, Korea, China, Japan, and Indonesia is used as edible oil. India is the second largest country producing rice in the world which helps in the growth of Indian economy if used for industrial purpose. There is not much research done on this oils yet hence for the present work this oils are considered.

The aim of the current study is to evaluate the physico-chemical, thermal, oxidative and tribological properties of RBO, KO and JO and study the feasibility of using them as base oil for industrial lubricant.

2. Materials and methods

The RBO, KO and JO are purchased from the market and preliminary tests like fatty acid composition, tribological, physico-chemical, thermal and oxidative properties were evaluated

2.1. *Materials*

The bran of the rice is squeezed and rice bran oil is extracted. The refined RBO was purchased from Kalady Rice Mill Consortium Pvt.Ltd, Kerala. Both KO and JO are extracted from the seeds.

2.2. *Fatty acid composition*

The fatty acid composition is very important in deciding the lubricating properties, oxidative stability and several other properties of oils. The fatty acid composition of the selected three vegetable oils was evaluated as per IS548 by Gas chromatography (GC).

2.3. *Tribological properties*

As per ASTMD 5183 standard the coefficient of friction for RBO, KO and JO was evaluated by using Four Ball Tester as shown in figure 1. The tests were carried out for 60 min at 75°C with 600 rpm. Chrome alloy steel balls with 12.7 mm diameter were used for the experiments. A rotating steel ball is pressed against three steel balls which are firmly held together and immersed in lubricant under test. All the parameters such as test load, duration, temperature and rotational speed will be set before starting the test. The wear scar diameter on the bottom three balls is evaluated by using optical microscope as shown in figure 2.

Figure1. Four ball tester

Figure 2. Optical microscope

IOP Conf. Series: Materials Science and Engineering **624** (2019) 012022 doi:10.1088/1757-899X/624/1/012022

2.4. *Physico-chemical properties*

The viscosity of RBO, KO and JO are evaluated by using Brookfield Rheometer as per standards and the equipment is shown in figure 3.

The chemical properties which indirectly explains the lubricating properties like acid number and iodine number were evaluated based on IS: 548 (Part 1) –1964. Acid number is defined as the measure of free fatty acids present in the oil. The amount of unsaturated bonds present in the oil is called as iodine number. The acid value is determined by chemical methods. Firstly, weigh 20 ± 0.1 grams of oil sample in a conical flask. Add 50 ml of methanol to the oil in the conical flask and drop 1 ml of the phenolphthalein indicator solution to the mixture in the conical flask. Then titrate the mixture against the potassium hydroxide in the burette. The break point is the point at which the colour of the mixture in the conical flask turns pink colour. Acid value = $\frac{V*N*56.1}{W}$, where V= volume of titrate sample, N= Normality of alkali solution (KOH), W=Weight of the sample

The iodine number indicates the amount of unsaturated bonds present in the sample. The material is treated, in carbon tetrachloride medium, with a known excess of iodine monochloride solution in glacial acetic acid (Wijs solution). The excess 'of iodine monochloride is treated with potassium iodide and. the liberated iodine estimated by titration with sodium thiosulphate solution.

Iodine value = $\frac{12.69N(B-S)}{w}$, where B = Volume in ml of standard sodium thiosulphate solution required for the blank, S = volume in ml of standard sodium thiosulphate solution required for the sample, N = normality of the standard sodium thiosulphate solution, and W = weight in g of the material taken for the test.

Figure 3. Brookfield Rheometer

Figure 4. Acid test stages a) before b) during c) after titration

2.5. *Thermal properties*

The flash point is the temperature at which the oil catches spark and gives out smoke whereas the fire point indicates the temperature at which the oil catches fire. Both of them gives the safe operating temperature of the oil. The flash point and fire point were estimated by Cleveland open cup equipment as per ASTMD92. The pour point is the temperature at which oil loses its flow ability and is evaluated by physical technique. The thermal degradation of the three oils are measured by using thermogravimetric analysis with an operating range of 0° C – 500° C.

2.6. *Oxidative stability*

The vegetable oils are having low oxidative stability because of the presence of poly unsaturated acids like linoleic and linolenic acids present. Hot Oil Oxidation Test (HOOT) as per American Oil Chemists Society (AOCSCd-12-57) was followed for evaluating the oxidative stability of the selected oils. The experiment is conducted in a dark oven with air vent at 100°C for 30 h.

3. Results and discussions

The fatty acid composition of RBO, JO and KO are shown in Table 1. It is clear that all the three oils are having higher percentages of oleic acid which is responsible for imparting good lubricating properties. Hence the selected oils can be potential alternate for mineral oil based lubricants.

Table 1. Fatty acid composition of rice bran oil, karanja oil and jatropha oil

Fatty acids	Rice bran oil %	Karanja oil %	Jatropha oil %
C18:0	2.12	7.2	6.8
C18:1	41.26	50.40	40.1
C18:2	32.63	15.54	33.9
C18:3	0.84	1.3	0.12

The viscosity of the oils under study is evaluated at different temperatures and are tabulated and shown in Table 2. From figure 5 it is clear that KO is having high viscosity when compared to other two oils but the change in viscosity with temperature is low for both RBO and JO. This helps in using the oils at wide range of temperatures. The thermogravimetric analysis (TGA) for the selected oils is shown in figure 6. RBO is stable up to a temperature of 320˚C. The mass degradation starts from 320 ˚C to 440 ˚C. Both JO and KO is stable till 200 ˚C. The rapid mass degradation of KO starts at 380 ˚C whereas rapid degradation of JO starts at 320 ˚C.

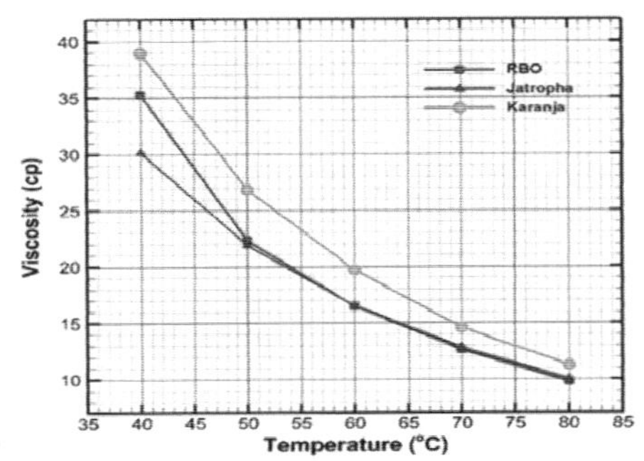

Figure 5. Viscosity plots for RBO, JO and KO

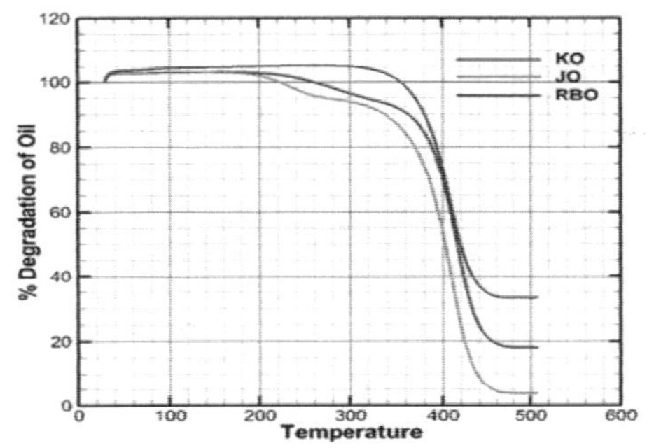

Figure 6. Thermogravimetric analysis of rice bran oil, karanja oil and Jatropha oil

Table 2. Viscosity values of RBO, JO and KO

Temperature (˚C)	Viscosity (cp)		
	Rice Bran Oil	Jatropha Oil	Karanja Oil
40	35.26	30.141	38.995
50	22.34	21.893	26.863
60	16.5	16.547	19.697
70	12.62	12.801	14.577
80	9.798	10.053	11.246

The coefficient of friction (COF) and wear scar diameter of the selected oils are evaluated by using four ball tester and are tabulated and shown in Table 3. Gamma oryzanol which is the natural anti-oxidant present in the RBO resulted in low wear scar. However the friction coefficient of RBO is high compared to the other oils.

Table 3. Coefficient of friction and wear scar diameter of RBO, KO and JO

Name of the oil	Coefficient of Friction	Wear Scar (mm)
Rice Bran Oil	0.0898	0.547
Karanja Oil	0.0624	0.584
Jatropha Oil	0.0673	0.571

The chemical properties of the selected oils are evaluated and tabulated in Table 4. The less amount of free fatty acids present in the RBO resulted in the low acid value. The iodine value indicates the presence of unsaturated fatty acids present in the oil. Among the 3 vegetable oils selected karanja has the lowest iodine value which indicates good oxidative stability

Table 4. Chemical properties of RBO, KO and JO

Name of the oil	Acid Value	Iodine Value
Rice Bran Oil	0.66	93.033
Karanja Oil	5.02	75.207
Jatropha Oil	19.5	91.833

The thermal properties of the selected oils are evaluated and tabulated in Table 5. The fire point of RBO is higher than KO and JO which implies its ability to work at higher temperatures. The pour point of RBO is lower than KO and JO which makes it able to flow at lower temperatures. The higher fire point and lowest pour point of RBO is due to the presence of unsaturated fatty acids present in it. It is clear that RBO can be worked at wide range of temperatures than JO and KO.

Table 5. Thermal properties of RBO, KO and JO

Name of the oil	Flash Point (˚C)	Fire Point (˚C)	Cloud Point (˚C)	Pour Point (˚C)
Rice Bran Oil	312	326	-5	-8
Karanja Oil	225	232	2	-6
Jatropha Oil	240	252	-6.5	-7.5

The hot oil oxidation test is conducted on all the selected oils and results are tabulated and shown in Table 6. The change in the acid value of RBO before and after heating is 1.24 whereas the change in acid value for KO and JO are 0.925 and 1.655 respectively. The increase in the acid value of JO due to heating indicates that JO has a very low oxidative stability and hence any type of chemical modifications are very much required for JO.

Table 6. Hot oil oxidation test for RBO, KO and JO

Name of the oil	Acid value at different heating hours					
	0	6	12	18	24	30
Rice Bran oil	0.66	1.11	1.22	1.45	1.68	1.9066
Karanja oil	5.02	5.12	5.5620	5.6717	5.805	5.945
Jatropha oil	19.5	20.18	20.4999	20.685	20.95	21.155

4. Conclusions

The lubricant properties of RBO, KO and JO are evaluated and the results were compared with each other. The following conclusions were made from this work:

i. The tribological properties of the oils were evaluated and coefficient of friction for KO is lowest than RBO and JO whereas wear scar of RBO is less when compared to the other selected oils

ii. Among all the selected oils KO has a higher viscosity at 40˚C and 80 ˚C than that of RBO and JO

iii. The acid value of RBO is lower than that of JO and KO and the iodine value of KO is lower than that of RBO and JO

iv. The fire point of RBO is higher than that of KO and JO and the pour point of RBO is lesser than that of KO and JO.

v. KO is showing better oxidative stability than RBO and JO.

vi. RBO is stable till 320 °C whereas JO and KO started degrading after 200 °C

From the feasibility study it is clear that the selected oils are suitable as base oil for industrial lubricants. However chemical modifications may be required for JO and RBO to improve their oxidative stability.

Acknowledgements: The authors sincerely acknowledge the financial contribution from Kerala State Council for Science, Technology and Environment [KSCSTE] for carrying out the research. Authors wishing to acknowledge the technical support provided by Dr. Anjan R Nair, Assistant Professor, Micro and Nano fluidics lab, Dept. of mechanical engineering, college of engineering, Trivandrum for the evaluation of viscosity on Brookfield rheometer. The refined RBO was purchased from Kalady Rice Mill Consortium Pvt.Ltd, Kerala. The crude karanja oil is purchased from Vijaya Agro Industries, Sangamner, Maharashtra. The crude jatropha oil is purchased from Jatropower Bio-Trading Private Limited, Coimbatore.

References

[1] Dowson D 1979 '*History of Tribology*'. New York: Longmans Green.125–254.
[2] Boyde S 2002 '*Green lubricants. Environmental benefits and impacts of lubrication*. Green Chem'. 4, 293–307.
[3] Asadauskas, S Erhan, S.Z 2000. '*Lubricant basestocks from vegetable oils*'. Ind.Crops Prod. 11, 277–282.
[4] Santos J, Santos IM, Conceiço MM, Porto SL, Trindade MF, Souza AG, Prasad S, Fernandes J, Araújo AS 2004. '*Thermoanalytical, kinetic and rheological parameters of commercial edible plant oils*' J. Therm. Anal. Calorim.75,419–428.
[5] Adhvaryu, A, Erhan, S.Z, Perez, J.M 2004. '*Tribological studies of thermally and chemically modified vegetable oils for use as environmentally friendly lubricants*'. Wear 257, 359–367.
[6] N.A. Zainala, N.W.M. Zulkiflia, M. Gulzarb, H.H. Masjukia 2018. '*A review on the chemistry, production, and technological potential of bio based lubricants*', Renewable and Sustainable Energy Reviews, 80–102.
[7] Fox NJ, Tyrer B, Stachowiak GW 2004. '*Boundary lubrication performance of free fatty acids in sunflower oil*', Tribol Lett.16,275–81.
[8] N.H. Jayadas, K. Prabhakaran Nair 2006. '*Coconut oil as base oil for industrial lubricants—evaluation and modification of thermal, oxidative and low temperature properties*', Tribology International. 39, 873–878.
[9] Matthew T. Siniawski, Nader Saniei, Bigyan Adhikari and Lambert A. Doezema 2007. '*Influence of fatty acid composition on the tribological performance of two vegetable-based lubricants*', Journal of synthetic lubrication. 24,101–110.
[10] S. Rani, M.L. Joy, K. Prabhakaran Nair 2015. '*Evaluation of physiochemical and tribological properties of rice bran oil – biodegradable and potential base stoke for industrial lubricants*', Industrial Crops and Products.65,328–333.
[11] Alessandro Ruggiero, Roberto D'Amato, Massimiliano Merolaa, Petr Valašek, Miroslav Müller 2017.'*Tribological characterization of vegetal lubricants: Comparative experimental investigation on Jatropha curcas L. oil, Rapeseed Methyl Ester oil, Hydrotreated Rapeseed oil*', Tribology International.109,529–540.
[12] S.A. Lawal, I.A.Choudhury, Y.Nukman 2017.'*Evaluation of vegetable and mineral oil-in-water emulsion cutting fluids in turning AISI 4340 steel with coated carbide tools*', Journal of Cleaner Production.52,1-12.
[13] Reeves CJ, Menezes PL 2016. '*Advancements in eco-friendly lubricants for tribological applications: past, present, and future*'. In: Davim JP, editor. Ecotribology. Switzerland: Springer International Publishing,41–61.

IOP Conf. Series: Materials Science and Engineering **624** (2019) 012023 doi:10.1088/1757-899X/624/1/012023

Design optimization of steering knuckle by adopting bionic design approach

Vikas Ghungarde[1], Sudhir Awachar[2], N K Vaidya[3], Jagadeesha T[4]

[1] Mechanical Department, National Institute of Technology, Calicut, India
[2] Deputy General Manager, Tata Motors Pune
[3] General Manager, Tata Motors Pune
[4] Assistant Professor, National Institute of Technology,Calicut,India

vikas.ghungarde@gmail.com jagdishsg@nitc.ac.in sudhir.awachar@tatamotors.com

Abstract. Weight optimization plays vital role in performance improvement for an automobile. In vehicle, steering knuckle is one of critical components as it is subjected to millions of varying cyclic loads and it contributes to unsprung mass of vehicle. Therefore, steering knuckle of passenger vehicle is selected for this research work. The aim of this research work is to relate optimized shape of the part with bionic shape, which will give the optimum design. Topology optimization is used as a tool for getting optimum material layout with given space and boundary conditions. The objective function for topology optimization is to maximize stiffness with minimum mass constraints as 30%. The optimization is done by Altair INSPIRE software in which the part is imported as cad model with SG 500/7 as part material. The final surface modeling is done by Polyworks software package for parametric model and result of that modelling has been applied for the standard analysis of the part. The proposed design of the parts meets the strength and stiffness as per requirement. For physical prototype the part is manufactured by additive manufacturing using AlSi10Mg as a material and required properties for same material is also generated. New proposed design of the part shows that there is reduction in weight by 35%.

1. Introduction

Mass manufacturing techniques evolved over past few decades have ensured economics of scale. However limitations of these techniques have invariably forced part designs to be actually inefficient. With the progress into a highly competitive market place further getting constrained to environment/ecology conservation, light weighting has become a focus area in automobile industry. While light weighting can be achieved in principal through extensive use of polymers; polymers offer three challenges namely cost, durability and end of life scrap handling have not allowed deep penetration of polymers in functional areas of automotive vehicles.

Evolution of 3D printing in past three decades has been offering this feasibility of creating most efficient part designs that can be realised physically which was impossible by conventional process. Bionic shape designs which are inspired through shapes of plants and animals has become a standard thought process for mechanical parts shape optimization. Nature is the best way to study for optimum material layout with best possible shape, because it has material only where it is required. Steering Knuckle is one of the main components used in the steering system of automobile which contains wheel hub and connects it to suspension and steering components. It undergoes millions of varying cyclic loading conditions which tends to fatigue failure of the part. Steering knuckle comes under unsprung mass so lighter the weight, greater the power saving and lesser the vibration (Less inertia). Topology optimization is the tool for the lightweight structure for the bionic design in a given design space [1]. shape optimization is carried out by performing static analysis as base, which intended to fulfill the weight reduction objective. Shape optimization aim is to modifying the part boundaries with given constraints and improve the performance of the structure [2]. Because of geometric complexity of the part after optimization were hardly convertible into manufactural parts due to the conventional manufacturing's inherent design restrictions. But Laser Additive Manufacturing is rising the grade of the design freedom and manufacturing possibilities [3]. Method for converting a optimize material layout into a STL file which is adaptive for additive manufacturing by using B-spline curve fitting as a tool [4]. V. Sivananth and S. Vijayarangan has done optimization of passenger car steering knuckle for the fatigue loading and he found that steering arm region is the critical part under the cyclic loading conditions. The failure region of the knuckle was optimized under operating load conditions and fatigue life increased [5]. Ti6Al4V alloy parts are widely used in this sector because of their good mechanical properties-to-weight ratio. For optimization Inspire software package is used as a tool. Further it is manufactured by direct metal laser sintering as an AM process [6].

2. Bionic design case study

Optimize structure inspire from nature are used as a basis to improve product design. an air suspension system was created using bionic approach to decrease the required air pressure and increase the suspension force. The bionic structure used for relating the air suspension system is albatross wings. Albatross wings have better flight mechanism that permits it to cooperate with complex environment. System consist of guide rail and a slider which is based on principle that the airflow is coming from the air outlet hole towards the guide rail. Using the bionic albatross wing's surface structure, the guide rail surface was designed into diverse structures and concludes that air pressure in air static rail is reduced by 37% in standard working condition [7].

3. Optimization

3.1. Process flow

The primary thing for the optimization is to define the objective function. Here the objective function is to maximize the stiffness with minimum mass constraint as 30%. Then optimization space have to identified and accordingly the given space is divide into 2 parts mainly design space and non-design space. Design Space is the volume where the optimization process carried away from, creating the most efficient material distribution capable of withstanding the forces applied to model and Non-Design Space is the region within the model in which material will not be removed. This area is still included in the simulation, but excluded from modification. The process flow for topology optimization is as shown in figure 1.

3.2. Load Estimation

The vehicle is run on the track for calculating the loads that are coming on the wheel which will then transfer to the steering knuckle and suspension system. For calculating the load Wheel Force Transducer is used which will give the six components as forces and torques in X, Y and Z directions on the rotating wheel. Data output referred to vehicle coordinate system. The forces are calculated is in the form of G force that is acceleration force. This loading data will be the input condition for the analysis of the part which further will give optimize material layout. Force data is collected in X, Y and Z directions at contact patches and wheel center.

3.3. Material

The material used for manufacturing the steering knuckle is SG 500/7 iron, the properties for the same is as shown in table1. Conventionally the steering knuckle is manufactured by sand casting process. The aim of this paper is to optimize the part with given material without giving any manufacturing constrain to obtain the optimum material layout.

IOP Conf. Series: Materials Science and Engineering **624** (2019) 012023 doi:10.1088/1757-899X/624/1/012023

Table 1. Material properties for SG 500/7.

Density (gm/cm3)	Yield strength (MPa)	Ultimate Strength (MPa)	Young's Modulus (GPa)
7.1	320	500	170

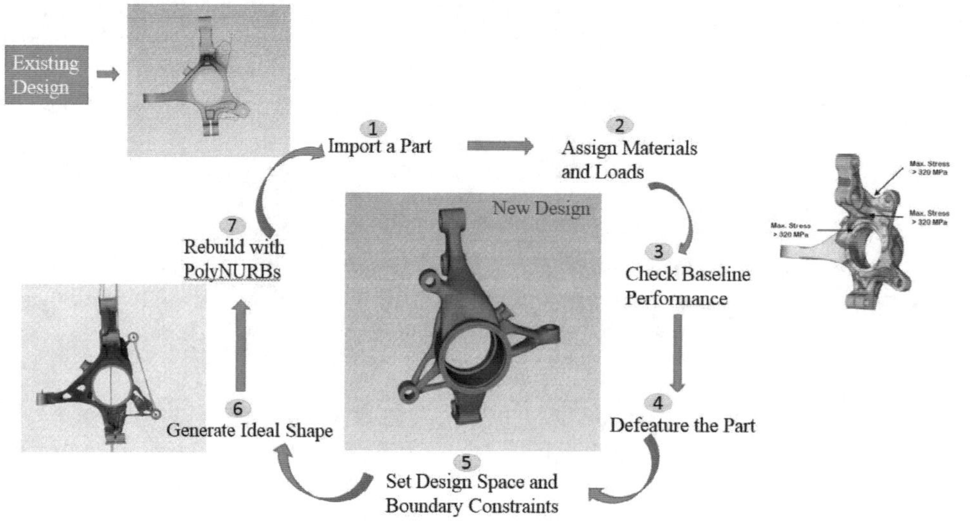

Figure 1. Steps and workflow in Topology optimization.

3.4. Load and Boundary conditions

The geometric model of steering knuckle is as shown in figure 2. The boundary conditions for optimization are as shown in figure 3. The lower ball joint and strut mounting points are constraints in all translation degree of freedom. The steering arm region is constraints in only y translation. The stiffness values for upper and lower caliper lug is also calculated based on unit load application along with the related displacement caused. Vehicle stability control is one of the major parameter for smooth operation of vehicle without any fatigue. For this toe and camber stiffness calculation becomes the primary concern and its value should be exceeding minimum permissible assessment. Toe and camber stiffness is determined by giving unit loading moment and analyzing the related displacement for stiffness calculation. This stiffness should match the particular standard for finalizing the part geometry.

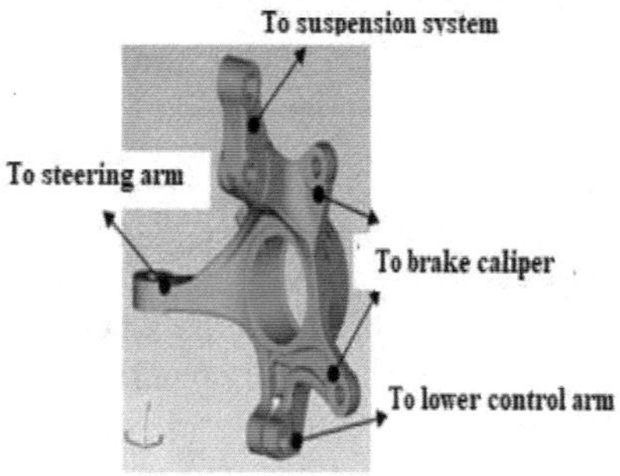

Figure 2. Steering knuckle.

Figure 3. Boundary conditions for steering knuckle.

3.5. Analysis

For Basic analysis of the part the boundary conditions and constraints are applied according to the loading cases. The von mises stress distribution plot for all loading conditions is as shown in figure 4. Four nod quadrilateral elements are considered for meshing. First the coarse meshing is done then in every subsequent iteration the local meshing is done at the critical section where the stress concentration is observed. Only critical load cases have been selected, because other cases are not having any considerable influence on the steering knuckle. The bump and kerb strike cases are influenced most due to sudden impact on the tire which will then transfer to the steering knuckle. During steering arm fatigue analysis or rack cycle force analysis the tie rod region is subjected to the cyclic loading of stress amplitude 6000N with mean stress as zero. For this analysis the wheel hub portion is constrain in all degree of freedom and load is applied in the tie rod direction. This will give the strength analysis of the steering arm.

3.6. Material layout

The optimum material layout is obtained as per the load cases, constraints and boundary conditions and is as shown in figure 5. Strut part and wheel hub bearing region is the most stress zone with given load cases because stress at part crosses the yield strength but it is within the permissible range of plastic strain.

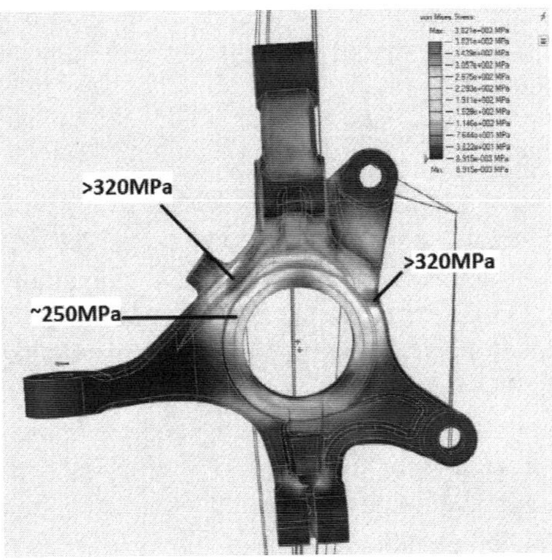

Figure 4. Von mises stress plot for static analysis.

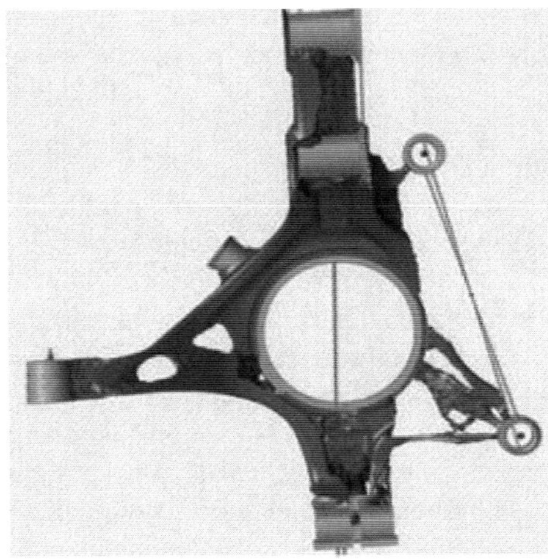

Figure 5. Optimum material layout.

The area of higher stress will get identified and according to optimization the area of higher stress will have more material. The regions where less stresses are there the pockets are made accordingly.

3.7. Modeling

The optimum material layout is considered as a base for the final modeling of the part. The material layout given by the software for modeling is infeasible according to the practical approach (under working condition of part) because cross-section at some points are having very less thickness and support for load carrying. Those portions are shown in figure 6a with red circles. Therefore, modification is done at steering arm region and brake caliper region. For modification n number of iterations followed by the analysis of the part have been carried out with all the functional requirement of the part remains the same. The final geometry of steering knuckle is as shown in figure 6b. For validation, the optimized part is again analyzed by using Hyper mesh for mesh generation, Abaqus as a solver and hyper work for result verification.

IOP Conf. Series: Materials Science and Engineering **624** (2019) 012023 doi:10.1088/1757-899X/624/1/012023

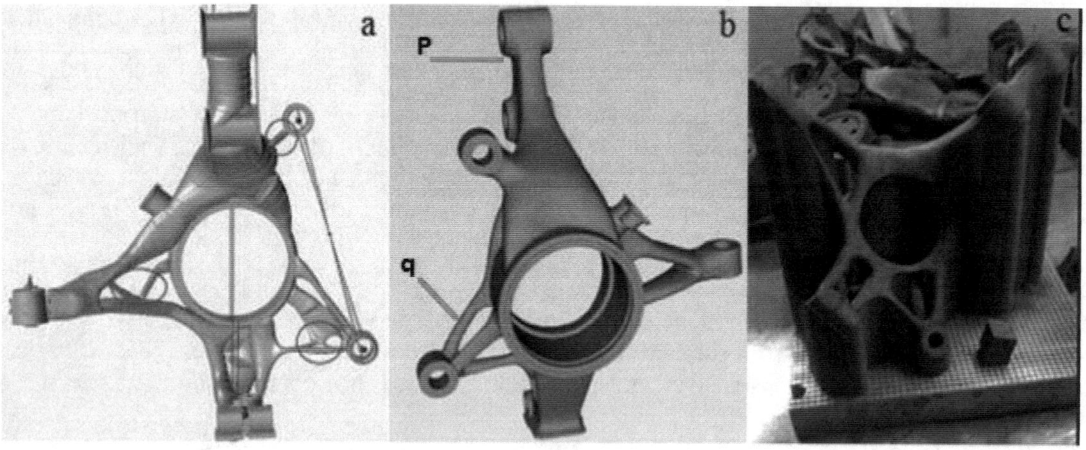

Figure 6.a Modeling as per material layout by software, **b** Final optimize design, **c** Metal 3d printed steering knuckle with support structures.

4. Bionic Design

The bionic approach methodology is used for transforming material layout to final design for steering knuckle, where the material layout is relating to the mimic nature and animals' anatomy. The calliper region is as shown in figure 6b (portion p) has relatable structure as a tridactyl bird feet anatomy. In birds this anatomy is for clamping and stiffness purpose. Same functional requirement for steering knuckle at the point of calliper side. The strut portion as shown in figure 6b (portion q) is structure similar to the horse anatomy. Nuchal ligament with the abdominal muscles will try to relieve the back muscle weight so that rider's weight can be contentedly carried out. By studying and implementing this bionic geometry the load bearing capacity of the part will get increased. This implementation phase is never ending and iterative process.

5. Manufacturing

5.1. Material

Considering the manufacturing feasibility of the final design Additive manufacturing is the only approach for physical part. This part for actual automotive application needs to be made in SG 500/7 from cost perspective. However we carried out an experiment using AlSi10Mg. Since SG500/7 printing technology does not exist today. Over the past few years, the AM industry has been using several types of aluminium alloy powders. A wide range of aluminium AM parts have been developed for a variety of application using AlSi10Mg and today it is the most popular aluminium alloy powder grade in the metal AM industry. So AlSi10Mg has been selected for the metal 3D printing of steering knuckle. The composition of AlSi10Mg is as shown in table 2.

Table 2. Composition of AlSi10Mg.

Aluminum (Al)	Copper (Cu)	Iron (Fe)	Magnesium (Mg)	Manganese (Mn)	Silicon (Si)	Titanium (Ti)	Zinc (Zn)
balance	≤ 0.05	≤ 0.55	0.2 - 0.45	≤ 0.45	9.0 - 11.0	≤ 0.15	≤ 0.10

The properties of the 3D printed materials depend on several factors based on which its value changes accordingly. The factors are particle size, powder production method, printing orientation etc. Due to above concern the properties for the AlSi10Mg has been generated.

5.2. Tensile Test

The standard specimens for tensile test are 3D printed as shown in the figure 7. Specimen tension tested as per ASTM E384-11. Stress relieved at 300°C±10°C for 2hours and it is air cooled. The tensile test results are as shown in table 3. AM material properties in Vertical and 45° are similar to 90% of the horizontal direction properties. AM material properties are in stress relieved condition and without any other heat treatment.

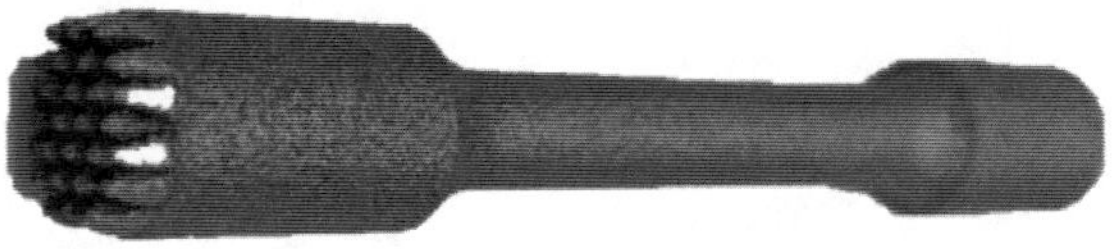

Figure 7. Standard Metal 3D printed part for testing.

Table 3. Tensile test result.

Orientation	Tensile strength (MPA)	Yield strength (MPA)	Average Elongation (%)
Horizontal	385.33	287.5	5.2
45° Angle	353.33	250	3
Vertical	353.33	275	5

The steering knuckle is 3D printed with Direct Metal Laser Sintering with AlSi10Mg as a printing material and is shown in figure 6c with support structures.

6. Conclusions

An overall idea about the material layout is identified by using topology optimization as a tool. This material layout is used for relating this shape to the bionic shapes and converts it to the solid model. The optimization process is accurate, quick and easy to implement. The quality of the product is matching the strength and stiffness criteria required for efficient functionality of the part. The bionic design weight optimization results in 35% reduction in weight. Additive manufacturing provides an option to print bionic shapes as per the actual design. In future we are exploring further same part by casting for production.

7. References

[1] Rahman, Kaysar, et al. "A New Bionic Topology Optimization Method Based Model of Bone Adaptation." *Applied Mechanics and Materials*. Vol. 433. Trans Tech Publications, 2013.

[2] Masching, Helmut, and Kai-Uwe Bletzinger. "Parameter Free Structural Optimization Applied to The Shape Optimization of Smart Structures." *Finite Elements in Analysis and Design* 111 (2016): 33-45.

[3] Emmelmann, Claus, et al. "Laser Additive Manufacturing and Bionics: Redefining Lightweight Design." *Physics Procedia* 12 (2011): 364-368.

[4] Liu, Shutian, et al. "A Realization Method for Transforming a Topology Optimization Design into Additive Manufacturing Structures." *Engineering* 4.2 (2018): 277-285.

[5] Sivananth, V., and S. Vijayarangan. "Fatigue Life Analysis and Optimization of a Passenger Car Steering Knuckle Under Operating Conditions." *International Journal of Automotive & Mechanical Engineering*11 (2015).

[6] López-Castro, J. D., et al. "Topological optimization and manufacturing by Direct Metal Laser Sintering of an aeronautical part in 15-5PH stainless steel." *Procedia Manufacturing* 13 (2017): 818-824.

[7] Gao, Siyang, Bangcheng Zhang, and Jianwei Sun. "Research on the Design Method of a Bionic Suspension Workpiece Based on the Wing Structure of an Albatross." *Applied bionics and biomechanics* 2019 (2019).

IOP Conf. Series: Materials Science and Engineering **624** (2019) 012024 doi:10.1088/1757-899X/624/1/012024

Optimal Design and Analyses of T-shaped rotor Magnetorheological Brake

S Ācharya [1,2], T R S Saini [1] and H Kumar [1]

[1] Department of Mechanical Engineering, National Institute of Technology Karnataka, Surathkal, Mangalore, India-575025.
[2] E-mail: subhashacharya12@gmail.com

Abstract. Magnetorheological (MR) brakes, belonging to the class of electromagnetic brakes, have a potential to replace conventional hydraulic brakes owing to reversible, rapidly controllable torque characteristics. In this study, T-shaped rotor MR brake was considered due to their higher braking torque capability and compactness compared to other configurations of brake. Optimal design of the brake was performed considering brake rotor radius, rotor thickness, flange length, casing thickness, coil height and coil width. Magnetostatic analyses were performed for different combinations of parameters of the brake dimensions to compute the magnetic flux density generated in the MR fluid region and the torque ratio and mass were calculated. The optimum dimensions of the brake were determined based on maximization of torque ratio and minimization of mass of the brake using multi-objective Genetic algorithm optimization technique. Further, magnetostatic analyses of the T-rotor brake with optimal dimensions were performed and torque characteristics were compared with those obtained for brake with simple disk rotor. It was concluded that T-rotor brake produces higher braking torque compared to simple disk rotor type MR brake for similar dimensions.

1. Introduction

A MR brake comprises of single or few revolving disks surrounded by MR fluid inside a casing embedded with electromagnet. By supplying current to the electromagnet coil, the iron particles in the MR fluid aligns with the generated magnetic field and its viscosity and yield stress increases in a fraction of a second. This yield stress which is controllable, generates torque due to shear friction on the revolving disk. Thus, the MR brakes when compared to conventional hydraulic brakes have quicker response, easier implementation of a new controller, lesser maintenance since there is no material wear and lower weight as it does not require additional auxiliary components. Based on the shape of the brake rotor, there are five designs of MR brake proposed by researchers namely drum, inverted drum, disk, T-shape rotor and multiple disks brake. T shaped rotor brake produces higher torque due to larger area of MR fluid subjected to MR effect at the periphery of rotor. However, it is complex and hence difficult to manufacture [1]. The performance of an MR brake is measured primarily in terms its torque characteristics.

A lot of research work has been performed by several researchers to design and enhance the torque characteristics of MR brake. Assadsangabi et al. [2] performed finite element analysis of MR brake with simple disk to analyse the magnetic field intensity distribution and then optimized braking torque capacity using Genetic Algorithm to obtain optimal dimensions. Nguyen and Choi [3] determined the optimized dimensions of several forms of brakes such as simple disk, drum, hybrid and T shaped types by maximizing the braking torque with the torque ratio higher than desired value. Further, the selection of MR brake based on radial and axial dimensions have been proposed. Nguyen and Choi [4] determined optimal dimensions of single and multiple disk MR brake for three types of commercial MR fluid supplied by the Lord Corporation for a passenger vehicle depending on the desired braking torque, temperature due to friction of MR fluid in the absence of magnetic field, mass of the brake and important dimensional parameters. Hung and Bok [5] determined optimal dimensions of a MR brake with T shaped rotor considering braking torque, availability of space, mass and heat generated by friction torque of the MR brake in the absence of magnetic field. Nguyen and Choi [6] designed three different configurations of MR brake with T shaped stationary drum embedded with two coils based on three different commercial Lord MR fluids and compared their size, mass and power consumption. Nguyen et al. [7] determined optimal dimensions of disk, drum, single-coil and double coil with and without combination of magnetic and non-magnetic materials for casing and T-type MR brakes with the aim of minimizing the mass of the MR brake needed to provide necessary braking torque. The selection of MR brake types was discussed depending on the mass of the brake, power consumption and braking torque. However, multi-objective optimization considering several objective functions such as mass of the brake, maximum braking torque and compactness have not been performed.

In this study, optimal dimensions of T-rotor MR brake were determined using multi-objective genetic algorithm optimization method. Initially, L-27 orthogonal array was used to analyse the effects of different parameters of the brake on the magnetic flux density generated in the MR fluid gap which was computed using magnetostatic analyses using ANSYS workbench software. The on state torques were calculated analytically based on the magnetic flux density in different portions of the MR fluid. The optimum dimensions of the brake were determined based on maximization of torque ratio and minimization of mass of the brake. Further, magnetostatic analyses of the T-rotor brake with optimal dimensions were performed and results were compared with those obtained for brake with simple disk rotor of similar dimensions.

2. Torque analyses of MR Brake

The MR brake consists of T-shaped rotor, stationary casing, MR fluid between rotor and casing and electromagnetic coils enclosed in the casing. Upon application of current to the electromagnet coil, the generated magnetic field lines increase the viscosity and hence the yield stress. This causes generation of braking torque due to resistance to the rotation of rotor. The schematic diagram of a T-rotor MR brake is shown in figure 1.

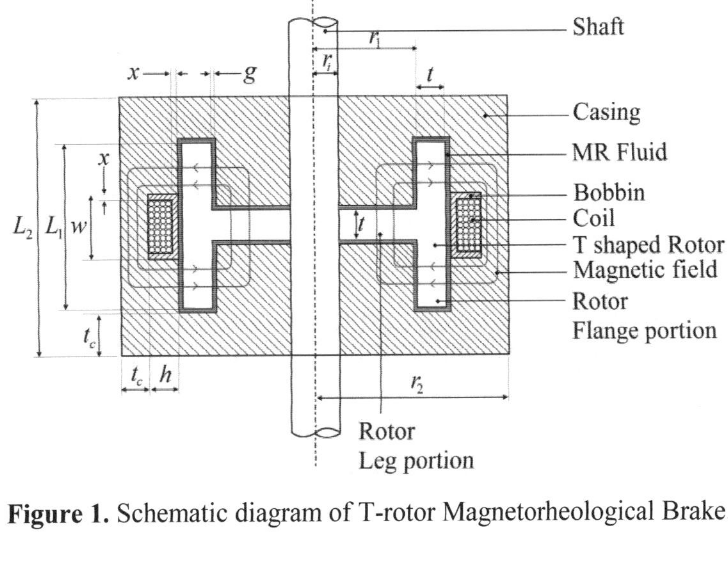

Figure 1. Schematic diagram of T-rotor Magnetorheological Brake.

The total torque generated in MR brake is the sum of on-state and off state torques. The braking torques were derived by modelling the rheological properties of the MR fluid using Herschel-Bulkley model. The off-state torque due to viscosity of MR fluid (T_η) is given by equation (1), [6]

$$T_\eta = \frac{4\pi\mu_{eq(r_1+t)}(r_1+t)^4}{(n_{r_1+t}+3)g}\left[1-\left(\frac{r_i}{r_1+t}\right)^{n_{r_1+t}+3}\right]\omega + \frac{4\pi\tau_{r_1+t}}{3}\left[(r_1+t)^3 - r_i^3\right] + 2\pi\tau_{r_1+t}\left[r_1^2 L_{if} + (r_1+t)^2 L_{of}\right]$$
$$+ 2\pi K_{r1+t}\left(\frac{\omega}{g}\right)^{n_{r1+t}}\left[r_1^{2+n_{r1+t}} L_{if} + (r_1+t)^{2+n_{r1+t}} L_{of}\right] \quad (1)$$

The on-state torque which is due to the application of magnetic field is the sum of the torque in the inner (T_{if}) and outer (T_{of}) portions of rotor flange and in the leg (T_1) of the rotor and is given by equations (2) to (6), [6]

$$T_h = 2T_1 + T_{if} + T_{of} \quad (2)$$

$$T_1 = \frac{2\pi\mu_{eql} r_1^4}{(n_1+3)g}\left[1-\left(\frac{r_i}{r_1}\right)^{n_1+3}\right]\omega + \frac{2\pi\tau_{yl}}{3}\left(r_1^3 - r_i^3\right) \quad (3)$$

$$T_{if} = 2\pi r_1^2 L_{if}\left[\tau_{yif} + K_{if}\left(\frac{\omega r_1}{g}\right)^{n_{if}}\right] \quad (4)$$

$$T_{of} = 2\pi(r_1+t)^2 L_{of}\left\{\tau_{yof} + K_{of}\left[\frac{\omega(r_1+t)}{g}\right]^{n_{of}}\right\} \quad (5)$$

$$\mu_{eql} = K_1\left(\frac{r_1\omega}{g}\right)^{n_1-1} \quad (6)$$

where, L_{of} is the effective length of outer portion of rotor flange, L_{if} is the effective length of the inner portion of the rotor flange, ω is the angular velocity and μ_{eq} is the equivalent viscosity at the required location of rotor. The rheological properties of Lord MRF 140CG fluid at different magnetic fields were calculated based on the equation (7), [6]

$$Y = Y_\infty + (Y_0 - Y_\infty)\left(2e^{-B\alpha_{sy}} - e^{-2B\alpha_{sy}}\right) \quad (7)$$

where, Y is the rheological parameter which could be yield stress (τ_y), consistency index (K) or flow behaviour index (n). Y ranges from value at zero magnetic field, Y_0 to that with magnetic field at saturation, Y_∞. α_{sy} is the saturation moment index and B is the magnetic flux density. The rheological parameters of MRF 140CG can be obtained by curve fitting the experimentally determined flow and viscosity curves of the fluid to Herschel-Bulkley model. The parameters obtained are $K_0 = 0.65\ \text{Pa}\cdot\text{s}^n$ $K_\infty = 5400\ \text{Pa}\cdot\text{s}^n$, $\alpha_{sk} = 5\ \text{T}^{-1}$, $\tau_{y0} = 25\ \text{Pa}$, $\tau_{y\infty} = 39000\ \text{Pa}$, $\alpha_{st_y} = 2\ \text{T}^{-1}$ $n_0 = 0.915$, $n_\infty = 0.24$, and $\alpha_{sn} = 35$ [6].

The torque ratio (K_T) is defined as ratio of on-state torque (T_h) to the off-state torque (T_η) and is given by equation (8),

$$K_T = \frac{T_h}{T_\eta} \quad (8)$$

IOP Conf. Series: Materials Science and Engineering **624** (2019) 012024 doi:10.1088/1757-899X/624/1/012024

3. Magnetostatic analyses of MR Brake

Magnetostatic analyses were carried out in ANSYS workbench to determine the magnetic flux density in the MR fluid region for different combination of dimensions of the brake whose values are listed in table 1.

Table 1. Design parameters of MR brake and their levels.

	Rotor radius (mm)	Rotor thickness (mm)	Flange length (mm)	Casing thickness (mm)	Coil height (mm)	Coil width (mm)
Lower level	30	5	50	10	10	10
Second level	50	10	75	15	15	15
Upper level	70	15	100	20	20	20

One fourth of MR brake was modelled and the material properties of their components were specified as input in the ANSYS software. Low carbon SAE1020 steel was chosen as material for rotor and casing and the B-H curve of this material was taken from ANSYS materials library. Bobbin is made of aluminium having a relative permeability of one. The B-H curve of Lord Corporation's MRF 140 CG is specified as per the material data sheet [8]. The analyses were performed at constant voltage of 12 V and current of 1 A. Angular velocity of 100 rad/s was considered for the analysis. The L27 orthogonal array was used for performing the magnetostatic analysis to examine the effects of variation in the dimensions of the brake on the magnetic flux density generated in the MR fluid gap. The response factors for the different combination of dimensions of the brake are shown in table 2. The values of magnetic flux densities in the MR fluid gap were substituted in equations (7) to calculate requisite properties of the fluid which are required to calculate on-state torque using equations (2) to (6).

Table 2. Response parameters for different combinations of MR Brake dimensions.

Sl. No.	Rotor Radius (mm)	Rotor Flange length (mm)	Rotor thickness (mm)	Casing thickness (mm)	Coil width (mm)	Coil height (mm)	On-State Torque (Nm)	Off-State Torque (Nm)	Torque ratio	Mass (Kg)
1	30	50	5	10	10	10	42.99	0.79	54.18	5.48
2	30	50	5	10	15	15	49.78	0.79	62.73	6.55
3	30	50	5	10	20	20	52.78	0.79	66.52	7.73
4	30	75	10	15	10	10	68.96	1.45	47.57	11.36
5	30	75	10	15	15	15	81.62	1.45	56.30	13.21
6	30	75	10	15	20	20	86.24	1.45	59.49	15.21
7	30	100	15	20	10	10	98.11	2.43	40.36	20.05
8	30	100	15	20	15	15	122.16	2.43	50.25	22.84
9	30	100	15	20	20	20	129.56	2.43	53.29	25.83
10	50	50	10	20	10	15	156.22	3.78	41.28	18.64
11	50	50	10	20	15	20	170.74	3.78	45.12	20.83
12	50	50	10	20	20	10	162.02	3.78	42.81	23.02
13	50	75	15	10	10	15	201.64	6.19	32.56	17.46
14	50	75	15	10	15	20	227.64	6.19	36.75	19.64
15	50	75	15	10	20	10	218.16	6.19	35.22	21.83
16	50	100	5	15	10	15	210.10	6.35	33.10	21.09
17	50	100	5	15	15	20	252.21	6.35	39.73	23.87
18	50	100	5	15	20	10	227.61	6.35	35.85	26.72
19	70	50	15	15	10	20	325.07	10.59	30.69	24.69
20	70	50	15	15	15	10	309.45	10.59	29.22	26.96
21	70	50	15	15	20	15	344.75	10.59	32.55	29.44
22	70	75	5	20	10	20	422.40	13.18	32.04	32.14
23	70	75	5	20	15	10	386.29	13.18	29.30	35.25
24	70	75	5	20	20	15	448.33	13.18	34.01	38.61
25	70	100	10	10	10	20	453.96	18.24	24.89	30.29
26	70	100	10	10	15	10	431.03	18.24	23.63	33.39
27	70	100	10	10	20	15	503.29	18.24	27.60	36.73

IOP Conf. Series: Materials Science and Engineering **624** (2019) 012024 doi:10.1088/1757-899X/624/1/012024

The mass of the brake for different configurations was calculated based on the product of volume and density of individual components. It can be observed that with increase in the dimensions of the brake there is significant increase in the on-state, off-state braking torques and mass though the torque ratio varies. The torque ratio is high for smaller dimensions of the brake which could be attributed to the smaller off-state torque.

4. Optimal Dimensions of MR brake

A magnetorheological brake should possess high torque ratio which means on-state torque should be high and off-state torque should be low. Also, mass of the brake is another important issue in the design of brake and should be as low as possible for given torque requirement. In this work, optimal dimensions were determined by maximization of torque ratio and minimization of mass using multi-objective Genetic algorithm optimization method using MATLAB and the Pareto front results is shown in figure 2. The selected optimized dimensions are mentioned in table 3.

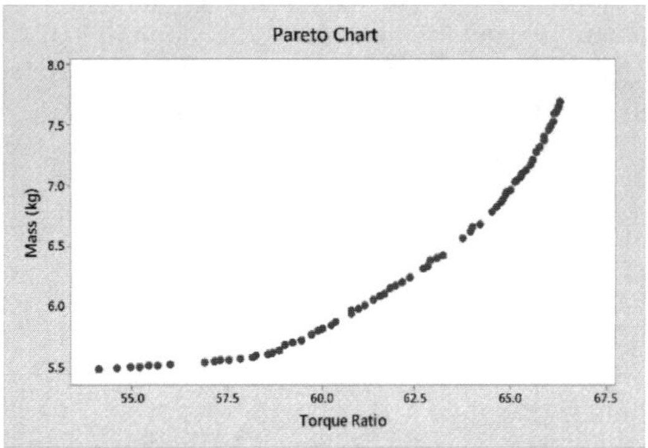

Figure 2. Pareto front of MOGA optimization.

Table 3. Optimized Parameters of MR brake.

Parameters	Optimized dimensions (mm)
Rotor radius r_1	30.01
Rotor thickness (t)	5.07
Flange length L_1	50.02
Casing thickness (t_c)	10.58
Coil height (h)	17.65
Coil width (w)	14.93

5. Comparative analyses of T-shaped and disk shaped rotor MR brakes

Magnetostatic analyses of T-rotor MR brake having optimized dimensions and simple disk shaped MR brake of same dimensions were performed in ANSYS workbench software. The magnetic flux density distribution in the T-shaped and simple disk MR brake are depicted in figures 3(a) and 3(b) respectively. The magnetic flux density is higher in case of T-rotor MR brake. At 1A current, the total braking torque of T-rotor and simple disk MR brakes are 44.1 Nm and 11.8 Nm respectively. Thus, the total braking torque of T-rotor brake is 3.74 times than that of simple disk brake. The mass of T-rotor and simple disk MR brakes are 6.78 kg and 6.39 kg respectively. The torque to mass ratio of T-rotor and simple disk MR brakes are 6.5 and 1.8 respectively.

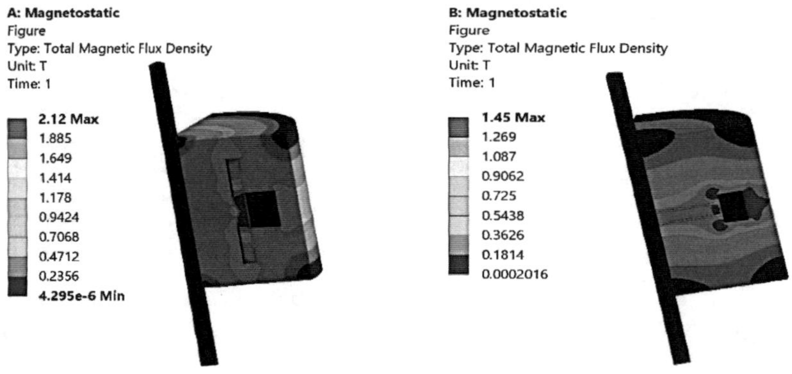

(a) (b)

Figure 3. Magnetic flux density distribution in the (a) T-rotor and (b) Simple Disk MR brakes.

Further, magnetic flux density and subsequently braking torques for both the MR brake configurations were determined at different current magnitudes of 0.1A, 0.2A, 0.5A, 1A and 1.5 A which are plotted as shown in figure 4. Beyond 1A current, there is not much increase in the magnitude of torque which is due to saturation of magnetic flux in the MR fluid region.

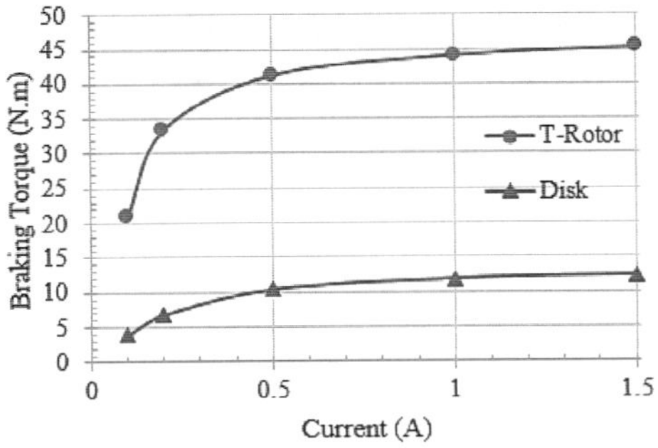

Figure 4. Comparison of torque of MR brakes at different applied currents.

6. Conclusions

In this study, magnetic flux generated in the MR fluid region was determined for different combinations of parameters of the brake dimensions to calculate on-state torque. The optimum dimensions of the brake were determined with the objective of maximization of torque ratio and minimization of mass of the brake using multi-objective Genetic algorithm optimization technique. Further, magnetostatic analyses of the T-rotor brake with optimal dimensions were performed and results were compared with those obtained for brake with simple disk rotor. It was observed that T-rotor MR brake produces significantly higher braking torque at all current magnitudes compared to simple disk rotor MR brake. Thus, T-rotor MR brakes are compact and have superior torque to mass characteristics for similar dimensions.

7. References

[1] Avraam M, Horodinca M, Romanescu I and Preumont A 2010 Computer controlled rotational MR-brake for wrist rehabilitation device *J. Intell. Mater. Syst. Struct.* **21** 1543 – 57

[2] Assadsangabi B, Daneshmand F, Vahdati N, Eghtesad M and Bazargan-Lari Y 2011 Optimization and design of disk-type MR brakes *Int. J. Automot. Technol.* **12** 921 – 32

[3] Nguyen Q H and Choi S B 2012 Optimal design of a novel hybrid MR brake for motorcycles considering axial and radial magnetic flux *Smart Mater. Struct.* **21** 055003

[4] Nguyen Q H and Choi S B 2010 Optimal design of an automotive magnetorheological brake considering geometric dimensions and zero-field friction heat *Smart Mater. Struct.* **19** 115024.

[5] Hung N Q and Bok C S 2012 Optimal design of a T-shaped drum-type brake for motorcycle utilizing magnetorheological fluid *Mech. Des. Struct. Mach.* **40** 153 – 62

[6] Nguyen Q H and Choi S B 2011 Selection of magnetorheological brake types via optimal design considering maximum torque and constrained volume *Smart Mater. Struct.* **21** 015012

[7] Nguyen Q H, Lang V T and Choi S B 2015 Optimal design and selection of magneto-rheological brake types based on braking torque and mass *Smart Mater. Struct.* **24** 067001

[8] Lord technical data, available at https://www.lordmrstore.com/ .../Data_Sheet_-_MRF140CG_Magneto -Rheological_Fluid. Accessed on 15th November 2018

Acknowledgements

The authors acknowledge IMPRINT Project No. IMPRINT/2016/7330 titled with ''Development of Cost Effective Magnetorheological (MR) Fluid Damper in Two wheelers and Four Wheelers Automobile to Improve Ride Comfort and Stability'' funded by Ministry of Human Resource Development and Ministry of Road Transfer and Highways, Govt. of India.

IOP Conf. Series: Materials Science and Engineering **624** (2019) 012025 doi:10.1088/1757-899X/624/1/012025

Multi-objective optimum design of an aero engine rotor system using hybrid genetic algorithm

Joseph Shibu K[1, 2], K Shankar[2], Ch. Kanna Babu[1] and Girish K Degaonkar[1]

[1]Aero Engine Research and Design Centre, HAL, Bengaluru, 560017
[2]Department of Mechanical Engineering, IITM, Chennai

Email ID: josephshk@gmail.com

Abstract. A working aero engine rotor system is subjected to multi-objective optimisation using Genetic Algorithm based optimisation. A Hybrid Genetic Algorithm (HGA) is introduced to reduce the weight and unbalance response of the rotor system with constrain on critical speed. The existing aero engine gear box casing vibration is found to be within the critical speed constraint and additional constraints are imposed to move the critical away from this zone. Bearing–pedestal model and Rayleigh damping model are used for accurate results. The optimisation resulted in Pareto optimal solutions and best solution selected using utopia point concept. The outcome of the paper is a comparative study which highlights the advantages of HGA over Controlled Elitist Genetic Algorithm (CEGA) and Goal Programming (GP).

1. Introduction

Aero engines are the backbone of an aircraft. Rotor dynamics analysis of the aero engine rotor system will ensure that unwanted vibrations are not hampering the satisfactory operation of the engine. Optimisation is introduced to rotor dynamics problems to improve the design of rotor systems. When more than one objective is to be addressed at a time, multi-objective optimisation is the preferred method. Initial studies on optimisation of the rotor system considered a single objective. Weight was optimized by programming the optimisation problem by converting it into a non-linear mathematical problem [1]. Further studies on weight optimisation introduced additional constraints and design variables [2]. Another weight optimisation problem discussed in detail the gyroscopic effects on optimisation [3]. All these studies employed weight as single objective and constraints are imposed for other potential design variables.

Simultaneously optimizing different design variables has its own advantage and researchers explored the same over time. The weighted function of vibration due to unbalance and stability limit is studied and support structure is modified to achieve positive results [4]. Pareto front is generated for the same problem to arrive at a set of alternate solutions for the designer [5]. Weight is combined with the force transmitted to bearing locations and multi-objective optimisation is carried out [6]. Immune Genetic Algorithm was applied to the same problem and resulted in better solutions [7]. This study explored the potential of the immune system and applied the same to the genetic algorithm. The initial convergence of the genetic algorithms is prevented and the search is extended to different optimal solutions. The emphasis of the algorithm was on the selection of the population. HGA used for the present problem concentrate on reducing the convergence time with improvements over the solutions obtained [8]. The novel elements of the present study are discussed here. An algorithm better equipped to reach optimum solutions in an efficient manner is introduced to multi-objective optimisation of an aero engine rotor system. The effectiveness of the algorithm is proved by comparing it with existing proven algorithms.

2. Optimisation Methods

Optimisation is the minimisation or maximisation of different functions by altering variables influencing the functions. Multi-objective optimisation results in Pareto optimal solutions which represent a trade-off between the objectives considered and are superior to all other possible solutions. Point close to the Utopia point, which represents minimum value of the objectives considering one at a time, is the optimum solution [9].

2.1. Controlled Elitist Genetic Algorithm(CEGA)

The theory about Genetic Algorithms is found in abundance in the literature and [10] is one of the best references among them. Non-dominated Genetic Algorithm (NSGA) is an algorithm to generate Pareto front for multi-objective optimisation and it works on the concept of non-domination of solutions [11]. The drawbacks such as difficulties associated with computation of solution, absence of elitism and the requirement of choosing an optimal parameter value for sharing parameter, led to the development of NSGA-II [12]. By incorporating control over the elite members, NSGA-II is modified and Controlled Elitist Genetic Algorithm is developed [8].

2.2. Goal Programming (GP)

The goal programming allows the user to set goals or targets for the objective functions and they are achieved satisfactorily [13]. Hence, this method is simpler compared to other traditional optimization algorithms, and results are attained quicker. Once the goals are set, the objective functions are transformed into constraints. The objective function is also restructured accordingly. Its restructuring methodology categorises the goal programming into scaled and unscaled versions. The present work uses the scaled version of the goal programming technique [14].

2.3. Hybrid Genetic Algorithm(HGA)

CEGA takes numerous function assessments to attain convergence although it reaches the area next to an optimal Pareto front comparatively fast. Hence CEGA is used for a diminutive number of generations to reach close to an optimum front and this solution is the initial point for goal programming solver that is quicker and proficient for a local search. The solutions obtained from HGA are combined with the existing population and a new Pareto front is obtained. Goals required for Goal Programming are provided by CEGA as the extreme points from the Pareto front established in previous run [8].

3. Rotor dynamics Model

The Rayleigh circular beam element is used for modeling the distributed shaft. Rigid disks representing the mass and inertia properties of compressor and turbines are attached to respective locations on shaft. Procedure for modeling is detailed in [15]. The single spool rotor system is made of stepped shaft carrying three disks and supported on two bearings. A segment of the turbines are selected as rotor and steel alloy hollow cylinder is used for connecting the two turbines. The schematic of the rotor system representing with the disks, bearings and their properties are shown in Figure-1. Initial configuration data and the material characteristics of the rotor are shown in Table-1.

Flexible pedestal of mass M_1 and stiffness of K_1 is connected to bearing of stiffness K_2.Stiffness of the bearing is in series with stiffness of pedestal. This forms four degrees of freedom system. This system is converted to two degrees of freedom system with stiffness K_{eq} as shown below:

$$K_{eq} = \begin{bmatrix} \dfrac{K_2(K_1 - \omega^2 M_1)}{K_2 + K_1 - \omega^2 M_1} & 0 \\ 0 & \dfrac{K_2(K_1 - \omega^2 M_1)}{K_2 + K_1 - \omega^2 M_1} \end{bmatrix} \quad (1)$$

ω is the angular velocity of the rotor system.

The proportional damping model known as Rayleigh damping [16] shown below is used.

$$[C] = \alpha[M] + \beta[K_{shaft} + K_{eq}] \quad (2)$$

[C] is damping matrix, α and β are Rayleigh damping coefficients,*[K_{shaft}]* is the stiffness due to shaft.

Table 1. Initial design for distributed rotor

Element	inside diameter (m)	outside diameter (m)	length (m)	Material	Density (kg/m^3)	Elastic modulus (N/m^2)
1	0	0.020	0.02	Steel alloy	7760	$2.1x10^{11}$
2	0	0.020	0.025	Steel alloy	7760	$2.1x10^{11}$
3	0	0.014	0.028	Steel alloy	7760	$2.1x10^{11}$
4	0	0.020	0.022	Steel alloy	7760	$2.1x10^{11}$
5	0	0.015	0.0055	Steel alloy	7760	$2.1x10^{11}$
6	0	0.015	0.0055	Steel alloy	7760	$2.1x10^{11}$
7	0	0.020	0.07	Steel alloy	7760	$2.1x10^{11}$
8	0	0.015	0.0055	Steel alloy	7760	$2.1x10^{11}$
9	0	0.015	0.0055	Steel alloy	7760	$2.1x10^{11}$
10	0	0.020	0.01	Steel alloy	7760	$2.1x10^{11}$
11	0	0.040	0.015	Nimonic alloy	8180	$2.05x10^{11}$
12	0	0.040	0.01	Nimonic alloy	8180	$2.05x10^{11}$
13	0.020	0.050	0.02	Steel alloy	7860	$2.0875x10^{11}$
14	0	0.040	0.01	Nimonic alloy	8180	$2.05x10^{11}$
15	0	0.040	0.01	Nimonic alloy	8180	$2.05x10^{11}$

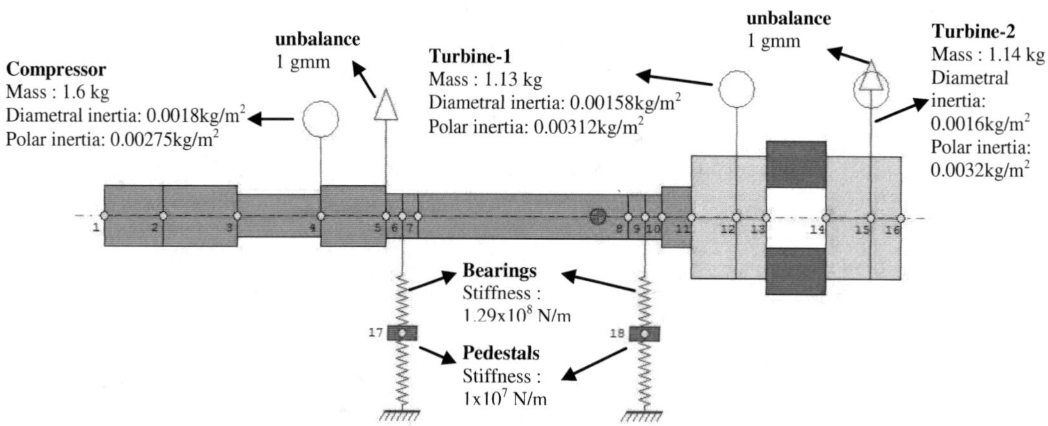

Figure 1. Rotor system schematic

Formulation of the multi-objective optimisation problem

3.1. Step-1: Selection of Pedestal stiffness of both bearing locations and diameters of the shaft as design variables

Design variables are selected based on numerical experiment. Diameters of the shaft at the bearing and disk locations are not considered as design variables to simplify the design modifications.

3.2. Step2: Selection of the second critical speed as constraint

$$\Omega_{low} \le \omega_2^c \le \Omega_{high} \tag{3}$$

ω_2^c is the second critical speed and Ω_{low}=1800 rad/s and Ω_{high}=3000 rad/s are the limiting frequencies due to self sustaining speed and idle speed. In this particular case the existing engine gear box casing vibration is found to be in at 420 Hz and it is decided to shift the critical away from this zone by keeping Ω_{high} at 2518 rad/s.

The equation of the rotor system in standard matrix form considering equations (1) and (2):

$$[M]\{\ddot{p}\}+(\alpha[M]+\beta[K]+\omega[G])\{\dot{p}\}+\lfloor K_{shaft}+K_{eq}\rfloor\{p\}=\{F\} \tag{4}$$

[M} is the mass matrix; *{p}* and *{F}* are the response and force vectors respectively.

Converting equation (4) into state space form,

$$\begin{bmatrix}(\alpha[M]+\beta\lfloor K_{shaft}+K_{eq}\rfloor+\omega[G]+\omega[M]) & [M] \\ [M] & 0\end{bmatrix}\frac{d}{dt}\begin{Bmatrix}p\\ \dot{p}\end{Bmatrix}+\begin{bmatrix}\lfloor K_{shaft}+K_{eq}\rfloor & 0 \\ 0 & -[M]\end{bmatrix}\begin{Bmatrix}p\\ \dot{p}\end{Bmatrix}=\begin{Bmatrix}0\\0\end{Bmatrix} \tag{5}$$

Solving equation (5) damped natural frequencies are obtained.

3.3. Step3:Selection of maximum amplitude of vibration due to unbalance and weight of shaft as the objective functions

Consider a solution of the form given below

$$\{p\}=\{p_s\}\cos\omega t+\{p_c\}\sin\omega t \tag{6}$$

Substituting in equation (6) in equation (4) yields

$$\begin{Bmatrix}p_c\\ p_s\end{Bmatrix}=\begin{bmatrix}(\lfloor K_{shaft}+K_{eq}\rfloor-\omega^2[M]) & \omega^2[G]-\omega\lfloor\alpha[M]+\beta\lfloor K_{shaft}+K_{eq}\rfloor\rfloor \\ -\omega^2[G]+\omega\lfloor\alpha[M]+\beta\lfloor K_{shaft}+K_{eq}\rfloor\rfloor & (\lfloor K_{shaft}+K_{eq}\rfloor-\omega^2[M])\end{bmatrix}\begin{Bmatrix}F_C\\ F_s\end{Bmatrix} \tag{7}$$

Using the above solution, vibration amplitude due to unbalance is written in the form below,

$$p=\sqrt{(p_c^2+p_s^2)} \tag{8}$$

Weight of the rotor system is calculated using equation (9)

$$W(x)=\sum_i^n \frac{\rho_i * \pi * (do_i^2 - di_i^2) * h_i}{4} \tag{9}$$

ρ is the density of shaft , *di* is the inner diameter, *do* the outer diameter and *h* the length of each element. *n* is the number of elements.

3.4. Step4: Deciding the bounding limits of design variables

$$\begin{aligned} &0.01 \le do_i \le 0.03 (i=1,2,4,7,10,13) \\ &0.01 \le di_i \le 0.03 (i=13) \\ &1\text{x}10^6 \le K_{eq} \le 1\text{x}10^9 \end{aligned} \tag{10}$$

4. Result and Discussions

The Pareto fronts generated by CEGA alone and by means of HGA are shown in Figure-2 and they are compared against each other using the spread and the average distance of the solutions. The average distance of the solutions on the Pareto front are enhanced by the use of hybrid function. The spread determines change in two fronts and that is superior for Pareto front generated by HGA. This indicates that the front has changed considerably from that obtained by CEGA with no hybrid function. It is clear that by using the HGA an optimal Pareto front is obtained but the diversity of the solution is reduced. This is pointed out by a higher value of the average distance measure and the spread of the front.

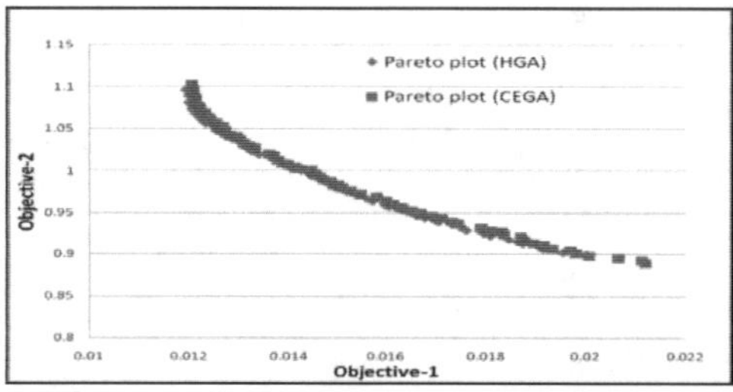

Figure 2. Pareto front for CEGA and HGA

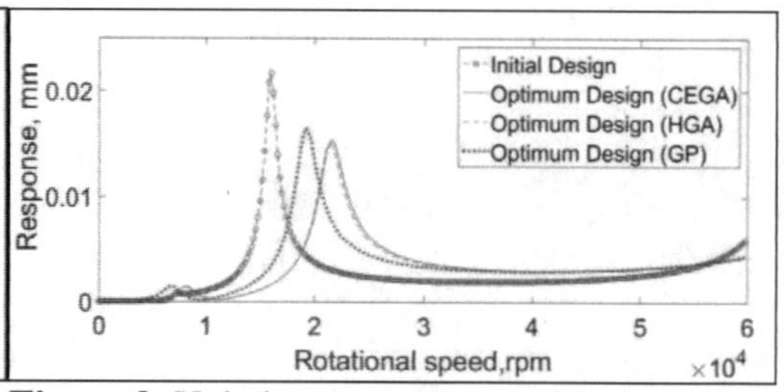

Figure 3. Unbalance responses of designs

The post optimal analysis is carried out to choose an optimal solution for implementing the results of optimisation. Utopia point method is used for the selection of a final solution. Amplitude of vibration for of initial design and optimum designs corresponding to HGA, CEGA and GP are shown in Figure-3. Table-2 and 3 represent the initial design and optimum designs corresponding to HGA, CEGA and GP objective function and constrain values.

Table 2. Different designs- Diameter of shaft element and bearing stiffness

Design variable	Element	Initial design	HGA	CEGA	GP
	1	0.020	0.0293	0.0280	0.0287
Outside diameter, m	2	0.020	0.0239	0.0252	0.0231
	4	0.020	0.0187	0.0179	0.0186
	7	0.020	0.0115	0.0115	0.0103
	10	0.020	0.0173	0.0169	0.0143
	13	0.050	0.0401	0.0401	0.0400
Inside diameter, m	13	0.020	0.0300	0.0299	0.0300
Pedestal Stiffness, N/m	Station 6	$1x10^7$	$4.73x10^8$	$4.73x10^8$	$4.0x10^8$
	Station 9	$1x10^7$	$3.37x10^8$	$2.57x10^8$	$2.5x10^8$

Table 3. Different designs- critical speed, shaft weight and amplitude of vibration

Parameter	Initial design	HGA	CEGA	GP
Critical speed, rpm	16500	19200	21600	21400
Amplitude of vibration, mm	$2.17x10^{-2}$	$1.50x10^{-2}$	$1.53x10^{-2}$	$1.64x10^{-2}$
Weight, kgf	1.120	0.978	0.975	0.950

Comparing with initial design,

- HGA optimum design: Amplitude reduced by 30.8% and the weight decreased by 12.6%.
- CEGA optimum design: Amplitude reduced by 29.4% and the weight decreased by 12.9%.
- GP optimum design: Amplitude reduced by 24.4% and the weight decreased by 15.6%.

HGA solution is found to be superior to CEGA as it showed better reduction in amplitude of vibration and comparable improvement in weight. Though GP optimum solution showed better improvement in weight, the corresponding reduction is amplitude was less. A better way to compare the results of three optimization methodologies is to identify the improvements of one objective for constant value of other objective. Table-4 shows such a comparison. For the same weight HGA has shown better reduction in amplitude of vibration compared to CEGA and GP.

Table 4. Different designs- shaft weight and amplitude of vibration

Parameter	HGA	CEGA	GP
Weight, kgf	0.95	0.95	0.95
Amplitude of vibration, mm	$1.62x10^{-2}$	$1.63x10^{-2}$	$1.64x10^{-2}$

5. Conclusions

Multi-objective optimisation of a rotor system for aero engine application is carried out through in-house MATLAB coding with additional constrain imposed by gear box casing vibration. The Pareto Front generated by HGA is found to be superior in terms of average distance measure and spread of solutions compared to CEGA. Optimum solutions are found out from HGA and CEGA Pareto Fronts separately using Utopia point method. HGA optimum solution resulted in better optimisation compared to initial design with amplitude of vibration reduced by 30.8% and the weight decreased by 12.6%. CEGA optimum design resulted in the amplitude of vibration reduction of 29.4% and the weight reduction of 12.9%. GP optimum design resulted in the amplitude of vibration reduction of 24.4% and the weight reduction of 15.6%. The optimal solution using HGA showed better improvement in amplitude of vibration compared to CEGA and GP optimisation optimum solutions. Weight reduction was more in GP solution but with less optimal reduction in amplitude of vibration. HGA and CEGA optimal solutions had similar weight reduction. Comparing the reduction in amplitude for same weight, it is confirmed that HGA solution is superior to CEGA and GP solutions.

6. Acknowledgments

The authors thank Chief Designer AERDC, HAL for his support to carry out the work and for his kind permission to publish this paper.

7. References

[1] T N Shiau and J L Hwang Minimum weight design of a rotor bearing system with multiple frequency constraints 1988 *Trans. ASME J. Eng. Gas Turb. Power* **110 (4)** 592-599
[2] T N Shiau and J L Hwang Optimum weight design of a rotor bearing system with dynamic behavior constraints 1990 *Trans. ASME J. Eng. Gas Turb. Power* **112 (4)** 454-462
[3] T Y Chen and B P Wang Optimum design of rotor-bearing system with eigenvalue constrains 1993 *Trans. ASME* **vol. 115 256**
[4] Panda K C and J K Dutt Design of optimum support parameters for minimum rotor response and maximum stability limit 1999 *Journal of Sound and Vibration* **223.1** 1- 21
[5] Babu, M Janardhana, A S Sekhar and K. Shankar Multi-objective optimisation of support characteristics of rotor bearing systems 2013 *International Journal of Structural Engineering* **4.4** 361-386
[6] T N Shiau and J.R. Chang Multiple-objective optimization of a rotor-bearing system with critical speed constraints 1993 *Trans. ASME J. Eng. Gas Turb. Power* **115** 246-255
[7] B K Choi and B S. Yang Multiobjective Optimum Design of Rotor-Bearing Systems with Dynamic Constraints Using Immune-Genetic Algorithm 2001 *Transactions of the ASME* 78 **/Vol. 123**
[8] Kalyanmoy Deb *Multi-Objective Optimization using Evolutionary Algorithms* John Wiley & Sons ISBN 047187339
[9] Arora J 2004 *Introduction to optimum design* Elsevier
[10] Goldberg D E 2006 *Genetic algorithms* Pearson Education India
[11] N Srinivas and Kalyanmoy Deb Multiobjective Optimization Using Nondominated Sorting in Genetic Algorithms 1994 *Evolutionary Computation 2* **no. 3** 221 – 248
[12] Kalyanmoy Deb, Amrit Pratap, Sameer Agarwal and T Meyarivan A Fast Elitist Multiobjective Genetic Algorithm: NSGA-II 2002 *IEEE Transactions on Evolutionary Computation 6* no. 2 182 – 197
[13] Charnes A and Cooper WW Programming with linear fractional functionals 1962 *Naval research logistics (NRL)* **9(314)** 181-6
[14] Gembicki F and Haimes Y Approach to performance and sensitivity multiobjective optimization: The goal attainment method 1975 *IEEE Transactions on Automatic control* **20(6)** 769-71
[15] Nelson H D and McVaugh J M The Dynamics of Rotor-Bearing Systems Using Finite Elements 1976 *Journal of Engineering for Industry* **98** 593-600
[16] Strutt, John William and Baron Rayleigh *The theory of sound* 1877 **vol. 1**

IOP Conf. Series: Materials Science and Engineering **624** (2019) 012026 doi:10.1088/1757-899X/624/1/012026

INFLUENCE OF S_0 TYPE OF ADDENDUM MODIFICATION ON SLIDING PERFORMANCE OF SPUR GEARS

Niranjan Y C[1], Prithvi C[2] and B K Sridhara[3]

[1]The National Institute of Engineering, Mysuru, India
[2, 3]Depart of Mechanical Engineering, The National Institute of Engineering, Mysuru, India

Email: cprith@gmail.com

Abstract. The effect of S_0 type of addendum modification on sliding velocity of the involute spur gear has been studied. For the analysis four critical points such as Beginning of Engagement, Low Point of Single Tooth Contact, High Point of Single Tooth Contact and Disengagement point corresponding to pinion are considered. Length of approach and length of recess for modified gears are evaluated. Effects of addendum modification on contact fatigue life and wear prediction are discussed. Product of contact stress and sliding velocity $(\sigma_h * V_s)$ at critical points for varied addendum modification factor X are evaluated, and its effect on sliding velocity has been analysed. Effect of change in value $(\sigma_h * V_s)$ at critical points of contact for modified gear pair on Hertzian fatigue is discussed.

Nomenclature

m	Module
$\emptyset$	Pressure angle
r_{bp}, r_{bg}	Base circle radius of pinion and gear respectively
r_p, r_g	Pitch circle radius of pinion and gear respectively
r_{ap}, r_{ag}	Addendum circle radius of pinion and gear respectively
r	Radius at any arbitrary point
C	Centre distance between gear pair
b	Face width
ρ_c	Radius of curvature at arbitrary point r
F	Tangential load
J	Number of teeth pair in contact
E_1, E_2	Elastic modulus of pinion and gear respectively

ϑ	Poisson's ratio
N_p, N_g	RPM of pinion and gear respectively
V_s	Sliding velocity
V_{rp}, V_{rg}	Rolling velocity of pinion and gear respectively
LPSTC	Low Point of Single Tooth Contact
HPSTC	High Point of Single Tooth Contact
BE	Beginning of Engagement
PP	Pitch Point
DP	Disengagement Point

1. Introduction

In mechanical systems transfer of power from source to application is very much necessary. Considering all other transmission systems, gears are having their own advantage of constant velocity ratio of transmission [1]. In a situation where pinion has to be designed with the number of teeth lesser than the minimum teeth required to avoid under cutting and interference, addendum modification universally adopted concept and amount of addendum modification is quantified using modification factor X [2]. Positively modified gear teeth will have increased bending strength and better contact and sliding relations [2]. There are two type of addendum modification, S type in which there is a scope for altering centre distance between the operating gear pair and other type is S_0, where the sum of modification provide on pinion (X_1) and gear (X_2) will be equal to zero [2]. One of the most significant research orientations in gear pair design is the selection of addendum modification factor (X) [3]. Transmitted load, contact stresses and sliding velocity determines specific wear rate on working surface of the gear pair [4]. Theoretically, the maximum value of contact stress will be on pitch point. Contact stress developed influences the wear rate of mating teeth surface. Product of contact stress and sliding velocity $(\sigma_h * V_s)$ is very much critical at beginning and end of path of contact on the line of action in wear prediction of mating teeth surface. Portion of gear teeth will have sliding velocity opposite to the rolling velocity. The crucial value $(\sigma_h * V_s)$ causes Hertzian fatigue [5]. Hassan [6] has carried out contact stress analysis at different points using Hertzian contact stress equation. H. K. Sachidananda [7] has carried sliding velocity analysis for profile modified gears with different amount of addendum modification factor and also by varying number of teeth of pinion and gear. Seok-Chul Hwang [8] carried out contact stress analysis of standard helical and spur gear in the region between LPSTC to HPSTC. From the literature it is evident that majority of the research considers the study of sliding velocity, contact stress for standard and addendum modified spur and helical gears. Also the influence of crucial value $(\sigma_h * V_s)$ on wear prediction for addendum modified spur gear. Whereas, research literature relating to influence of S_0 type of addendum modification on contact stress, sliding velocity and wear behaviour of involute spur gears is scare. In this work, an attempt has been made to study the sliding velocity, contact stress and influence of S_0 type of addendum modification on sliding velocity.

2. S_0 Type of Addendum Modification

S_0 type of addendum modification is a type of profile modification in which addendum and dedendum of the pinion and gear are disposition with respect to base circle. In some of the gear pair design procedure, the centre distance between the rotating gear pair is fixed by some other design features. In such cases, improved performance through modification cannot be obtained by varying the centre distance. In S_0 type of addendum modification even though the addendum and dedendum of the gear and pinion are disposition, pitch and base circle are not affected. The amount of profile modification is decided by the modification factor "X". In most general case pinion is provided with positive modification and gear with negative modification. Provided positive modification for pinion and negative modification for gear will increase the bending strength of the pinion and protects pinion teeth from bending fatigue [2]. Weight of pinion gets increased with the increase in modification factor X and weight of gear decreases with the increase in modification factor X. Therefor overall weight of meshing gear pair in contact decreases with the increase in modification factor X.

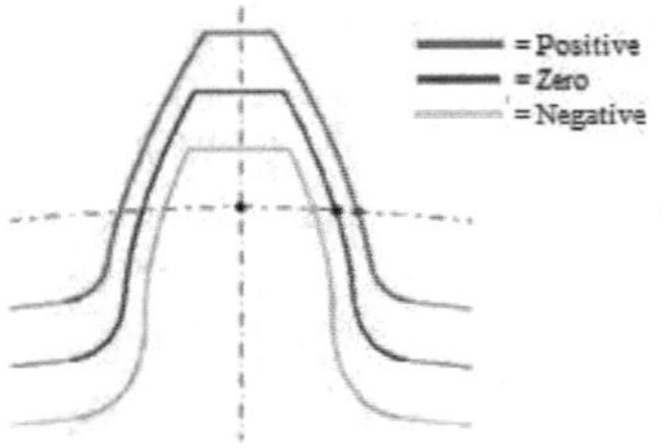

Figure 1. Schematic representation of S_0 type of addendum modification.

3. Location of BE, LPSTC, HPSTC and DP

Beginning of Engagement and Disengagement Point are crucial in lubrication and wear point of view. For a gear pair contact ratio 1.55 indicates, about 55% of the revolution two Pair of teeth will be in contact and for remaining 45% of revolution one pair of teeth will be in contact. When contact ratio of gear pair lies in between 1 and 2, LPSTC and HPSTC are the lowest point and highest point on the tooth surface where there will be only one pair of teeth in contact. And one pair of teeth in contact carries entire load at any point in between LPSTC to HPSTC on the gear teeth surface. As load carried by single teeth in contact, wear rate will be maximum in the region between LPSTC and HPSTC compared to any other region on the gear surface.

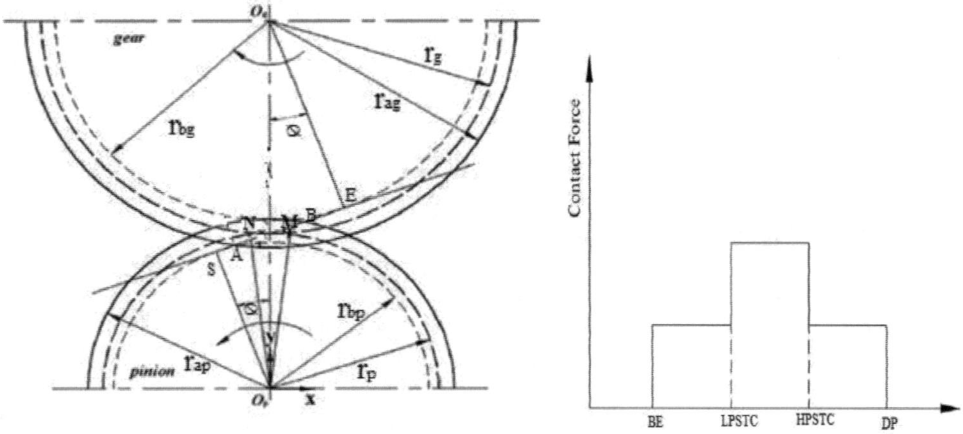

Figure 2. Representation of BE, LPSTC, HPSTC and DP and contact force on line of action.

In the fig shown above,
SE is the line of action.
AB is the length of contact.
B and A are the BE and DP of the pinion on line of action respectively.
M and N are the LPSTC and HPSTC of the pinion on line of action respectively.
At any other region other below LPSTC and above HPSTC, the applied load will be shared between two pair of teeth in contact. Therefore LPSTC and HPSTC are considered as the critical points considered for analysis [6] [8].

Using mathematical equations presented in [6], radius at critical points and radius of curvature of both gear and pinion can be calculated.

Theoretically, the maximum contact stress is on pitch point. For a pair of standard gear pair, pitch points lies between LPSTC and HPSTC. For the case of S_0 type of addendum modified gear pair, the location of pitch point depends on modification factor "X". The contact fatigue life of the gear pair can be increased considerably by locating the pitch point in the double tooth contact region [7]. In modified gear pair as the modification factor varies the position of LPSTC and HPSTC gets altered not the pitch point. In case of S_0 gearing, as modification factor increases, pitch point turns towards double tooth contact region. When modification factor reaches X = 0.4 to 0.5 the pitch point lies in double tooth contact region. With this modification the contact fatigue life has been increased.

Table 1. Data of mating spur gear considered for analysis.

Parameters	Pinion	Gear
Number of teeth	18	72
Module (mm)	5	5
Pressure angle (^{0})	20	20
Pitch diameter (mm)	90	360
Base diameter (mm)	84.5723	338.289
Number of rotation (RPM)	1440	360
Tangential load (N)	1473.8393	
Modulus of elasticity (Mpa)	$2*10^5$	
Poissons ratio	0.29	

4. Length of Approach (LA) and Length of Recess (LR)

Length of approach and length of recess for a pair of modified gear pair differs with modification factor compared to standard gear pair in contact. Length of approach and length of recess can be tabulated using the formulas 1 and 2.

$$\text{Length of Approach (LA)} = \sqrt{{r_{ag}}^2 - {r_{bg}}^2} - r_g \sin \emptyset \quad (1)$$

$$\text{Length of Recess (LR)} = \sqrt{{r_{ap}}^2 - {r_{bp}}^2} - r_p \sin \emptyset \quad (2)$$

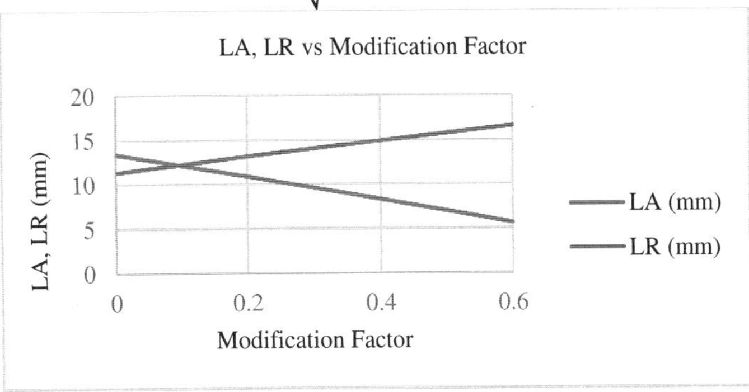

Figure 3. Variation of LA and LR with modification factor.

In wear prediction of gear pair LA and LR plays a major role. LA reduces and LR increases with the increase in modification factor X. The approach action of gear teeth is like pushing a piece of chalk across a blackboard, the chalk screeches. Recess action is like pulling the piece of chalk across the blackboard, it glides smoothly. Thus, recess action is always preferable because of lower frictional forces and smoothness of operation [10]. In recess action the abrasive wear of material (scoring) takes place as one tooth slides over the other. The scoring limits the value of LR in turn modification factor.

IOP Conf. Series: Materials Science and Engineering **624** (2019) 012026 doi:10.1088/1757-899X/624/1/012026

5. Contact Pressure * Sliding Velocity ($\sigma_h * V_s$)

Involute gear pair in contact, the meshing action is a combination of both sliding and rolling. This is because of difference in radius of curvature of mating pairs. At any given point the difference in rolling velocity of driving and driven gear defines the sliding velocity at that point. Gear tooth surface wear rate depends on the contact stresses, sliding velocity and applied load. So the product of contact stresses and sliding velocity ($\sigma_h * V_s$) at any point of contact on the gear teeth mating surface is one of the most crucial value which defines the wear behaviour of the gear pair teeth in contact.
Sliding velocity = Rolling velocity of pinion - Rolling velocity of gear.
Sliding velocity at any point can be written as [1]

$$V_s = V_{rp} - V_{rg} \tag{3}$$

$$V_{rp} = \frac{2*\pi*N_p*\rho_p}{60} \tag{4}$$

$$V_{rg} = \frac{2*\pi*N_g*\rho_g}{60} \tag{5}$$

In some points on the gear teeth sliding velocity is negative. The negative sliding velocity indicates direction of sliding velocity is opposite to the rolling velocity of gear. The frictional forces due to sliding are high in the negative velocity region.
Gear pair mating surface is assumed to be of Hertzian contact. This means two mating surfaces of gear are assumed as two cylinders in contact. For a gear pair in contact, contact stresses developed due to Hertzian contact depends on the radius of curvature of mating surfaces in contact point.
As the pinion is provided with positive addendum modification factor, radius of curvature (ρ_c) of pinion gets increased. The increased radius of curvature reduces the contact stresses of the mating gear pairs. The reduced contact stresses retard the pit initiation.
Equation for Hertzian contact stresses will be,

$$\sigma_h = \sqrt{\frac{F[(1/r_1)+(1/r_2)]}{jb\pi[(1-\vartheta_1{}^2)/E_1+(1-\vartheta_2{}^2)/E_2]}} \tag{6}$$

The value ($\sigma_h * V_s$) will be crucial at beginning and end of engagement of the gear tooth in contact. The negative value of ($\sigma_h * V_s$) will allow the lubricating oil to enter into surface cracks of mating teeth and by the hydraulic pressure propagation mechanism the surface crack growth gets accelerated which will cause Hertzian fatigue [7].
Increases radius of curvature due to addendum modification will increase the value($\sigma_h * V_s$) in recess region of gear pair. The increased value ($\sigma_h * V_s$)can result in squeezing out of lubricating oil from the tooth surface, which will cause metal-to-metal contact of gear teeth. The increased value ($\sigma_h * V_s$)will cause scoring (abrasive wear) in the gear teeth. Squeezing of lubricant and scoring in recess of gear teeth will limit the amount of addendum modification [7].

Table 2. Value of ($\sigma_h * V_s$) at crucial points for different modification factor.

Contact Region / ($\sigma_h * V_s$)	Addendum modification factor				
	0	0.2	0.4	0.5	0.6
BE	-949673.01	-524723.50	-325125.52	-253386.94	-191851.33
LPSTC	-217911.29	-97287.87	6281.11	53617.99	98111.34
PP	0.00	0.00	0.00	0.00	0.00
HPSTC	76631.90	204106.15	326036.66	385851.77	445370.03
DP	269404.69	308529.44	345758.58	363855.88	381659.29

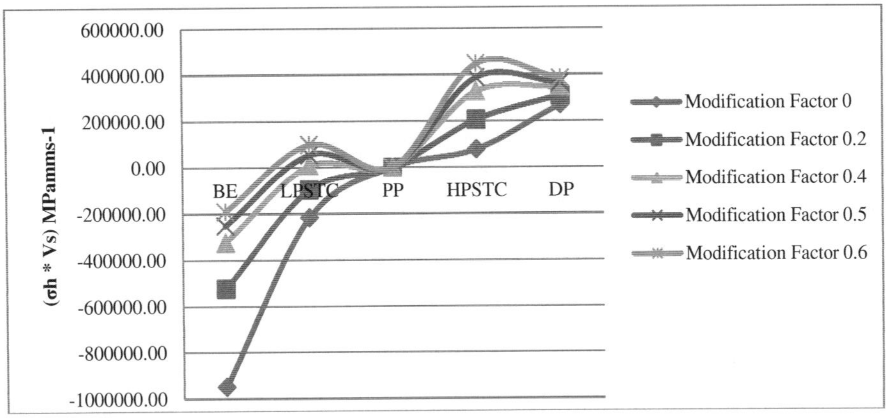

Figure 4. Graph of $(\sigma_h * V_s)$ at different points of pinion.

6. Conclusion

Theoretically the contact stress will be maximum at pitch point other than any other point on the teeth surface. By S_0 type of addendum modification, the position of LPSTC and HPSTC gets altered corresponding to modification factor X. Thus pitch points lies in the region of double tooth contact when modification reaches 0.4 and above for considered gear pair. Thus modification will increase the contact fatigue life of the gear pair. Modification also increases the length of recess by reducing length of approach. Gears are smoother in operation in the recess region than approach region. The critical value $(\sigma_h * V_s)$ is negative in BE and LPSTC. Through modification, the value $(\sigma_h * V_s)$ comes towards the positive region in BE and LPSTC. This increases Hertzian fatigue life. Meanwhile the positive value $(\sigma_h * V_s)$ in HPSTC and DP (recess region) gets increased with increasing modification factor. The increased positive value $(\sigma_h * V_s)$ leads to scoring. Thus scoring limits the magnitude of modification factor. For the case of S_0 type of addendum modification considering above mentioned parameters the optimal value of modification factor can be considered in between 0.4 to 0.5.

References

[1] Hsiang His Lin and Chuen-Huei Liou 1998 A Parametric Study of Spur Gear Dynamics, *NASA / CR-1998-206598, U.S. ARMY,* **ARL-CR-419**.

[2] Gitin M Mithra 2001 *Handbook of Gear Design, Tata McGraw-Hill Publication Company Limited – Second Edition.*

[3] Singh D, Suhaib M 2014 Kinematic considerations in gear drives — a review *International Journal of Innovative Research in Science, Engineering and Technology* **3(1)**:8204–8214.

[4] Shivamurthy B, Bhat K U and Anandhan S 2013 Mechanical and sliding wear properties of multi layered laminates from glass fabric/graphite/epoxy composites *Materials & Design* **44**:136–143.

[5] He S, Gunda R and Singh R 2007 Effect of sliding friction on the dynamics of spur gear pair with realistic time-varying stiffness *Journal of Sound and Vibration Elsevier* **301**:927–949.

[6] Ali Raad Hassan 2009 Contact Stress Analysis of Spur Gear Teeth Pair, *International Journal of Mechanical and Mechatronics Engineering* **3(10)**:1279-1284

[7] H. K. Sachidananda, K. Raghunandana and Joseph Gonsalvis 2017 Sliding velocity in profile corrected gears, *Lubrication Science* **29**:43–58

[8] Seok-Chul Hwang, Jin-Hwan Lee, Dong-Hyung Lee, Seung-Ho Han and Kwon-Hee Lee 2013 Contact stress analysis for a pair of mating gears *Mathematical and Computer Modelling* **57**: 40-49.

[9] Stephen P. Radzevich 2012 *Dudley's hand book of Practical Gear Design and Manufacture, 2nd Edition, CRC press*, New York.

[10] John J. Uicker,Jr, Gordan R. Pennock and Joseph E. Shigley 2014 *Theory of Machines and Mechanisms, 4th Edition, OXFORD PRESS.*

IOP Conf. Series: Materials Science and Engineering **624** (2019) 012027 doi:10.1088/1757-899X/624/1/012027

Vibration based real time brake health monitoring system – A machine learning approach

Alamelu Manghai T M and Jegadeeshwaran R

[1]School of Mechanical and Building Sciences, VIT University, Chennai, Tamil Nadu, India

Email: alamelumangai.m@vit.ac.in

Abstract. In an automobile, the brake system is the most important control component which ensures the safety of both passengers and vehicles. The continuous application of the brake causes the system gets faulty due to reasons like wear, mechanical fade, oil leak, etc., These faults or failures need to be monitored using proper monitoring techniques in order to avoid the incidents that may lead to accidents. Thus, the continuous monitoring of the brake system is very much essential for the safety of the vehicle. In this study, an experimental investigation was carried out for monitoring the brake system using vibration signals. An experimental setup which resembles the brake system was fabricated. The vibration signals were acquired under various brake condition such as good and faulty. From the acquired vibration signals, the features were extracted using statistical feature extraction techniques and feature selection was carried out. The selected features were then classifieds using a set of tree family classifiers such as random forest, random tree, LMT and decision tree. The classification accuracy of all the algorithms was compared and discussed.

1. Introduction

In an automobile, the brake system is one of the vital control element which is responsible for reducing the vehicle speed and also to stopping distance. Any failure in the brake system makes an adverse effect on vehicle stability and the traveler's safety. The National Motor Vehicle Causation Survey (NMVCCS) indicated 22 percentages of crashes were occurred because of brake related failure [1]. In 2007, Bayerische Motoren Werke (BMW), recalled over 1, 60,000 SUVs, due to the brake fluid failures. Chrysler took back 60,000 vehicles which were sold out in 2007 because of brake failure [2]. Monitoring of such system for faults will reduce the number of accidents and increase the degree of safety. There is no specific approach for monitoring the failure of the brake system. Hence it is necessary to develop an approach to monitor the brake related faults. The hydraulic brake system may get faulty due to reasons like mechanical fade, pad wear, the insufficient fluid pressure in the master cylinder, oil leak, etc. If these faults were gone unnoticed, the accidents will occur. Hence, the preventive maintenance of the brake system is suggested in order to avoid damage. This is achieved through the continuous Condition Monitoring (CM) process. CM emphases on numerous parameters in which the substantive changes indicating the development of a fault [3,4]. Fault detection is a subfield of condition monitoring which anxieties itself with observing a framework, identifying when a fault has emerged and seen the sort of fault. Fault detection techniques are mainly classified into two types namely signal processing based fault detection (SPBFD) and model-based approach (MBA). In this study SPBFD approach has been focused on the fault diagnosis of hydraulic brake study. In recent years, the vibration signal has been focused on many fault diagnostic study. The hydraulic brake system is producing the vibration under different operating conditions. The produced vibration signatures are directly related with the fault conditions. Hence, the vibration signal under various fault conditions has been considered for the learning process in this study.

The statistical feature analysis has been studied for many applications. The features can be extracted from the raw signals using various techniques like visual basic code, LabVIEW graphical program, and MATLAB. In this study, a novel LabVIEW graphical program was used for quarrying the features. Feature selection is required so as to diminish the computational unpredictability. The selection of feature was done with the new approach called Incremental Wrapper-based Subset Selection (IWSS). In this selection method first, the ranking was done with the filter method followed by a wrapper approach [5]. The suggested features were then used for feature classification. Classification is a feature based supervised learning. Many learning algorithms such as decision tree (Gearbox fault diagnosis) [6], best first tree (Brake fault diagnosis) [7], Random Forest (Engine) [8], K Star (Brake fault diagnosis and tool condition monitoring) [9,10], Fuzzy (Centrifugal pump) [11], Logit Boost (Brake fault diagnosis) [12] were reported for many fault diagnosis applications including brake failures. In these studies, the possibility of implementing in real time with real-time experiments is a huge challenge for the researchers. In this study, a real time brake health monitoring has been initiated for monitoring the brake conditions. Figure 1 shows the methodology for monitoring the condition of the brake.

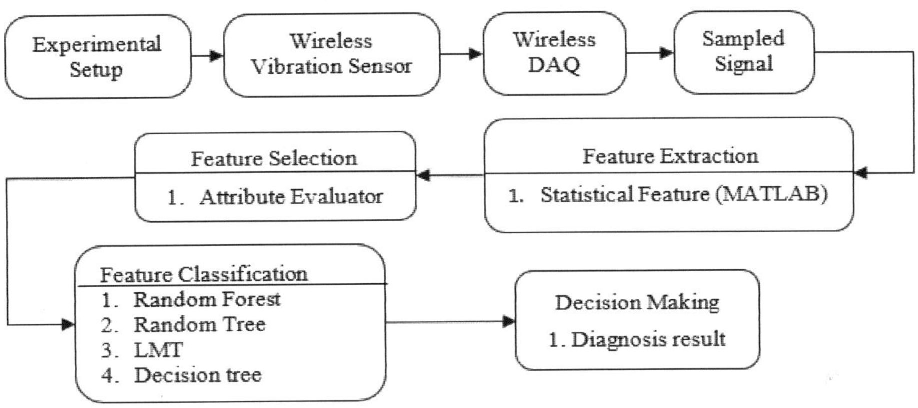

Figure 1. Flow Chart – Brake fault diagnosis procedure

2. Experimental Study

A commercial passenger vehicle (Maruti Zen) brake system was considered for the experimental study (Figure. 2). The real vehicle was used for the test. A fabricated setup with a free roller was used to drive the wheel. The drive shafts were allowed to run at a constant speed (331 rpm) over the free wheel on the test setup. Piezoelectric type shear uniaxial accelerometer (500 g range, 10.26 mV/g sensitivity) was used for acquiring the vibration signals and a wireless data acquisition hardware (DAQ - Model NI 9234, 4 channel, 51.2 kilo Samples /sec) through a LabVIEW graphical program. The following frequently occurring faults namely, Disc Brake Pad Wear Even Inner (DBPEI), Disc Brake Pad Wear Even Inner and Outer (DBPEIO), Disc Brake Pad Wear Uneven Inner(DBPUEI), Disc Break Pad Wear Uneven Inner and Outer(DBPUEIO), Reservoir Leak (RL), Brake Oil Spill on the Disc (BOS) and a Brake without any fault (GOOD) were considered for the simulation. Once the faults were simulated, the vibration signals were captured from the LMV brake system with the following settings:

1) *Sample length:* 12000 chosen arbitrarily
2) *Observation frequency:* 25 *kHz* (As per Nyquist sampling theorem)
3) *Number of samples:* 65

IOP Conf. Series: Materials Science and Engineering **624** (2019) 012027 doi:10.1088/1757-899X/624/1/012027

4) *Wheel speed:* 331 rpm

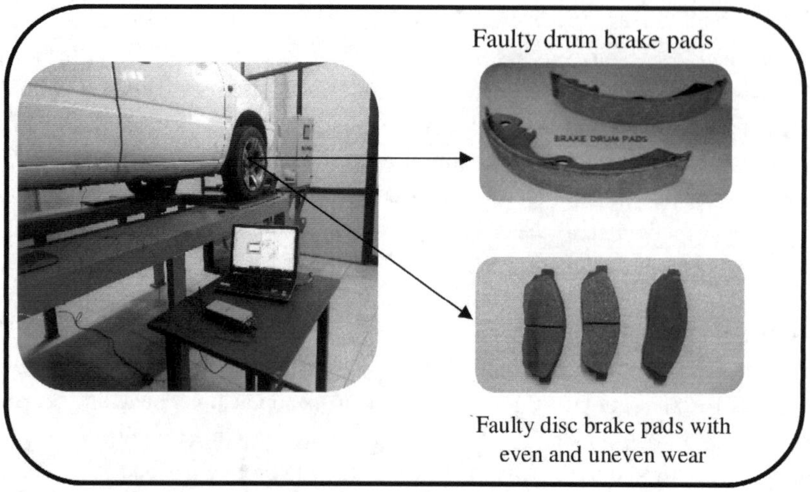

Figure. 2. Experimental Setup – Hydraulic Brake system

3. Result and Discussion

The vibration signal was acquired under different fault condition from the experimental test setup. The acquired vibration signals were extracted using various statistical feature extraction techniques. Table 1 shows the extracted sampled statistical features from the acquired vibration signal. After the feature extraction, the IWSS attribute evaluator and ranker search feature selection method was carried out. And the selected features were classified using different classification techniques such as random forest, random tree, LMT and decision tree.

Table 1. Sampled extracted statistical feature from the acquired vibration signal

Feature	Values	Feature	Values
Mean	0.05223	Maximum	0.141155
Standard Deviation	0.412191	Minimum	1.836311
Variance	0.169901	Range	-1.39755
Kurtosis	3.233003	RMS	3.233864
Median	0.046706	Impulse factor	0.41547
Mode	0.138532	Shape factor	35.15817
Skewness	0.05223	K factor	7.954618
Standard Error	0.762931	Total	12000
Sum	626.7599		

3.1. Classification using Random Forest

By using the random forest, the selected statistical features were classified. A random forest is a collaborative approach for structure an expansive number of decision trees at the preparation time and yielding a strategy for ordering the individual tree. The random forest starts with a standard machine learning method, a decision tree in which information is entered. As the information brings down, this tree gets assigned to smaller and smaller subsets. The overall classification accuracy of the random forest algorithm is 93.19%.

Misclassification details are presented as misperception matrix. Table 2 demonstrates the misperception matrix for the random forest. In the misperception matrix, the primary line first component (A) speaks to the absolute number of data points relating toward ''GOOD(GD)'' condition of the brake system. The primary component in the first column (A) speaks how many are effectively classified as 'A' condition. Among the 65 data points, 56 are effectively classified. Nine data point was misclassified as "Disc Pad Wear Even Inner (DBPEI)" belongs to *'B'* condition. Similarly, the third element in the third column (C) represents "Disc Pad Wear Even Inner & Outer (DBPEIO)" condition. Among the 65 data, all are effectively classified and there are no misclassified condition data points. Similarly, the other elements in the third row represent the miss-classification details. In the third row other than the third element, all other values are zero so there is no misclassification. The same way the classification accuracy can be calculated.

Table 2. Misperception Matrix for random forest

Category	*a*	*b*	*c*	*d*	*e*	*f*	*g*
a	*56*	*9*	*0*	*0*	*0*	*0*	*0*
b	*9*	*56*	*0*	*0*	*0*	*0*	*0*
c	*0*	*0*	*65*	*0*	*0*	*0*	*0*
d	*0*	*0*	*0*	*57*	*0*	*6*	*2*
e	*0*	*0*	*0*	*0*	*64*	*0*	*1*
f	*0*	*0*	*0*	*2*	*0*	*63*	*0*
g	*0*	*0*	*0*	*2*	*0*	*0*	*63*

Table 3. Misperception Matrix for random tree

Category	*a*	*b*	*c*	*d*	*e*	*f*	*g*
a	*51*	*14*	*0*	*0*	*0*	*0*	*0*
b	*11*	*54*	*0*	*0*	*0*	*0*	*0*
c	*0*	*0*	*65*	*0*	*0*	*0*	*0*
d	*0*	*0*	*0*	*58*	*0*	*5*	*2*
e	*0*	*0*	*0*	*0*	*65*	*0*	*0*
f	*0*	*0*	*0*	*8*	*0*	*57*	*0*
g	*0*	*0*	*0*	*1*	*0*	*0*	*64*

a - GOOD; b- DBPEI; c – DBPEIO; d – DBPUEI; e – DBPUEIO; f – BOS; g - RL.

3.2. Classification using Random tree

The random tree algorithm can manage both arrangement and regression issues. The random tree is an accumulation (outfit) of tree indicators that is called a forest. Using feature selection method all the sixteen contributing features were selected for the random tree classification. The selected features were classified using a random tree. Table 3 shows the misperception matrix of a random tree. The maximum classification accuracy of a random tree is 90.99%.

3.3. Classification using Logistic Model tree (LMT)

Logistic Model tree (LMT) is a well-known tree induction and linear logistic regression approach for classification. LMT made up of the target variables (both binary and multiclass), Missing and numeric values. The outcome of LMT in the form of a tree which encompass with binary leaves on numerical attributes. Using feature selection method top eleven features were contributed for the LMT classification. The selected features were classified using LMT. Table 4 shows the misperception matrix of LMT. The LMT produces the maximum classification accuracy of 93.85 %.

3.4. Classification using Decision tree

A decision tree is a tree-based knowledge structure used to select features through the classification rules. J48 is an classification algorithm used to generate a decision tree. Using feature selection method top nine features were contributed for the decision tree classification. The selected features were classified using a decision tree. Table 5 shows the misperception matrix of the decision tree. The decision tree produces the maximum classification accuracy of 92.09 %.

Table 4. Misperception Matrix for LMT.

Category	*a*	*b*	*c*	*d*	*e*	*f*	*g*
a	*58*	*7*	*0*	*0*	*0*	*0*	*0*
b	*10*	*55*	*0*	*0*	*0*	*0*	*0*
c	*0*	*0*	*65*	*0*	*0*	*0*	*0*
d	*0*	*0*	*0*	*58*	*0*	*5*	*2*
e	*0*	*0*	*0*	*0*	*64*	*0*	*1*
f	*0*	*0*	*0*	*1*	*0*	*64*	*0*
g	*0*	*0*	*0*	*2*	*0*	*0*	*63*

Table 5. Misperception Matrix for Decision tree

Category	*a*	*b*	*c*	*d*	*e*	*f*	*g*
a	*56*	*9*	*0*	*0*	*0*	*0*	*0*
b	*8*	*57*	*0*	*0*	*0*	*0*	*0*
c	*0*	*1*	*64*	*0*	*0*	*0*	*0*
d	*0*	*0*	*0*	*57*	*0*	*6*	*2*
e	*0*	*0*	*0*	*0*	*64*	*0*	*1*
f	*0*	*0*	*1*	*4*	*0*	*59*	*1*
g	*0*	*0*	*0*	*3*	*0*	*0*	*62*

4. Comparative study

In this study, the acquired vibration signals were extracted using the statistical technique. The extracted features were classified using different classifier such as a Random Tree, Random Forest, LMT and Decision Tree classification algorithm. Figure 3 shows the overall classification accuracy of various feature classifiers. The comparative study also reveals the maximum classification accuracy among the considered classifiers. The LMT produced the maximum accuracy of 93.85% (427 out of 455 data points) in just 0.07 seconds. The random forest produced the accuracy of 93.19% (424 out of 455 data points) in 0.09 seconds. The decision tree produced classification accuracy as 92.088% (419out of 455) and the random tree produced classification accuracy as 90.99 % (414out of 455). The computational time of best first tree and the random forest is high compare to LMT. The LMT produced maximum classification accuracy as 93.85%.

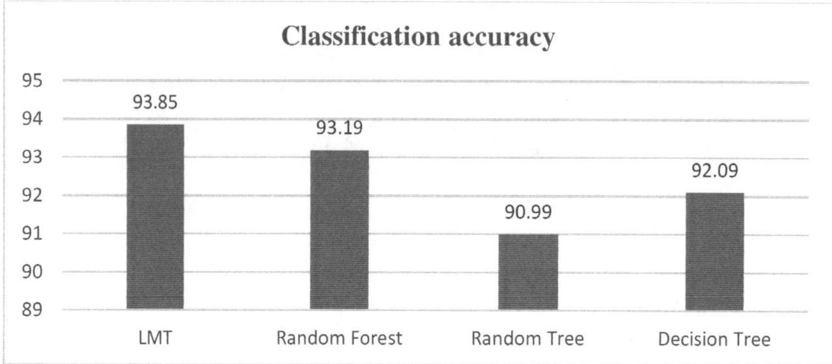

Figure. 3. Overall classification accuracy

5. Conclusion

In the paper, the statistical feature extraction techniques were discussed for the hydraulic brake fault diagnosis study through the vibration signal. Random forest, random tree, LMT and decision tree classifiers were used for classifying the faults. In this study, six fault conditions along with the good condition of a brake system were considered. The vibration signals occurred under all simulated conditions was acquired using a piezoelectric type accelerometer. Using statistical feature extraction techniques the acquired vibration signals were extracted. The attribute evaluator and effect of number of features methods used for feature selection. Then the selected features were classified using Random forest, random tree, LMT and decision tree classifiers. The experimental study shows that the LMT produced better classification accuracy of 93.85% compare to the other classifiers. This study can be extended on a real road condition for monitoring the health status of the hydraulic brake system. This detailed investigation will also provide a pathway for making and on-board diagnostic (OBD) module for finding the brake related faults well in advance.

Acknowledgment

The authors would like to acknowledge the research funding from DST – Science and Engineering Research Board (SERB Grant) File No. ECR/2016/002068.

References

[1] Singh S. Critical reasons for crashes investigated in the National Motor Vehicle Crash Causation Survey. NHTSA's National Center for Statistics and Analysis. 2015.

[2] NickJohnson. ⟨http://www.articles3k.com/article/144/176520/brake_failure_accidents⟩.

[3] Thomas Edward Wallace, Richard J Rink, Mark David Chandler. Brake Monitoring System and Method. US patent. US 2010/0188203A1. 2010.

[4] Sudhanshu Goel, Ripul Ghosh, Satish Kumar and Aparna Akula. A Methodical Review of Condition Monitoring Techniques for Electrical Equipment. National Seminar & Exhibition on Non-Destructive Evaluation, NDE. 2014, 1 – 8.

[5] Bermejo P, Gamez JA, Puerta JM. Improving incremental wrapper-based subset selection via replacement and early stopping. International Journal of Pattern Recognition and Artificial Intelligence. 2011; 25(05): 605-625.

[6] Saravanan N, Kumar Siddabattuni VNS, Ramachandran KI. A comparative study on classification of features by SVM and PSVM extracted using Morlet wavelet for fault diagnosis of spur bevel gear box, Expert Systems with Applications. 2008; 35: 1351–1366.

[7] Jegadeeshwaran R, Sugumaran V. Comparative study of decision tree classifier and best first tree classifier for fault diagnosis of automobile hydraulic brake system using statistical features. Measurement. 2013; 46(9): 3247-3260.

[8] Devasenapati SB, Ramachandran KI. Random forest based misfire detection using kononenko discretiser. ICTACT Journal on Soft Computing. 2012; 2(2): 270-275.

[9] Jegadeeshwaran R, Sugumaran V. Vibration Based Fault Diagnosis Study of an Automobile Brake System Using K Star (K*) Algorithm–A Statistical Approach. Recent Patents on Signal Processing. 2014; 4(1): 44-56.

[10] Madhusudana CK, Hemantha Kumar, Narendranath S. Condition monitoring of face milling tool using K-star algorithm and histogram features of vibration signal. Engineering Science and Technology, an International Journal. 2016; 19(3): 1543 – 1551.

[11] Wang HQ, Chen P. Fault diagnosis of centrifugal pump using symptom parameters in frequency domain. Agricultural Engineering International: CIGR Journal. 2007.

[12] Alamelu Manghai TM, Jegadeeshwaran R, Sugumaran V. Vibration based Condition Monitoring of a Brake System using Statistical Features with Logit Boost and Simple Logistic Algorithm. International Journal of Performability Enginnering. 2018; 14(1): 1-

IOP Conf. Series: Materials Science and Engineering **624** (2019) 012028 doi:10.1088/1757-899X/624/1/012028

Real time condition monitoring of hydraulic brake system using naïve bayes and bayes net algorithms

Alamelu Manghai T M*, Jegadeeshwaran R and Sakthivel G

[1]School of Mechanical and Building Sciences, VIT University, Chennai, Tamil Nadu, India
Email: alamelumangai.m@vit.ac.in

Abstract. The vehicles usage is increasing day by day due to the recent, technological development in the automotive field. In a competitive global market in order to survive, the reliability needs to be ensured, through a proper monitoring system. Brake system is one such control component, in which much focus is very much essential. An efficient brake system should provide reliable and effective performance in order to ensure the safety. If it is not properly monitored, it may lead to a serious catastrophic effect such as accidents, brake down, etc. Hence, the brake system needs to be monitored continuously. In this study, an experimental investigation was carried out for monitoring the brake system using vibration signals. An experimental setup which resembles the brake system was fabricated. The vibration signals were acquired under various brake condition such as good and faulty. From the acquired vibration signals, the features were extracted using statistical and histogram feature extraction techniques and feature selection was carried out. The selected features were then classifieds using a Naive Bayes and Bayes Net a classifier. The classification accuracy of all the algorithms were compared for finding the best feature classifier model for monitoring the brake condition.

1. Introduction

The brake system of an automobile is one of the most important safety systems which is used to stop at moments and slow down the vehicle. in order to avoid an accidents. Hence, brakes should be regularly inspected, through the trusted repair facility, test drive, visual inspection, measuring pad thickness for finding pad wear etc. Unless otherwise there is a warning or a monitoring system the people will not mind about the condition nowadays. Hence, it is very important that the brake components must be monitored for faults. Many researchers are performing the condition monitoring (CM) study on the systems for diagnosing faults. CM emphases on numerous parameters in which the substantive changes indicating the development of a fault. This CM increase extends the useful lifespan of equipment by identifying troublesome areas before they cause major issues. Fault detection is a subfield of condition monitoring which anxieties itself with monitoring a system, identifying when a fault has arisen and perceiving the type of fault. There are two fault detection techniques namely signal processing based fault detection (SPBFD) and model based approach (MBA) which are mostly used for the fault diagnosis study. In the MBA some knowledge based models are used to decide about the fault manifestation. In SPBFD, some statistical processes are used to abstract the information about the fault. SPBFD is a field concerned with finding faults arising on rotating devices and machinery such as presses, pumps, electric motors, brakes, and internal combustion engines, in which the vibration is inevitable. These vibrations can be cast-off for finding the condition of the mechanical components. In this study SPBFD approach has been focused for the brake fault diagnosis study.

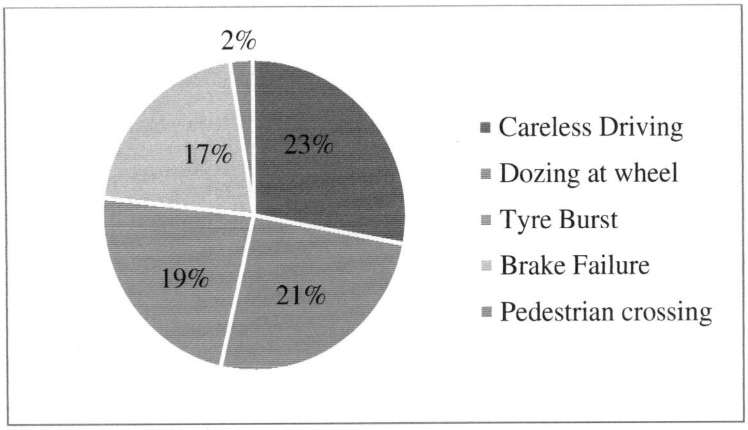

Figure 1. Details of major accident causes

The frequent application of brake causes faults in nature [1,2]. In recent years accident due to brake failures is increasing. Figure 1 shows the details of major causes. According to the analysis report from figure 1 the monitoring of brake failures no such automated systems are available in low and medium level LMVs. In safety point of view, the brake health should be ensured. In this study, an effort has been made to develop a monitoring model for diagnosing the brake faults. In recent years, the vibration signal has been focused in many fault diagnostic study. The hydraulic brake system is producing the vibration under different operating conditions. The produced vibration signatures are directly related with the various fault conditions has been considered for the learning process in this study.

The statistical and histogram feature analysis has been studied for many applications. The features can be excavated from the raw signals using various techniques like visual basic code, LabVIEW graphical program, and MATLAB. In this study, a novel LabVIEW graphical program was used for quarrying the features. Feature selection is necessitated in order to reduce the computational complexity. Suggested features were then used for feature classification. Classification is a feature based supervised learning.

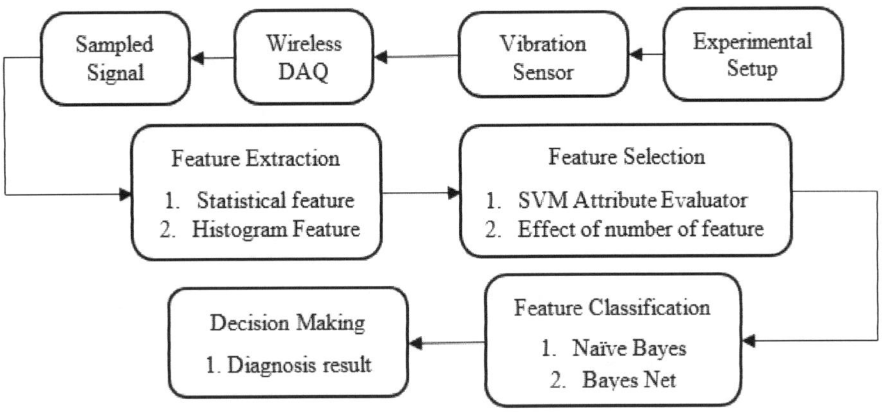

Figure 2. Flow Chart – Brake fault diagnosis procedure

Many learning algorithms such as decision tree (Gearbox fault diagnosis) [3], best first tree (Brake fault diagnosis) [4], Random Forest (Engine) [5], Bayes and Naïve Bayes (Pump fault diagnosis) [6], Support Vector Machine (Rolling Element Bearing and brake fault diagnosis) [7, 8], K Star (Brake fault diagnosis and tool condition monitoring) [9, 10], Fuzzy (Centrifugal pump) [11], Logit Boost (Brake fault diagnosis) [12] were reported for many fault diagnosis applications including brake failures. In these studies, the possibility for implementing in real time with real time experiments is a huge challenge for the researchers. In this study, a real time brake health monitoring has been initiated for monitoring the brake conditions. Figure 2 shows the methodology for monitoring the condition of the hydraulic brake system.

2. Experimental Study

A commercial passenger vehicle (Maruti Zen) brake system was considered for the experimental study (Figure. 3). The real vehicle was used for the test. A fabricated setup with free roller was used to drive the wheel. The drive shafts were allowed to run at a constant speed (331 rpm) over the free wheel on the

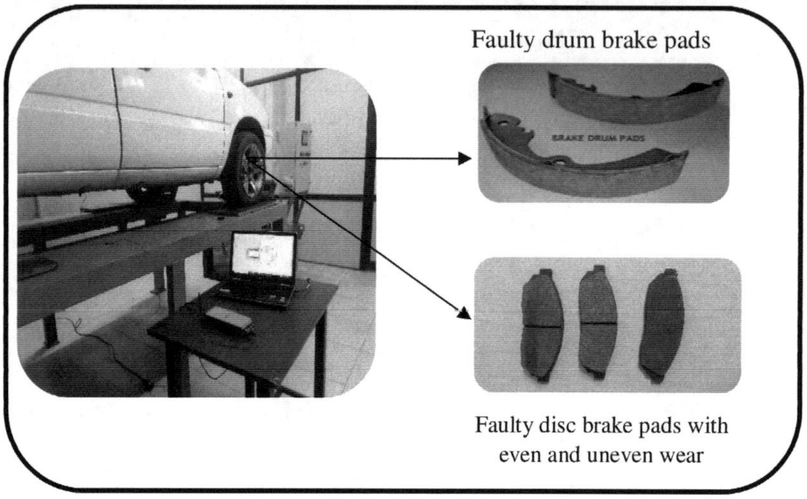

Figure. 3. Experimental Setup – Hydraulic Brake system

test setup. Vibration signals were acquired using a piezoelectric type shear uniaxial accelerometer (500 g range, 10.26 mV/g sensitivity) and a wireless data acquisition hardware (DAQ - Model NI 9234, 4 channel, 51.2 kilo Samples /sec) through a LabVIEW graphical program. The following frequently occurring faults namely, Inner brake pad wear in disc brake, both inner and outer pad wear in disc brake, uneven wear in inner brake pad, uneven wear in inner and outer brake pad, reservoir leak, oil spill on the disc and a good condition brake pad were considered for the simulation. Once the faults were simulated, the vibration signals were captured from the LMV brake system with the following settings:

1) *Sample length:* 12000 chosen arbitrarily
2) *Observation frequency:* 25 *kHz*
3) *Number of samples:* 65
4) *Wheel speed:* 331 rpm

3. Result and Discussion

3.1. Feature Extraction

IOP Conf. Series: Materials Science and Engineering **624** (2019) 012028 doi:10.1088/1757-899X/624/1/012028

3.1.1 Statistical Feature Extraction

The vibration signal was acquired under different fault condition from the experimental test setup. The acquired vibration signals were extracted using various statistical feature extraction techniques. Table 1 shows the sampled statistical features extracted from raw vibration signal. After the feature extraction, the IWSS attribute evaluator and ranker search feature selection method was carried out. And the selected features were classified using different classification techniques such as Naive Bayes and Bayes Net a classifier.

Table 1: Sampled Statistical Feature extracted from raw vibration signal

Feature	Values	Feature	Values
Mean	0.05223	Maximum	0.141155
Standard Deviation	0.412191	Minimum	1.836311
Variance	0.169901	Range	-1.39755
Kurtosis	3.233003	RMS	3.233864
Median	0.046706	Impulse factor	0.41547
Mode	0.138532	Shape factor	35.15817
Skewness	0.05223	K factor	7.954618
Standard Error	0.762931	Total	12000
Sum	626.7599		

3.1.2. Histogram Feature Extraction

A histogram shows plot the range of variation in a better way. The information derived from a histogram plot can be used as features in the fault diagnosis. The vibration from the brake system will give better information for the histogram analysis. The histogram features can also be extracted using visual basic macro code and Micro soft office excel**.** Figure. 4 shows the sample extracted histogram feature from the test setup under various fault and good conditions of the hydraulic brake system. The bin range for the histogram plot was calculated from the minimum maximum and value of the signal. From the acquired vibration signal totally 44 bins were extracted.

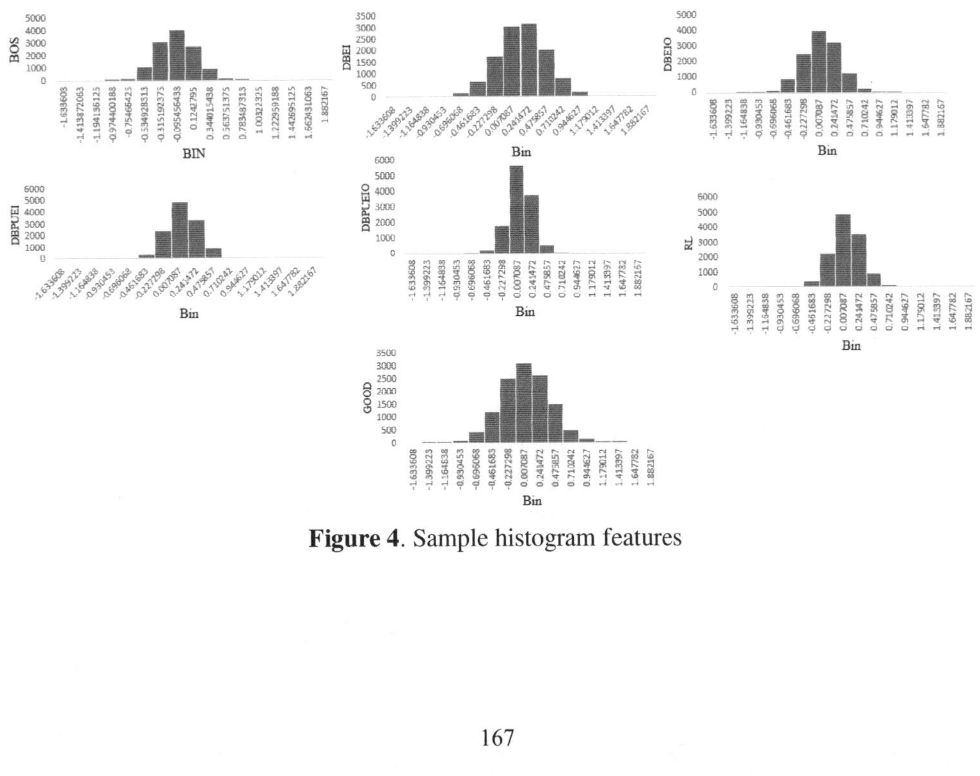

Figure 4. Sample histogram features

3.2. Feature Classification

3.2.1 *Bayes Net* (BN)

Bayesian network comprises of a set of variables, $C = \{B_1, B_2, . . ., B_N\}$ and a set of coordinated edge A, between variables, which frame a Directed Acyclic Graph (DAG) D = (C, A) where a joint distribution of factors is spoken by the result of contingent dispersions of every variable given its folks. The selected histogram and statistical features were classified using Bayes Net. Table 2 shows the overall classification accuracy of BN classifier. Based on table 2 statistical feature gives maximum classification accuracy 90.55% (549 out of 550 data points) compare to histogram. The histogram produced classification accuracy as 78.85% (425 out of 455 data points).

Table 2 Overall Classification accuracy

Features	Classification accuracy (%)	
	Bayes Net	Naive Bayes
Statistical	90.55	89.23
Histogram	78.85	71.06

Table 3(a) Misperception Matrix for Bayes Net (Statistical)

Category	***a***	***b***	***c***	***d***	***e***	***f***	***g***
a	51	14	0	0	0	0	0
b	15	50	0	0	0	0	0
c	0	0	64	1	0	0	0
d	0	0	0	57	0	5	2
e	0	0	0	0	64	0	1
f	0	0	0	1	0	64	0
g	0	0	0	4	0	0	61

Table 3(b) Misperception Matrix for Naïve Bayes (Statistical)

Category	***a***	***b***	***c***	***d***	***e***	***f***	***g***
a	50	15	0	0	0	0	0
b	16	49	0	0	0	0	0
c	0	0	64	0	0	0	0
d	0	0	1	54	0	8	2
e	0	0	0	0	65	0	0
f	0	0	0	4	0	61	0
g	0	0	0	2	0	0	63

a - GOOD (Brake without any fault); b- DBPEI (Disc Pad Wear (Even) Inner); c – DBPEIO (Disc Pad Wear (Even) Inner & Outer); d – DBPUEI (Disc Pad Wear (Uneven) Inner); e – DBPUEIO (Disc Pad Wear (Uneven) Inner & Outer); f – BOS (Brake Oil Spill); g - RL (Reservoir Leak).

Misperception details are presented as misperception matrix. Table 3(a) shows the misperception matrix for Bayes Net. In the misperception matrix, the first-row first element (A) represents the total number of data points corresponding to ''GOOD(GD)'' condition of the brake system. The first element in the first column(A) represents how many are correctly classified as *'A'* condition. Among the 65 data points 51 are correctly classified. 14 data point was misclassified as "Disc Pad Wear Even Inner (DBPEI)" belongs to *'B'* condition. The same way the classification accuracy can be calculated

3.2.2. Naive Bayes

The Naïve Bayes (NB) algorithm is a classification algorithm in light of Bayes rule, which accepts the independent attributes. The estimation of this presumption is that it significantly rearranges the representation of *P(X/Y)* and the issue of assessing it from the training data. From the acquired vibration signal 60% of data can be used for training and 40% of data can be used for testing. The selected histogram and statistical features were classified using Bayes Net. Table 2 shows the overall classification accuracy of BN classifier. Based on table 2 statistical feature gives maximum classification

IOP Conf. Series: Materials Science and Engineering **624** (2019) 012030 doi:10.1088/1757-899X/624/1/012030

Milling cutter condition monitoring using machine learning approach

A D Patange[1*], Jegadeeshwaran R[1] and N C Dhobale[1]

[1]School of Mechanical and Building Sciences, VIT University, Chennai, Tamil Nadu, India

Email: abhipatange93@gmail.com

Abstract. The cutting tool condition drives the economy of machining processes in manufacturing industry. The failures in cutting tool are unbearable and affect the drive of machine tool which reduces life. Hence it necessitates reducing power consumption using monitoring cutting tool condition and hence requires an efficient supervision to monitor and predict faults. Simply stated, the condition which curtails cutting tool life highlighted before it turns into a tool wear, breakage and failure. This ensures optimized and effective use of a cutting tool, saves maintenance/repair time, enhances constancy in a process etc. The recent development in Machine Learning (ML) and its applicability for condition monitoring approach has drawn attention of researchers. Machine learning examines existing and past indications to predict conditions in future. This paper presents machine learning based condition monitoring of milling cutter of vertical machining centre (VMC). The vibration signals acquisition of 4 insert milling cutter is carried out with healthy and various fault conditions. The Visual Basic (VB) code and script is used to extract statistical features and decision tree algorithm is used to select relevant features. The different conditions of milling cutter are classified using tree family classifiers. The effort made in this work is to check applicability of ML approach for milling cutter fault diagnosis for reducing power consumption of drive of machine tool.

1. Introduction

The health monitoring of cutting tool is inspired and originated due to necessity of developing unmanned machining systems which are self-sustained and intelligent [1,2]. One of the most important monitoring requirements in an unmanned manufacturing system is in-process detection of tool breakages and execution of prompt action [3]. This has evolved the need of 'Tool condition monitoring' or 'Health monitoring of tool' in past several years. The tool condition monitoring methodologies are usually classified into two broad types such as 'Offline/Direct methods' and 'Online/Indirect methods'. The direct methods are mostly appropriate for inspecting and analysing of complex failures (hard faults) which usually are unpredictable and hence found unsuitable for machine learning domain [4]. On the other hand indirect methods are based most appropriate method for monitoring soft faults which evolve gradually with respect to time leading to continuous deterioration of the tool [5]. The recent development in Machine Learning (ML) and its applicability for condition monitoring approach has drawn attention of researchers. Machine learning examines existing and past indications to predict conditions in future [6]. In this paper, the vibration signals acquisition of 4 insert milling cutter is carried out with healthy and various fault conditions. The Visual Basic (VB) code and script is used to extract statistical features and decision tree algorithm is used to select relevant features. The different conditions of milling cutter are classified using tree family classifiers i.e. J48, Logistic Model Tree and Random Forest algorithms.

IOP Conf. Series: Materials Science and Engineering **624** (2019) 012030 doi:10.1088/1757-899X/624/1/012030

2. Literature review

The research related to tool condition monitoring through machine learning approach is presented here to highlight the research gap. This research is specifically confined to the study of vibration involved in machining and its relation with cutting tool condition. Vibration monitoring is useful to identify primary symptoms about tool condition, dimensional variations and surface roughness in machining applications than others. The customized regulation of vibrations for corresponding removal rate and avoidance of tool breakage at the shank, in process tool breakage, wear and chatter recognition is a vital task [7]. It is recommended that vibration signals are more sensitive, robust, reliable, requires less accompanying instrumentation and can be employed for real time application [8]. The cutting forces change dynamically for cutter exit which leads to vibration especially when forces change their directions due to the variation of dynamic chip thickness [9]. In the experiment carried by [10] it is observed that if nose wear is greater than 0.2 mm lead to the possibility of cataclysmic breakdown. Moreover he found that the notch wear conversely rose progressively in the beginning of machining, and continued to be unchanged for the rest tool life. Second step, these signals are extracted as pertinent features using statistical, wavelet or histogram approaches etc. In the comparison of seven statistical features amongst which root mean square found most significant [11]. In the statistical analysis of acquired signals using I-kaz technique where correlation of I-kaz coefficients and Z^{∞} values with resultant flank wear width data were presented. It has been observed that Z^{∞} values reduce when VB rises and reduces when machining speed rises [12]. In experimentation of [13] obtained classification accuracy of 72% with thirteen extracted features. But when only standard error, standard deviation, variance, kurtosis and mode were selected accuracy increased to 76.9%. In the comparison of both histogram and statistical features, it is demonstrated that classification accuracy of 86.34% is found when statistical approach is used than histogram (73.61%) i.e. statistical technique extracts features more accurately than histogram technique [14]. The J48 algorithm was utilised for selecting features based on the information gain of the features which helps in reduction of domain knowledge [15]. Lastly, the fault diagnosis of tool condition is accomplished by analysing extracted features applying machine learning techniques. The features classification carried using clonal selection classification algorithm by analysing diverse population with high fitness antibodies instead of a single antibody solution [16]. When Naïve Bayes and Bayes Net classifiers was used and found accuracy of 85.28% and 86.34% respectively [14,17]. It is found that the use of statistical technique for feature extraction is most favourable as compared to others. The decision trees are suitable for feature selection as it reduces dimensionality. It is also found that use of tree family classifiers for tool condition monitoring is not reported till now.

3. Experimental study

The experimentation for signal acquisition was conducted on Vertical Machining Centre (VTEC Model CCS00605) in Axis Metalcut Technologies, Bhosari-Pune.

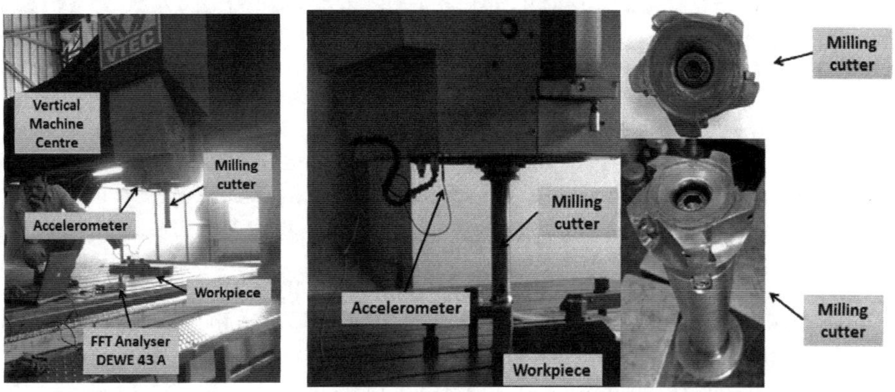

Figure 1. Experimental setup

Figure 1 shows the experimental setup. The piezoelectric accelerometer with sensitivity 9 mV/g was mounted on near to spindle frame. The 8 channel FFT analyser (DEWE 43 A) was used for

IOP Conf. Series: Materials Science and Engineering **624** (2019) 012030 doi:10.1088/1757-899X/624/1/012030

acceleration signal acquisition from accelerometer. Vibration measurements are thus usually taken near to the rotating component of a machine. Here when the machining is carried out on VMC, the vibration generated due to variation in parameters will directly affect cutting tool first and hence spindle. As the spindle revolving in vertical direction, orientation of accelerometer is kept vertical. The literature review [7,10] provides the suggestion for deciding orientation of accelerometer for tool condition monitoring. Based on the basis mentioned above and literature, the vertical orientation is incorporated. The time domain graph of acceleration (g) Vs time (sec) was recorded using DEWE software in real time. The sampling frequency of 20 kHz was used to convert analog acceleration into equivalent digital acceleration value. The vibration signals are acquired for face milling operation which was performed on Mild Steel C shaped hollow cuboid workpiece with dimensions 650mm*250mm*100mm using milling cutter of diameter 63 mm with 4 carbide inserts. The machining parameters recommended by KomGuide – Technical Manual was selected for all machining operations such as spindle speed 900 rpm, feed 2000 mm/min and depth of cut of 0.25 mm. The 6 machining operations were performed with 6 different insert conditions where speed, feed and depth of cut was kept constant as mentioned above and vibration signal was captured. The 6 different conditions are described as below in Table 1:

Table 1. Insert conditions for machining

Operation No.	Insert 1	Insert 2	Insert 3	Insert 4	Insert condition Category
1	New	New	New	New	A
2	New	New	New	Flank wear	B
3	New	New	New	Nose wear	C
4	New	New	New	Notch wear	D
5	New	New	Flank wear	Nose wear	E
6	New	New	Flank wear	Notch wear	F

3.1. Signal acquisition: The vibration signal was captured for all 6 operations and respective time domain plots acuqired in DEWE software as shown in figure 2. The sampling frequency of 20 kHz was selected according to Nyquist theorem [8]. Each operation is performed for time interval of 1 second. Initially rough machining was performed to eliminate oxide layer and unevenness of the workpiece and allowed to get stabilized. It is observed in figure 2 that the acceleration amplitude varies and rises as the fault condition changes.

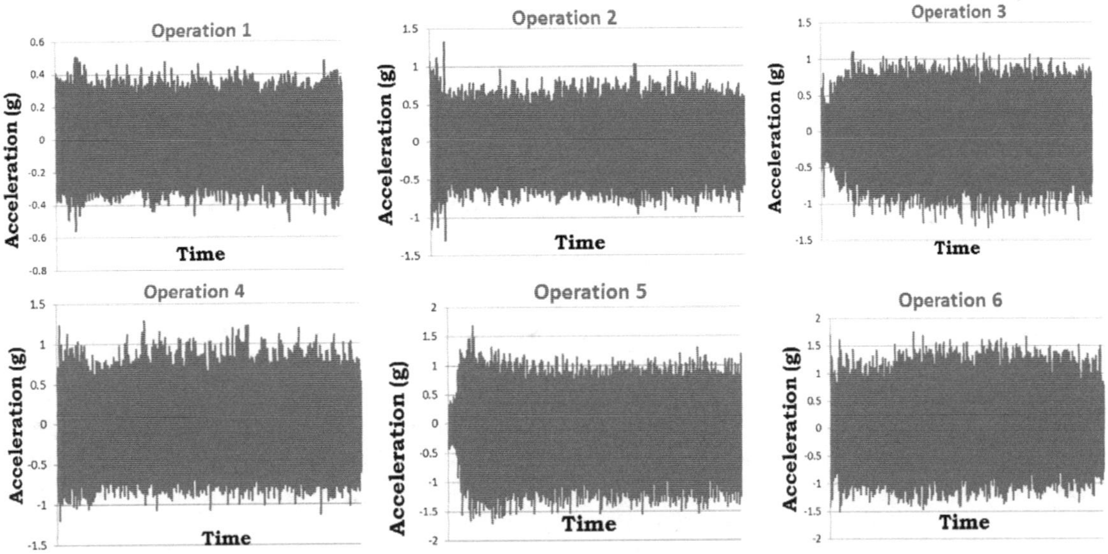

Figure 2. Signal acquisition in time domain for all machining operations

4. Result, discussion & validation using machine leaning approach

4.1. Feature extraction: The time domain features are nothing but the distinct characteristics observed directly from the acquired time domain plots. The VB code described by [14] was used in Microsoft Excel to extract statistical features from all 6 time domain plots of respective operations. Total 13 statistical features such as mean, standard error, median, mode, standard deviation, variance, kurtosis, skewness, range, minimum, maximum, summation, count were computed to serve as features. Here total 600 samples are considered (i.e.100 samples for each condition * 6 conditions = 600 samples).

4.2. Feature selection: The feature selection is the most significant step in machine learning approach as it selects most significant features amongst all. It reduces dimensionality; speed up the classification algorithm; increases classification accuracy and makes the results more comprehensible [15]. The all 13 features are fed to J48 algorithm and generated decision tree structure is shown in figure 3. It is observed that only 9 feature such as mean, standard error, median, standard deviation, kurtosis, skewness, range, minimum, maximum, found to be selected by J48 algorithm will serve to be most significant. Hence these 9 features are only used for feature classification.

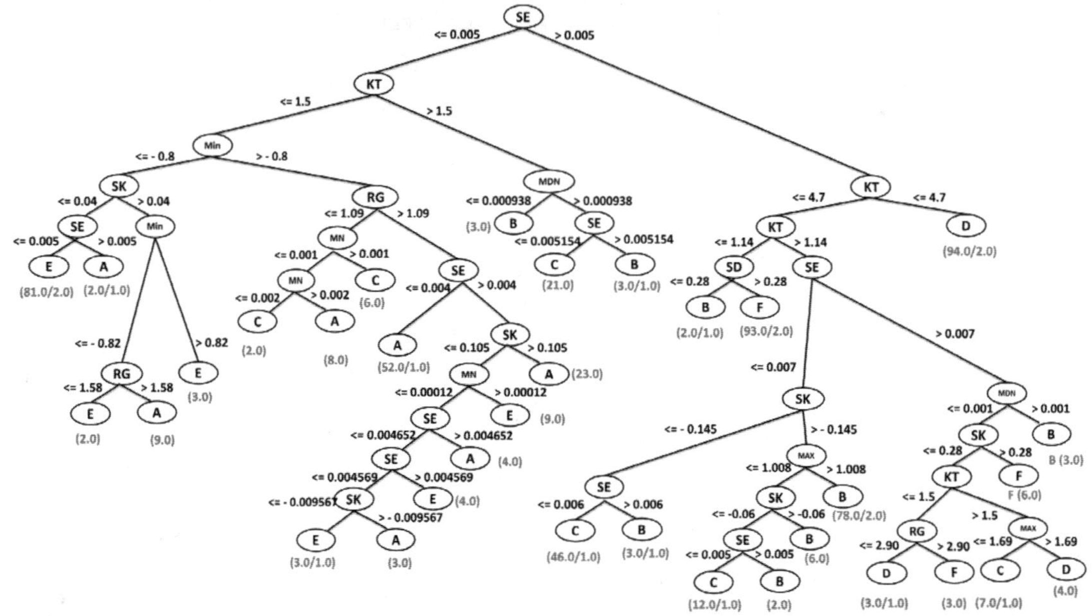

Figure 3. Result of decision tree J48 for feature selection

Nomenclature: MN: Mean, SE: Standard Error, MDN: Median, SD: Standard Deviation, KT: Kurtosis, SK: Skewness, RG: Range, MIN: Minimum, MAX: Maximum and A, B, C, D, E, F are the tool conditions as already

4.3. Feature classification: A data mining freeware named 'WEKA' was used for feature classification. The tree family classifiers give best classification accuracy amongst all other classifiers. That why comparison of tree family classifiers is advocated in this study to obtain best between them. We can use some other algorithms like K star algorithm, Naïve Bayes, LWL provided it should provide better classification accuracy. Here, the different conditions of milling cutter are classified using tree family classifiers i.e. J48, Logistic Model Tree and Random Forest algorithms based on 10-fold cross validation. Here classification validation is presented using confusion matrix as it describes categorization of distinctive tool conditions.

4.3.1 J48 classifier: In this experimentation, the classification accuracy for J48 algorithm is obtained as 85.5%. It is observed from confusion matrix shown in figure 4(a) that the diagonal element represents the correctly classified samples, while non diagonal element represents misclassified samples. The confusion matrix for J48 tree shows that for 'A' condition (all inserts are new) of tool, 86 samples were correctly classified as 'All New'. For condition 'B' (three inserts new and one with flank wear), 81 samples were correctly classified as condition 'flank wear', whereas 12 samples were misclassified as 'C' (i.e. nose) condition and so on.

4.3.2 Random Forest Classifier: Random forest classifier creates assemblies of decision trees and then integrates them with each other to obtain improved, steady and truthful prediction. It is considered as combined approach as it facilitates classification and regression both. It put extra randomness to training model at time of developing the tree structure. It examines the most significant feature amongst a random subset than finding it at time of separating the node. Hence this gives better model because of broaden assortment. The classification accuracy for random forest is obtained as 88.16%. It is observed from confusion matrix shown in figure 4(b) that for 'A' condition (all inserts are new) of tool, 94 samples were correctly classified as 'All New'. For condition 'B' (three inserts new and one with flank wear), 84 samples were correctly classified as condition 'flank wear', whereas 10 samples were misclassified as 'C' (i.e. nose) condition and so on.

4.3.3 Logistic Model Tree Classifier: The principle of this classifier is based on collaborative approach, as it uses decision tree and logistic regression both classifiers. A piecewise linear regression model is obtained by this decision tree using linear regression models at its leaves. An LR model at every node in the tree is created using LogitBoost algorithm. After this the node is separated using the C4.5 criterion. A number of LogitBoost iterations that does not overfit the training data is identified from cross-validation. The classification accuracy for LMT is obtained as 88.33%. It is observed from confusion matrix shown in figure 4(c) that for 'A' condition (all inserts are new) of tool, 88 samples were correctly classified as 'All New'. For condition 'B' (three inserts new and one with flank wear), 84 samples were correctly classified as condition 'flank wear', whereas 10 samples were misclassified as 'C' (i.e. nose) condition and so on.

```
=== Confusion Matrix ===

  a  b  c  d  e  f   <-- classified as
 86  2  3  0  8  1 |  a = A
  0 81 12  3  1  3 |  b = B
  5 13 75  2  2  3 |  c = C
  1  2  2 94  0  1 |  d = D
 13  1  2  0 83  1 |  e = E
  3  1  1  1  0 94 |  f = F
```

(a) Decision tree J48

```
=== Confusion Matrix ===

  a  b  c  d  e  f   <-- classified as
 94  0  3  0  3  0 |  a = A
  0 84 10  3  0  3 |  b = B
  4 12 76  3  3  2 |  c = C
  0  4  2 92  0  2 |  d = D
 12  0  3  0 85  0 |  e = E
  0  2  0  0  0 98 |  f = F
```

(b) Random Forest

```
=== Confusion Matrix ===

  a  b  c  d  e  f   <-- classified as
 88  1  1  0 10  0 |  a = A
  1 84 10  3  2  0 |  b = B
  8  8 80  2  1  1 |  c = C
  1  5  2 90  0  2 |  d = D
  8  0  2  0 90  0 |  e = E
  0  1  0  1  0 98 |  f = F
```

(c) LMT

Figure 4. Classification result for (a) J48 algorithm (b) Random Forest (c) LMT

4.3.4 Discussion on result: The table 2 represents comparison of various algorithms used for feature classification. It has been observed that logistic model and random forest algorithms yields very close accuracy. The table 2 shows that Logistic model tree algorithms performs best amongst all but when considered real time implantation of these algorithms for future prediction, the time taken to build model is a concern. In comparison of LMT and random forest, the time required to build the model is really less for random forest i.e. 0.6 seconds with the accuracy 88.16%. If accuracy is only constrain irrespective of time, then LMT served to be best. One can consider J48 algorithm too, if the time is constrain, as it required lesser time amongst all.

Table 2. Comparison of various algorithms used for feature classification

Sr. No.	Constrains used for classification (9 constrains selected by J48 tree)	Algorithm type → Parameter ↓	J48	LMT	Random Forest
1	Mean + Stand. Error + Median + SD + Kurtosis + Skewness + Range + Min + Max + Condition	Accuracy (%)	85.5	88.3333	88.1667
		Time for building model (seconds)	0.2	8.6	0.6

IOP Conf. Series: Materials Science and Engineering **624** (2019) 012030 doi:10.1088/1757-899X/624/1/012030

5. Conclusion

The machine learning based condition monitoring of milling cutter of vertical machining centre (VMC) was successfully demonstrated. The vibration signals acquisition for milling cutter was carried out with fault tool conditions. The Visual Basic (VB) code and script is used to extract statistical features and decision tree algorithm is used to select relevant features. The different conditions of tool are classified using tree family classifiers i.e. J48, LMT and random forest algorithms. Considering the classification accuracy and time required to build the model, the random forest observed to be best classifier. The LMT classifier yields slightly more accuracy than random forest but time required is considerable more than the random forest classifier. This approach is most suitable for real time implementation of tool fault diagnosis which ultimately reduces power consumption of drive system. This classification data can be used as history of particular instance and compared with same instant in present to predict future of the tool condition i.e. to build and incorporate a model using data fusion approach will be incorporated for future correspondence.

References

1. Daneshmand L K and Pak H A 1986 *Journal of dynamic systems measurement and control* Adaptive Control System using economic performance index in turning vol 108 pp 215-222
2. Altintas Y and Peng J 1990 *Computers in Industry* Design and Analysis of a Modular CNC System vol 13 pp 305-316
3. Altintas Y, Newell N and M. Ito 1993 *Manufacturing Science and Engineering* A Hierarchical Open Architecture Multi-Processor CNC System for Motion and Process Control vol 64 pp 195-205
4. Botsaris P N and Tsanakas J A 2008 *The Proceedings of The International Conference of COMADEM* State-of-the-art in methods applied to tool condition monitoring (TCM) in unmanned machining operations: A review
5. Cho S, Binsaeid S and Asfour S 2009 *International Journal of Advanced Manufacturing Technology* Design of multisensory fusion-based tool condition monitoring system in end milling pp 1-14
6. Nebot J A and Subiro R 2010 *International Journal of Advanced Manufacturing Technology* A Review of Machining Monitoring Systems Based on Artificial Intelligence Process Models vol 47 pp 237–257
7. Dimla D and Lister P 2000 *International Journal of Machine Tools Manufactures* On-line metal cutting tool condition monitoring: force and vibration analyses vol 40 pp 739–768
8. Shewale M S Mulik S S Deshmukh S P Patange A D 2018 *Proc. of 2nd Inter. Conf. on Data Engineering and Communication Technology* Advances in Intelligent Systems and Computing Springer Singapore A novel health monitoring system vol 828 pp 461-468
9. Ming L, Jiawei M and Dinghua Z 2016 *Chinese Journal of Aeronautics* Time-domain modelling of a cutter exiting a workpiece in the slot milling process vol 29 pp 1852–1858
10. Dimla D E 2002 *The International Journal of Advanced Manufacturing Technology* The correlation of vibration signal features to cutting tool wear in a metal turning operation vol 19 pp 705–713
11. Jantunen E and Jokinen H 1996 *International Journal of Flexible Automation and Integrated Manufacturing* Automated On-line Diagnosis of Cutting Tool Condition vol 4 pp 273–287
12. Ahmada M, Nuawia M, Abdullaha S, Wahida Z, Karimb Z and Dirhamsyahc M 2015 *Procedia Engineering* Development of tool wear machining monitoring using novel statistical analysis method, I-kaz vol. 101 pp 355 – 362
13. Painuli S, Elangovan M and Sugumaran V 2014 *Expert Systems with Applications* Tool condition monitoring using K-star algorithm vol 41 pp 2638–2643
14. Elangovan M, Ramachandran K and Sugumaran V 2010 *Expert Systems with Applications* Studies on Bayes classifier for condition monitoring of single point carbide tipped tool based on statistical and histogram features vol 37 pp 2059–2065
15. Sugumaran V, Muralidharan V and Ramachandran K 2007 *Mechanical Systems and Signal Processing* Feature selection using Decision Tree and classification through Proximal Support Vector Machine for fault diagnostics of roller bearing vol 21 pp 930–942
16. Jegadeeshwaran R and Sugumaran V 2015 *Engineering Science and Technology, An International Journal* Brake fault diagnosis using Clonal Selection Classification Algorithm (CSCA): A statistical learning approach vol 18 pp 14-23
17. Nalavade S P, Patange A D, Prabhune C L and Mulik S S 2018 *Lecture Notes in Mechanical Engineering, Springer* Development of 12 Channel Temperature Acquisition System for Heat Exchanger Using MAX6675 and Arduino Interface Vol 1 pp 119-125

Impact of Adaboost and Support Vector Machine Classifier in Automotive Sector

Graceson Jeriel Frederick [1], Sakthivel G [2] and Jagadeeshwaran [2]

[1] Student, SMBS, VIT University, Chennai Campus.
[2] Associate Professor, SMBS, VIT University, Chennai Campus.

E-mail: graceson.frederick2017@vitstudent.ac.in

Abstract. The automobile industry is moving towards complete automation of vehicles. While the automation of vehicles can reduce the workload to human beings, we also need to make sure that safety is not compromised in the process. One of the integral components of the automation of vehicles is the detection of several obstacles. Obstacles may include lamp post, pedestrians, animals, vehicles, etc. This project involves detection of commercial vehicles that are present in front of the host vehicle and is limited only to far and middle region. The classification is done in two steps: Adaboost Training and Support Vector Machine Training. Adaboost being a weak classifier is used to reduce the reaction time of the entire system. Even though the classifier does not produce more accurate results it will reduce the time taken for classification considerably. Support Vector Machine algorithm will produce better classification result as it makes use of a hyperplane to separate between the classes. Support Vector Machine being a strong classifier gives better result but increases the reaction time taken for classification. In order to counter this problem Adaboost Training is done. Adaboost is a set of weak classifiers which will classify at a better speed but the results may not be accurate. Therefore, this Adaboost classification is done at first and only the positive classification from the Adaboost classifier are sent for Support Vector Machine classification. This will make sure that more time will not be used for the Support Vector Machine classification. Based on the classification if the result specifies that there is an object, the distance will be computed. Based on the distance and the speed of the host vehicle, the time to avoid collision can be calculated. The needs to be a lookup table to match different time frames with reduction of gears and the speed respectively. The ECU will manage the speed variation and the gear shift required through the control system that will give the time to avoid collision as the output. Therefore, the collision can be avoided by obstacle detection and computation of distance of the object.

1. Introduction

The automobile sector is going towards automation at various levels. Even though the level of automation may vary from application to application the significance. In the automotive field, automation plays a very important role. In self driving sector there are several applications such as parking, lane changing, smart energy systems, obstacle detection and so on. Machine learning also plays an important role in the field of marketing the vehicles and in predicting the needs of the customers and the type of vehicles that will create high impact in the environment of the customers. Artificial intelligence is another blooming field that helps in creating a setup where the human beings can be mimicked and the decision making can also be monitored. When the artificial intelligence and the Machine Learning techniques are combined, we can improve the decision making. There is a requirement to produce a system that will be able to train itself automatically. Thereby, we can create a model that will train itself and improve the detection with any particular change that is brought to the dataset. The prediction accuracy was estimated to be 94.8 percentage. Once the deep learning is done and the data is further transferred to ECU. The data of the presence of the obstacle along with the distance of the obstacle is sent to the ECU. Before sending the data to the ECU there are certain computations such as computation of the time taken for collision, speed of the host vehicle and the speed of the movement of the obstacle needs to be done. Thus, the speed reduction can be achieved in the stipulated amount of time.

2. Literature Survey

Ye et al.,(2019) developed a decentralized resource allocation mechanism that can be used to improve the communication in between the vehicles. It is based in a reinforcement learning technique that will make use of a deep learning technique. This paper shows an autonomous decentralized mechanism to communicate between the vehicle and at the same time increase the efficiency of the communication between the systems without creating any drastic changes to the input or the output that is produced by the system. There is a mechanism that is used to produce a link so that the transmission data can be transferred easily through the system. The communication needs to be done without creating any interference with the environment. There are certain simulation results that are also produced in this paper. [1]

Moghimi et al.,(2011) has mentioned that vehicles are detected based on a boosting technique. Their proposed system are the ones that have been tested in real time life like situations and in scenarios that were taken from the surveillance scenarios. There are different conditions with various lighting techniques that are tested in several conditions. The obtained results in the real surveillance videos present in different lighting conditions show us the efficiency and the extent of practicality of the method. [3]

Chintalacheruvu (2012), has proposed a vehicle detection system that is based on certain pre-existing videos which can help in reducing the complexity involved in the calibration of the entire system. This will help in increasing the performance of the system even if the resolution of the available video is very low. Based on the distance differences, there are a different set of resolutions that need to be used. The accuracy is increased in terms of the number of vehicles that are detected and the speed of those vehicles that have been detected.[4]

Hadi et al.,(2014) has presented briefly about that various image processing techniques and the several analysis tools that were used in the several methods that were used for traffic light monitoring and surveillance techniques. The processing methids have been classified into three different categories in order to precisely explain in clarity about the traffic systems. The tracking method also takes into consideration the various motion constraints that are involved as well when it comes to detection in traffic surveillance systems. [5]

Beymer et al.,(1998) developed a tracking system that can help in detection in several challenging conditions based on certain features that have been extracted from the total set of features that are present with the vehicular data. The major advantage is that it works without showing much change in efficiency when it comes to different lighting conditions. After the tracking of the vehicles is done they are segregated and labelled as separate vehicles thereby helping in the process of tracking the vehicles continuously. [6]

Avery et al.,(2002) has presented a method to detect the width of the vehicle based on a width based distance calculation algorithm. This algorithm can be useful in detecting the distance at which an object is present from the the host vehicle and at the same time aids in accurate prediction of the presence of obstacles. This method is of greater use when it comes to detection any obstacle present.[9]

Based on the above mentioned literature survey it is learnt that adaboost and SVM have been used for detection in several fields, but have not been used together for the application of reducing the reaction time of classification. Using a classification technique used for detection of obstacles to vary the speed of the vehicle to avoid collision is something that has not been in any of the aforementioned literature work.

3. Impact in Obstacle Detection

In the automotive industry automation is becoming a field of at most impportance because there seems to be a need for the automation of several parts in the vehicle. There are several levels of automation, which can vary from assisting the driver in the completion of any task to the extent of automatically driving the entire vehicle. The major constrain in the automation of vehicles is the ability to extract necessary data from the computer vision system that already does exist. Therefore, the field of obstacle detection becomes one of at most importance. Obstacles will include the lamps, sidewalks, vehicles of all kinds, pedestrians and several other things. Therefore, it is required to create a model where the various obstacles can be detected using a computer vision based machine learning technique.
Obstacle detection is about detecting different things that come in the way of the host vehicle. The obstacles may include lampstands, dividers, pedestrians and several types of vehicles. Pedestrian detection is a long-standing problem not only in automotive sector in ADAS but also in robotics. Nowadays, there is a lot of improvement in the field of vehicle detection. It has further been taken into dividing further into several other types of vehicles. Thus, the machine learning will have an effect on the detection of several objects. Sensor fusion is another field that will have greater impact on this.

Frame capture is done from the existing vehicle videos. The images required for the dataset need to be generated by this process only. Based on the annotation that has been done on the videos, a csv file is created. Using this csv file the images are cropped out of the respective videos. It is required to separate the positives and negatives in these images. These images are used as positive and negative images while training the adaboost or the SVM model. The positive and negative images need to be generated for both the far region and middle region based on the requirement. The far region needs to be in HD as the clarity needs to be high when the distance is more as there is a need for more pixels in the same screen size. Whereas for the middle region we need only VGA images as the IHOG classification can be done for lesser pixels as the distance is less in comparison with far region.

4. Pre-processing and Feature Extraction

There are certain pre-processing steps that need to be taken into account whenever there is a need for classification. These pre-processing steps need to done only once and need not be done every time the code is run for the classification of vehicles and in the real time detection. The three main pre-processing steps that are used before the implementation of the training process using the given set of images are mentioned below. The cropping of images needs to be done in the respective regions in order to generate the required images in the required resolution and the correct detection. Next the resizing of the cropped images needs to be done in order to make all the images in the same region of the same region. This will help in creating a particular uniformity within the images, thereby making the classification region. The Contrast Limited Adaptive Histogram of Variants (CLAHE) needs to be done in order to increase the contrast but within a limit in order to make the images seem similar in both the day and night times. The contrast variation using CLAHE will help in making the edges more prominent in order to make the IHOG feature extraction easier. The above-mentioned pre-processing steps are more or less same in both the far and middle regions.

5. Feature Classification

Classification is done in order to help any system to know which particular class the given sample falls into. Therefore, while using certain machine learning techniques the classification of data can be done with a greater extent of ease. In this method we are using mainly two classifications they are Adaboost and Support Vector Machine.

5.1. AdaBoost

Adaboost is a week classifier where it is having 100 iteration. The output of the adaboost is having 5 set of values they are alpha, direction, dimension, threshold and error. The output of the Adaboost is given as the input to the Support Vector Machine classifier. SVM being a strong classifier where it will actually classify the window which contains vehicle and non-vehicle. AdaBoost is the best used algorithm in boosting the performance of any decision tree when implemented on any binary classification related problems. Adaboost is used to increase the efficiency of the system by reducing the reaction time of the system. It can be used to boost any particular machine learning algorithm. In our case we are using adaboost along with Support Vector Machine algorithm.

Error is calculated by calculating the difference between the correct classification and the total number of classifications and then normalizing it with respect to the total sample.

$$\text{error} = (\text{correct} - N) / N \tag{1}$$

Error can also be called as misclassification rate. This misclassification rate needs to be reduced as much as possible. If we take for example the number of correct classifications to be 72, then the misclassification rate can be calculated as: ((72-100))/100 which will come to 0.28 or 28 percentage. This is how the misclassification rate is calculated.

There are certain changes that are carried over to make it feasible for all the available training instances. The equation after making the changes are mentioned below:

$$\text{error} = \text{sum}(w(i) * \text{terror}(i)) / \text{sum}(w) \tag{2}$$

The adaboost model file that was used after repeated alterations done to the dataset and after balancing of the positives and negatives so that the detection has increased is mentioned below:

{1.024641, 120, 0.015100, -1, 0.114125}, {0.892237, 1472, 0.004014, -1, 0.114125}, {0.678667, 1411, 0.135415, 1, 0.059877}, {0.693005, 1224, 0.006608, -1, 0.054248}, {0.651721, 527, 0.045992, -1, 0.041965}, {0.681770, 1403, 0.289748, 1, 0.038895}, {0.581958, 1014, 0.001396, -1, 0.031218}, {0.611232, 1327, 0.050055, 1, 0.030194}, {0.547301, 113, 0.034173, -1, 0.024565}, {0.559930, 1488, 0.003952, -1, 0.020471}, {0.593717, 25, 0.096921, 1, 0.015353}, {0.565087, 424, 0.118433, -1, 0.014841}, {0.598132, 989, 0.030135, -1, 0.013306}, {0.597178, 344, 0.021429, -1, 0.011771}, {0.504489, 332, 0.098198, 1, 0.009724}, {0.591283, 1432, 0.003367, -1, 0.007165}, {0.472304, 604, 0.162644, 1, 0.006141}, {0.496130, 395, 0.173487, -1, 0.005118}, {0.547841, 1192, 0.001335, -1, 0.003582}, {0.534296, 1344, 0.029758, 1, 0.002559}, {0.484472, 935, 0.133996, 1, 0.002047}, {0.545644, 1215, 0.028317, -1, 0.001024}, {0.529149, 595, 0.282335, 1, 0.000512}, {0.515515, 200, 0.142500, -1, 0.000000}, {0.480180, 1495, 0.000677, -1, 0.000512}, {0.443694, 1542, 0.009537, 1, 0.000512}, {0.520398, 180, 0.181511, 1, 0.000512}, {0.480547, 222, 0.071252, 1, 0.000000}, {0.483433, 426, 0.015580, -1, 0.000000}, {0.531209, 993, 0.037307, -1, 0.000000}, {0.484859, 1351, 0.017728, 1, 0.000000},{0.501244, 866, 0.096911, 1, 0.000000}

5.2. Support Vector Machine

The Support Vector Machine is a machine learning technique that is used to classify any particular object based on the attributes that are extracted. There will be certain features that are extracted, these features need to be passed into the SVM module in order to detect the necessary object. Support Vector Machine algorithm makes use of a hyperplane in order to segregate or classify the given sample into its particular class. The number features and the respective attributes need to be specified in order to increase the value of each feature and to make the detection process more precise.

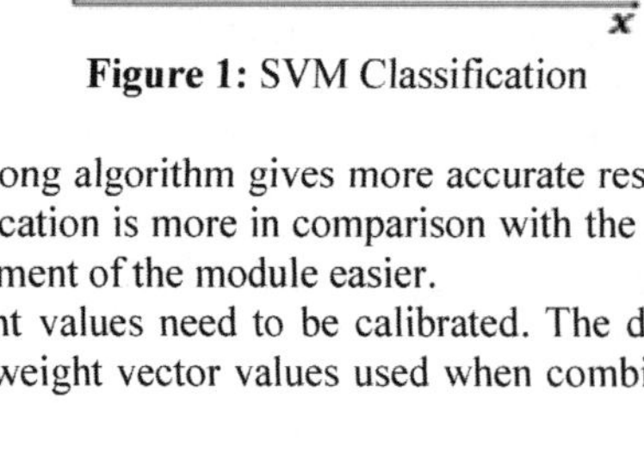

Figure 1: SVM Classification

Support Vector Machine being a strong algorithm gives more accurate results, but the limitation of this process is that the time taken for the classification is more in comparison with the other weaker classifiers. SVM library is used in order to make the development of the module easier.

The Support Vector Machine weight values need to be calibrated. The detection accuracy will vary based on how these weight vectors are. The weight vector values used when combining both adaboost and SVM for the optimal dataset is mentioned below:

IOP Conf. Series: Materials Science and Engineering **624** (2019) 012031 doi:10.1088/1757-899X/624/1/012031

{28351,441,13439,3901,15736,27209,8210,7333,4248,9666,4414,-1813,1768,11620,15631,-980,1386,8228,1227,1666,-5103,8305,7365,-8560,130,-301,5997,-1069,-5555,-2748,8111,7042,3125,10926,13769,-430,305,11273,8902,-1431,-475,-3536,1509,-8180,375,-2570,6844,-9395,579,2650,6104,-1660,-5967,-2827,-8678,-9503,6116,1476,-2477,-1748,-7548,-2552,10781,5,186,-606,5546,-5848,3427,1636,-628,-11708,-3299,9602,8666,-1761,-3118,4890,-3821,-7936,7684,2986,-42,-536,-6725,2834,717,4583,13106,12981,6155,10378,7652,8799,-3811,-12364,-2228,7166,7797,-2501,-1742,6693,4391,-3990,-1540,-1112,15775,1489,511,10270,-5335,2833,10703,14321,11953,10069,5311,1245}

6. Merging of Adaboost and Support Vector Machine

In order to reduce the reaction time taken by the entire system adaboost and SVM are combined together. When adaboost is used the classification accuracy is reduced. Therefore, the number of iterations that are used in adaboost are altered and the detection varies accordingly.

The number of iterations is calculated by keeping track of the error percentage. The number of iterations will freeze at a point where the error percentage is as low as possible and is saturated at the same time.Once the adaboost training is done, only the positive classification from the adaboost model is sent for Support Vector Machine classification. Since Support Vector Machine is a strong classifier it takes more time. Adaboost will take less time, therefore when we limit the number of inputs sent for Support Vector Machine classification the reaction time will reduce to a greater extent.

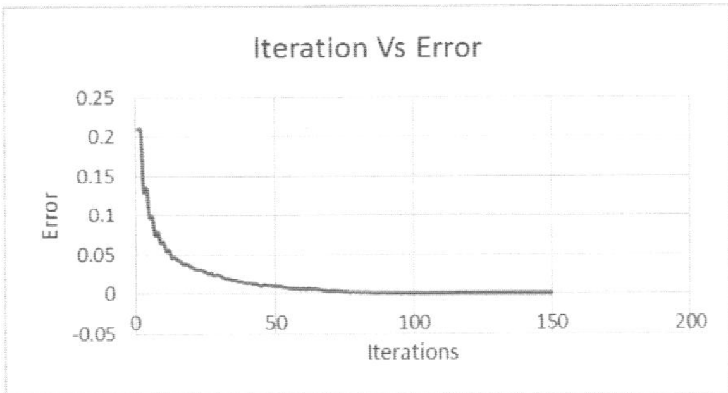

Figure 2: Iteration Vs Error Graph for Adaboost

7. Multiple Box Reduction and Evaluation

In the segmentation window every box that has the commercial vehicle in it will be displayed. In this there will be multiple boxes and will increase the density of boxes in the areas where there are vehicles present. When the density of boxes is more in one particular region, the merging of bounding boxes is done. This process is known as non-maximal suppression. This non-maximal suppression reduces the boxes if the box density in a particular region is high. Thus, the multiple boxes in the same region can be reduced to one single box that is around the vehicle.

It is required to evaluate the model after creation of it to get the desired results. Accuracy and precision are calculated in order to find out the extent to which the positives and negatives are being classified by the system. Accuracy is calculated with respect to false positives and precision is calculated with respect to false negatives.

8. Impact on Power Transmission of Vehicles

At the most basic level, automatic transmission is a simple computer. They will take in inputs such as the gas pedal position, vehicle speed and road speed, and use a basic lookup table to decide which gear to be in. Some of the older transmissions work on just two variables- generally it will shift into a higher gear as the pedal position is reduced or speed increases.

More modern, ECU-controlled transmissions being advanced are more sophisticated. More variables are being used to decide when to shift gears, including engine temperature, whether the vehicle is towing or not, and often the manual override inputs from the driver. Many of these systems are also capable of 'learning' the driver's style of driving the vehicle, by remembering the general trend of how vigorously the driver normally drives the vehicle. If the driver is calm and composed, it will tend to shift up earlier to optimize fuel economy, at the expense of performance. A rasher driving style will make the transmission to tend to hold lower gears for longer periods of time, thereby improving responsiveness.

Therefore, the data from the power transmission needs to be transmitted to a server and it needs to be monitored continuously over a larger period of time. The dta needs to be transferred in packets in order to make sure that the transmission monitoring process is carried out in a systematic manner. The speed variation needs to be achieved based on the detection of vehicles and the distance at which the vehicle is present from the host vehicle. Another parameter that needs to be taken into consideration is the relative speed of the vehicle. With the variation in the relative speed the amount of speed reduction that is required will also get affected drastically. A threshold value needs to be set in order to maintain the safety involved in the system on the whole. Based on the threshold value the transmission rate also needs to vary. Therefore, the prediction from the SVM classifier can be directly related to the transmission rate.

9. Conclusion

Thus the detection of various obstacles is achieved using a combination of both adaboost and support vector machine algorithm, based on which there has been a reduction in the reaction time taken for classification of various objects. Based on the data of the distance that has been computed in the classification process and the speed of the vehicle and the obstacle, the estimated time for collision can be calculated. This time is made use of to create a speed reduction system in order to reduce the speed of the power transmission through the electronic control unit. A dedicated controller module is used to facilitate this process in order to make sure that the speed reduction takes place faster. Thus the speed reduction of the power transmission can be achieved at a faster rate by making use of the above mentioned classification methods.

References

[1] Hao Ye ; Ye Geoffrey Li ; Biing-Hwang Fred Juang, "Deep Reinforcement Learning for Resource Allocation in V2V Communications", IEEE Vehicular Technology Society, 04 February 2019.

[2] Dietrich KnodlerWerner BoesAnton OttKurt O. Venger, "Torque-control system for automatic speed changer shiftable under load",Google Patents, 14 March, 2019.

[3] Mohammad Mahdi Moghimi, Maryam Nayeri, Majid Pourahmadi and Mohammad Kazem Moghimi, "Moving Vehicle Detection Using AdaBoost and Haar-Like Feature in Surveillance Videos", Journal of Transportation Technologies, 2011.

[4] Naveen Chintalacheruvu and Venkatesan Muthukumar "Video Based Vehicle Detection and Its Application in Intelligent Transportation Systems", Journal of Transportation Technologies, 2012, 2, 305-314.

[5] Raad Ahmed Hadi, Ghazali Sulong and Loay Edwar George, "Vehicle detection and tracking techniques: A concise review", An International Journal (SIPIJ) Vol.5, No.1, February 2014.

[6] David Beymer, Philip McLauchlan, and Jitendra Malik, "A Real-Time Computer Vision System for Vehicle Tracking and Traffic Surveillance", Transportation Research-C, 1998.

[7] Swathy M., Nirmala P. S., Geethu P. C., "Survey on Vehicle Detection and Tracking Techniques in Video Surveillance", International Journal of Computer Applications, February 2017.

[8] Nachwa Abou Bakr, James Crowley, "Histogram of Oriented Depth Gradients for Action Recognition", INRIA, 29 Jan 2018.

[9] Ryan P. Avery, Yinhai Wang, and G. Scott Rutherford, "Length-Based Vehicle Classification Using Images from Un-calibrated Video Cameras", U.S. Department of Transportation, 2002.

[10] Volkswagen Ag, "Automated friction clutch controlling method for motor vehicle, involves adapting final set point for rotational speed gradient of drive-side element to speed gradient of output side element for completion of closing process", 25-10-2010.

[11] Zf Friedrichshafen, "Method for operating a transmission device with a plurality of friction-locking and positive-locking shifting elements", Google Patents, 02 Feb, 2012.

IOP Conf. Series: Materials Science and Engineering **624** (2019) 012032 doi:10.1088/1757-899X/624/1/012032

Vibration Analysis of Heterogeneous Gearbox Faults using EMD Features and SVM Classifier

Setti Suresh[1] **and VPS Naidu**[2]
[1]School of Avionics, Institute of Science and Technology, JNTUK, Kakinada, India.
[2]Multi Sensor Data Fusion Lab, National Aerospace Laboratories, Bangalore, India.

Email: settisuresh1@gmail.com

Abstract. Gearbox is one of the important mechanical power transmission device most commonly used in automobiles and industries to get the desired change in speed and torque. The gearbox fault diagnosis has given utmost importance for its significance in preventing halts of a mechanical system and guaranteeing an advantage of sufficient maintenance. This paper presents the vibration analysis of heterogeneous gearbox faults using EMD features and SVM classifier. The vibration signal is converted into intrinsic mode functions (IMF) with decreasing order of frequencies using empirical mode decomposition (EMD) method. Feature vector consisting of information theoretic features have been computed for each IMF and concatenated to form a feature set. By using random permutations, the feature set has been divided into training and testing sets. The support vector machine (SVM) algorithm has been used as a classification technique to diagnose the gearbox faults, which consists of five-class classification. The accuracy of the developed algorithm has been validated using 100 Monte Carlo runs. A comparative study has been carried between computed features and varying IMF components. The observations made were - clear discrimination of the gearbox faults and improved classification accuracy, which contain - chipped tooth, missing tooth, root fault, surface fault and healthy working state of the gear.

1. Introduction

Gearbox is a rotary complex system, which aims at effective mechanical power transmission in terms of speed and torque. It consists of different gear mechanisms embedded together to achieve desired mechanical benefits. As it is such a complex mechanical system, it is common to experience different faults associated with gear. Several qualitative and quantitative research has carried out from decades to bring the abnormalities in the gear, which lead to failure of the gearbox. Still, there is an ambiguity in the fault detection and diagnosis of gearbox due to its diverse applications and innovative gear designs. Therefore, it is important to develop different algorithms using trending mathematical computations to detect and diagnose the gearbox faults.

The condition monitoring of rotating machines was attained by measurements of vibratory, thermal, acoustics and oil-based signals. Among them, vibration analysis is the most widely used method, as the measured signal consists of information regarding the cause of vibration and characteristic vibrations of rotating components [1]. When there is a fault in the rotating machinery, the characteristics of acquired vibration signal changes. The vibrations measured from the gearbox are complicated that contain composite signal information, which needs to be analysed through different fault diagnosis methods to state the condition of the gearbox. Over the past decades, many researchers and scholars have done a pivoting role in implementing signal-processing techniques to diagnose the different types of gearbox faults. The ideology of dividing the original signal into local windows, then computing the characteristic features of tri-domain that include time, frequency and time-frequency features followed by local feature-based gated recurrent unit (LFBGRU) model has been proposed in [2]. The methodology to localize the faulty gear using EMD and the estimation of two new condition indicators (features) on the angular pitch rather than using the entire vibration signal have asserted in [3]. The analogy of pre-processing the signal using differentiation to enhance the frequency and fault diagnosis of bearing using wavelet decomposition through mean and standard deviation features has presented in [4]. There are several other proposed ideas and experimental analyses [6-8] carried out by distinguished researchers with a common aim to detect and diagnose the fault of rotating machinery.

This paper presents a comparative study of vibration analysis of heterogeneous gearbox faults using EMD features and SVM classifier. The proposed workflow sequence contains, dividing the original vibration signal into segments/windows/epochs, pre-processing, application of empirical mode decomposition in the fault diagnosis of rotating machinery and SVM classifier. A comparative study of individual features behaviour and feature combination behaviour with respect to varying IMF components has been presented.

2. Methodology

The proposed workflow sequence involved in the gearbox fault diagnosis is shown in Figure 1. The workflow starts with data acquisition from the gearbox followed by pre-processing the vibration signal. The segmented signal is decomposed into intrinsic mode functions (IMF's) using the EMD method. Further diagnostic features are computed for each IMF to get the discriminative information from the faulty signals. All the computed feature vectors from each IMF append together to form a feature set. To advance the diagnosis of the gearbox from fault detection to fault classification, the feature set is divided into training and testing sets to classify the fault between measured gearbox vibration signals. The training set is used to train the SVM model along with labels and the testing set is used to test the developed SVM model to get the final decision of fault classification. The application of each block in the fault diagnosis of gearbox is illustrated in the following subsections.

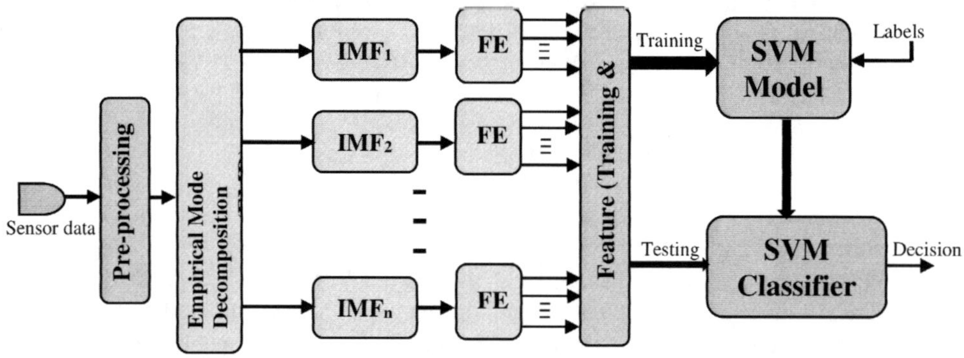

Figure 1. Workflow diagram of gearbox fault diagnosis.

2.1 Data Acquisition

The vibration signals of the gearbox dataset [5] have been collected from Drivetrain Dynamic Simulator (DDS). The dataset consists of two operating conditions with a combination of rotating speed and load configuration set as 20Hz-0V and 30Hz-2V. Each configuration has five-class classification, which include one healthy state and four fault states (chipped tooth, missing tooth, root fault and surface fault) of the gearbox. Each state composed of seven vibration signals (1-motor vibration, 3-planetary gearbox vibration along x, y, z direction, 3-parallel gearbox vibration along x, y, z direction) and one torque sensor data.

2.2 Pre-processing

Pre-processing is an important part in the analysis of complex vibration signals. The original signal is differentiated to enhance the frequency of the signal [4]. In this paper, the acquired signals are segmented over a length of 32768 samples [3]. All the sequence of steps (as shown in Figure 1) involved in the gearbox fault diagnosis after pre-processing are applied to the segmented signal and the information from all such segments are concatenated to get the final decision in fault diagnosis.

2.3 EMD

Huang et al [9] introduced Empirical Mode Decomposition (EMD) for the first time as a data-driven technique and used effectively with non-linear and non-stationary time series data. EMD is a self-adaptive technique, which helps to analyse the signal in both the time and frequency domain. It has applied to decompose the segmented signal into a set of intrinsic mode functions (IMF) based on local time scale characteristics with a decreasing order of frequencies [10].

2.4 Feature Extraction

In general, feature extraction is an act of representing the original signal in terms of reduced dimensionality. In place of analysing the huge length of observations (cause of time lapse), it has converted into features that describe the properties of the original signal. Table 1 shows the expression of computed features for gearbox fault diagnosis and gear wear prediction. Here, each IMF component is converted into a feature vector comprising of estimated features. The significance of each feature in the fault diagnosis of the gearbox has illustrated briefly in the results and discussion section.

Table 1. Computed feature vectors for gearbox fault diagnosis and gear wear prediction.

Features	Mathematical formula
RMS [11]	$X_{rms} = \left[\frac{1}{N}\sum_{i=1}^{N} x_i^2\right]^{\frac{1}{2}}$
Variance [11]	$\sigma^2 = \frac{1}{N-1}\sum_{i=1}^{N}(x_i - \mu)^2$
Energy [7]	$X_{energy} = \sum_{i=1}^{N} x_i^2$
Entropy [7, 11]	$X_h = -\sum_{i=1}^{n} p(z_i)\, log_2 p(z_i)$
Skewness [11]	$X_{skew} = \frac{1}{N}\sum_{i=1}^{N}(x_i - \mu)^3 \cdot \left[\frac{1}{N}\sum_{i=1}^{N}(x_i - \mu)^2\right]^{-\frac{3}{2}}$
Kurtosis [11]	$X_{kurt} = \frac{1}{N}\sum_{i=1}^{N}(x_i - \mu)^4 \cdot \left[\frac{1}{N}\sum_{i=1}^{N}(x_i - \mu)^2\right]^{-2}$

Note: x_i = i^{th} value of IMF; N = length of IMF; $\mu = \frac{1}{N}\sum_{i=1}^{N} x_i$; n = $p_1, p_2, \ldots p_n$ = set of probabilities

2.5 SVM (Feature Classification)

Support vector machine - A modern classification or regression technique which defines a plane and divide the feature space into subspaces to attain minimum classification error has used in the feature classification of gearbox vibration signals. The feature set comprising of feature vector observations of the healthy signal, chipped tooth, missing tooth, root fault and surface fault, of planetary gearbox has divided into 80% of training and 20% as testing sets. The training set is used to train the SVM model and the testing set is used to test the developed SVM model with five-class classification.

IOP Conf. Series: Materials Science and Engineering **624** (2019) 012032 doi:10.1088/1757-899X/624/1/012032

3. Results and Discussions

The proposed workflow was carried under both the configurations (20-0 & 30-2) and random permutations were used, to prepare the feature set into training and testing sets. The SVM model along with the classifier is set under 100 Monte Carlo runs to achieve the consistency in the classification accuracy of gearbox fault diagnosis. Table 2 shows the classification accuracy of the developed SVM model with five-class classification under 20-0 configuration for the planetary gearbox considering original signal. A comparative study of each computed feature, with the variation in the level of IMF components (k) to achieve a theoretical conclusion of gearbox fault diagnosis, is tabulated exclusively. The numerical values in percentage (mean and standard deviation of 100 Monte Carlo runs) denote the SVM classification accuracy with individual features considered and with different feature combinations (FC).

The feature 'energy', which indicates the strength of the signal, alone classifying the gearbox fault signals with a maximum accuracy of 98.18% at k = 6. It is observed that each feature has its own fault classification accuracy at different levels of IMF components i.e., 'entropy', which is a measure of non-linearity of the signal is with a maximum classification accuracy of 98.20% at k=4, 'variance', which denotes the squared deviation of the signal from its mean value having an accuracy of 93.27% at k=7, 'root mean square (RMS)', which aims at measuring the effective value of power content in the vibration signal having a maximum accuracy of 98.18% at k=4, 'kurtosis', which determines the tailedness of the probability distribution (the damaged vibration signal of gearbox contain sharp peaks with higher value, indicating sharp probability distribution) having a better accuracy of 87.18% at k=3, 'skewness', which describes the asymmetry in the distribution of vibration signal having an accuracy of 27.57% at k=2 as highest. It indicates that the skewness feature is unable to classify the gearbox dataset consisting of five-class classification. Further, to improve the classification accuracy, feature combinations has considered. FC_1 denotes the combination of all features considered having a maximum accuracy of 91.00% at k=1, which is ineffective. FC_2 is equal to FC_1 minus skewness feature having an accuracy of 96.64% at k=1 as highest, which is ineffective comparatively with individual features considered. FC_3 is equal to FC_2 minus kurtosis feature showing the highest classification accuracy of 99.03% at k=3, which is considered as achieved maximum successful classification accuracy. FC_4 equal to FC_3 minus variance having an accuracy of 96.62% at k=1. FC_5 equal to FC_4 minus RMS having an accuracy of 98.95% at k=4. Therefore, it can be observed that, with feature combination FC_3 the fault classification accuracy of SVM is improved to 99.03%. Similar case study as explained in Table 2 is carried out for planetary gearbox under 30-2 configuration and observed that, feature combination FC_3 has brought a maximum classification accuracy of 93.56% at k=3 with 100 Monte Carlo runs. Further, to improve the fault classification accuracy, time-derivative of the original signal is carried out and similar case studies as explained in Table 2 were exercised. However, the results shown were not satisfactory for the considered gearbox dataset, assuring a maximum fault classification accuracy of 91.44% at k=5 under 20-0 configuration. Figure 2(a) & 2(b) shows the classification plots of gearbox multi-class faults using SVM algorithm. It can be observed that, skewness and kurtosis features are ineffective in classifying the faults (poor).

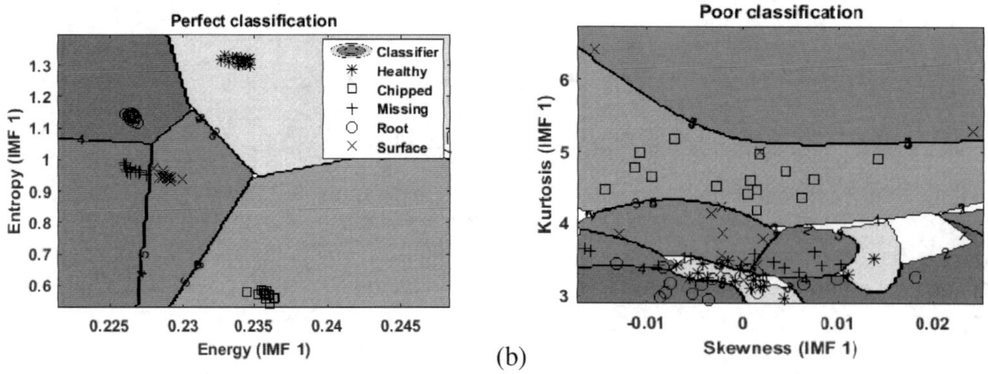

(a) (b)

Figure 2. Gearbox multi-class classification using SVM (a) Perfect, (b) Poor classifications.

Figure 3(a) & 3(b) shows the confusion matrix of the SVM fault classification accuracy under both 20-0 & 30-2 configurations. The five-class classification of known test labels (93 in total) of 20% testing set were mapped with output decision labels of SVM classifier. In confusion matrix 1 it is observed that in the missing tooth, 1-label is showing as healthy signal. In confusion matrix 2, in the root fault 3-labels are observed as a surface fault, 1-label as chipped tooth and 1-label as healthy, in the surface fault 1-label is predicted as root fault. These incorrect classification lead to a fault classification accuracy of 99.9% for 20-0 and 93.5% for 30-2 configuration (with single Monte Carlo run).

Table 2. SVM classification accuracy (%) with 100 Monte Caro runs for planetary gearbox.

Features	Number of IMF's (k)								
	k=1	k=2	k=3	k=4	k=5	k=6	k=7	k=10	k=12
Energy	37.52 ± 4.97	87.05 ± 3.65	89.58 ± 3.91	97.47 ± 1.59	97.40 ± 1.61	**98.18 ± 1.37**	97.28 ± 1.58	82.72 ± 3.69	47.49 ± 5.49
Entropy	66.98 ± 4.29	84.55 ± 4.57	97.95 ± 1.46	**98.20 ± 1.35**	95.31 ± 2.51	94.13 ± 2.50	89.41 ± 3.23	67.01 ± 5.37	49.47 ± 4.60
Variance	53.32 ± 4.77	65.47 ± 5.47	76.76 ± 3.46	84.44 ± 3.93	86.92 ± 4.72	91.49 ± 2.65	**93.27 ± 3.25**	81.37 ± 3.98	70.90 ± 4.52
RMS	53.78 ± 4.51	77.52 ± 4.52	96.67 ± 2.62	**98.18 ± 1.93**	97.98 ± 1.56	97.17 ± 1.51	96.56 ± 1.78	82.65 ± 3.38	68.97 ± 5.32
Kurtosis	59.69 ± 4.53	79.85 ± 4.32	**87.17 ± 3.26**	86.14 ± 3.67	81.98 ± 3.78	78.03 ± 4.15	72.60 ± 4.50	45.56 ± 5.04	30.34 ± 4.67
Skewness	24.26 ± 3.64	**27.57 ± 4.47**	25.00 ± 4.16	24.01 ± 4.23	22.88 ± 4.11	20.94 ± 3.61	19.94 ± 3.76	19.73 ± 3.54	19.44 ± 3.52
FC_1	**91.00 ± 3.63**	83.61 ± 5.22	62.40 ± 6.56	44.14 ± 5.60	24.50 ± 4.33	21.01 ± 3.82	20.03 ± 4.09	19.96 ± 4.30	20.57 ± 3.63
FC_2	**96.64 ± 1.89**	95.97 ± 2.67	93.04 ± 2.74	89.20 ± 3.28	80.22 ± 4.07	73.61 ± 4.70	55.41 ± 5.22	19.74 ± 3.34	19.88 ± 3.45
FC_3	84.38 ± 4.07	98.65 ± 0.99	**99.03 ± 0.98**	98.22 ± 1.33	97.99 ± 1.18	96.89 ± 1.51	86.87 ± 3.40	23.96 ± 3.91	23.24 ± 3.96
FC_4	**96.62 ± 1.67**	96.40 ± 2.19	93.67 ± 3.14	88.95 ± 3.12	81.82 ± 4.27	74.71 ± 4.97	59.18 ± 4.92	40.47 ± 4.15	21.25 ± 3.57
FC_5	75.61 ± 4.62	98.36 ± 1.27	98.45 ± 1.14	**98.95 ± 1.02**	97.94 ± 1.46	97.61 ± 1.23	94.43 ± 2.37	49.30 ± 4.54	26.03 ± 4.48

Note: FC = Feature combination;

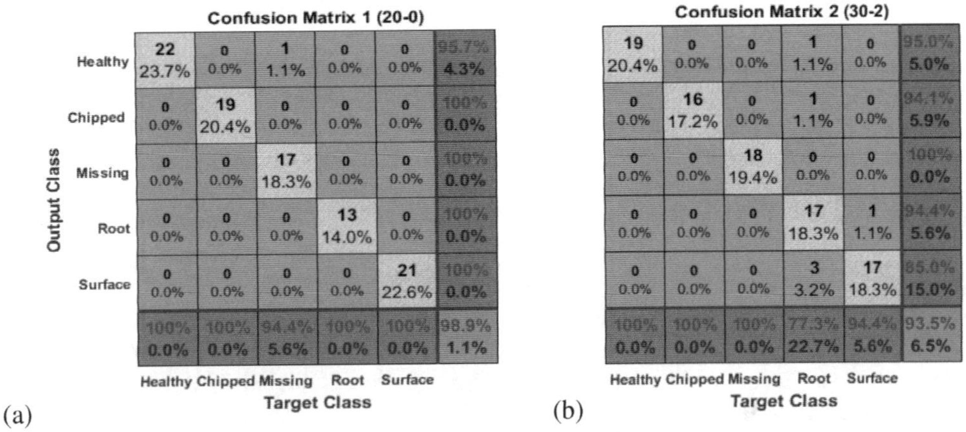

Figure 3. Confusion matrix of SVM classification accuracy for (a) 20-0, (b) 30-2 rotating speed and load configurations.

Experimental results are presented in Table 3. From the results, it has observed that the proposed method shows the improved fault classification accuracy compared with LFBGRU method in [2]. Even though the achieved classification accuracy is less for 30-2 configuration, it is elevated as a consistent result tested with 100 Monte Carlo runs (*).

Table 3. Classification results of the gearbox dataset.

Fault Diagnosis Method	Gearbox (Rotating speed – Load configuration)	
	20-0	30-2
LFBGRU [2]	**94.8%**	**95.8%**
Proposed method	**99.09 ± 0.98***	**93.56 ± 2.33***

4. Conclusion

A new method based on EMD features and SVM classifier has been proposed to classify the heterogeneous gearbox faults. The EMD technique was implemented to decompose the segmented signal into IMF components, which contain different characteristic frequencies. The statistical features have been computed on each IMF to provide discriminative information to the classifier in the diagnosis of gearbox faults. All the feature vectors were concatenated to form a feature set and divided into training and testing set. The developed SVM model using the training set has tested using the testing set with 100 Monte Carlo runs. A comparative study has been carried out with the computed features (individually and selected feature combinations) to improve the fault classification accuracy. The result shows that the proposed method is effective in classifying the heterogeneous gearbox faults.

5. References

[1] Ilhan Asilturk ,Halit Aslanci ,Ugur Ozmen , (2017) " Machinery Monitoring Using Vibration Signal Analysis " , International Journal of Mechanical and Production Engineering (IJMPE) , pp. 21-25, Volume-**5**, Issue-2.

[2] R. Zhao, D. Wang, R. Yan, K. Mao, F. Shen and J. Wang, "Machine Health Monitoring Using Local Feature-Based Gated Recurrent Unit Networks," in IEEE Transactions on Industrial Electronics, vol. **65**, no. 2, pp. 1539-1548, Feb. 2018.

[3] Buzzoni, M., Mucchi, E., D'Elia, G., & Dalpiaz, G. (2017). "Diagnosis of Localized Faults in Multistage Gearboxes: A Vibrational Approach by Means of Automatic EMD-Based Algorithm", *Shock and Vibration*, 2017, 1–22.

[4] Shanmukha Priya, V, Mahalakshmi P & VPS Naidu, (2015). "Bearing health condition monitoring: Wavelet decomposition". *Indian Journal of Science and Technology,* 8.10.17485/ijst/2015/v8i26/81712.

[5] https://github.com/cathysiyu/Mechanical-datasets [Accessed on 20-03-2019].

[6] Junsheng, Cheng & Dejie, YU & Tang, Jiashi & Yang, Yu. (2009). Application of SVM and SVD Technique Based on EMD to the Fault Diagnosis of the Rotating Machinery. *Shock and Vibration*, **16**. 89-98.

[7] Yu, Y., YuDejie, & Junsheng, C. (2006). A roller bearing fault diagnosis method based on EMD energy entropy and ANN. *Journal of Sound and Vibration*, **294(1-2)**, 269–277.

[8] L. Guan, Y. Shao, Fengshou Gu, B. Fazenda and A. Ball, "Gearbox fault diagnosis under different operating conditions based on time synchronous average and ensemble empirical mode decomposition," *2009 ICCAS-SICE*, Fukuoka, 2009, pp. 383-388.

[9] N. E. Huang, Z. Shen, S. R. Long et al., "The empirical mode decomposition and the Hilbert spectrum for nonlinear and non-stationary time series analysis," The Royal Society of London. Proceedings. Series A. *Mathematical, Physical and Engineering Sciences*, vol. **454**, no. 1971, pp. 903–995, 1998.

[10] Klionski, D. M., Oreshko, N. I., Geppener, V. V., & Vasiljev, A. V. (2008). "Applications of empirical mode decomposition for processing nonstationary signals", Pattern Recognition and Image Analysis, **18(3)**, 390–399.

[11] Caesarendra, Wahyu & Tjahjowidodo, Tegoeh. (2017). A Review of Feature Extraction Methods in Vibration-Based Condition Monitoring and Its Application for Degradation Trend Estimation of Low-Speed Slew Bearing. *Machines*. 5. 21. 10.3390/machines5040021.

IOP Conf. Series: Materials Science and Engineering **624** (2019) 012033 doi:10.1088/1757-899X/624/1/012033

Contact analysis of a planetary gear train using linear complementarity

M Pathak and S Rakshit

Mechanical Engineering Department, IIT Madras, Chennai, India - 600036

E-mail: srakshit@iitm.ac.in

Abstract. The current computation models for gear contact analysis and wear prediction are mostly based on finite element principles which consumes a lot of computation time and effort. In this paper, an alternate approach of linear complementarity is used for gear contact analysis. This approach was successfully applied for a pair of rigid spur gears in our previous paper. In this paper, we are extending this to a planetary gear train. The linear complementarity problem formulation is revised to take bilateral constraints into consideration. A linear complementarity solver computes the contact forces between meshing teeth of gears in gear train. From the contact forces, sliding wear in gear teeth is predicted. Archard's wear model is used for the wear prediction.

1. Introduction

Gear trains are one of the effective and popular means for power and motion transmission between shafts. Contact analysis is an essential part to evaluate life expectancy, safety parameters and integrity of gear trains. Currently, two types of analysis models are being followed : 1) Lumped Parameter Model [1] 2) Finite Element Model [2]. Gear train is a multi-body system with multiple contact configurations that cannot be solved using equations of motion alone. Hence, we have used linear complementarity [3] for doing multiple contact analysis in gear trains. This approach was successfully applied to a pair of rigid spur gears in our previous work and the results were compared with those from finite element model [4]. The contribution of current work is twofold. First, we have extended our linear complementarity analysis of a pair of meshing spur gears to a more complicated case of planetary gear train. Second, we have incorporated the reaction forces at bearings as bilateral constraints in linear complementarity problem (LCP) formulation. From linear complementarity analysis, we have calculated the contact forces at meshing teeth in gear train. From the contact forces, sliding wear in gear teeth is predicted using Archard's wear model [5].

2. Method

The first step in contact analysis is to create kinematic model for determining the contact points along path of contact of meshing gears and second step is the linear complementarity analysis to get the contact forces at the contact points.

IOP Conf. Series: Materials Science and Engineering **624** (2019) 012033 doi:10.1088/1757-899X/624/1/012033

2.1. Kinematic Model

For a pair of mating spur gears, contact points always lie on the line of action. This line of action is fixed for a given gear pair owing to the involute profile of gear tooth. Sometimes there is only one tooth in contact and other times there are two teeth in contact. Thus, the contact ratio is in between 1 and 2.

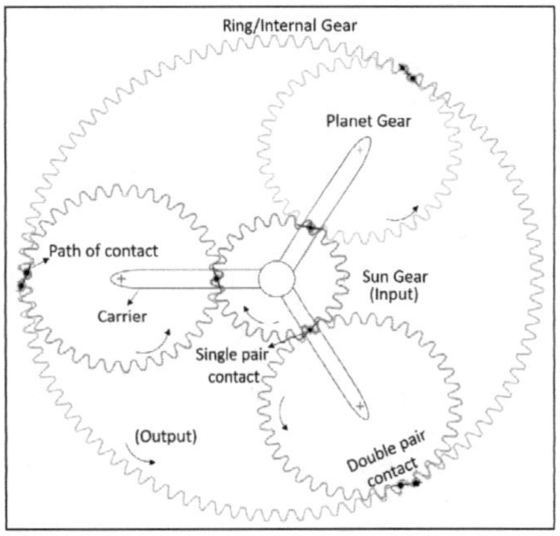

Figure 1. Kinematic model of planetary gear train

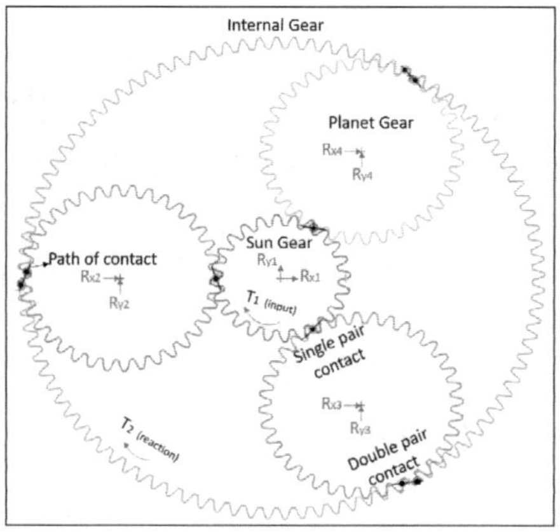

Figure 2. Input and output torques for linear complementarity model

The kinematic model for planetray gear train is shown in Fig. 1. The input motion is given to sun gear. The carrier arm is fixed and the output is ring gear. The equations for the involute profile of spur gears and internal gear are used for the generation of exact gear tooth geometry [6].

2.2. Linear Complementarity Model

LCP consists of a set of non-negative variables that are pairwise complementary (i.e. either one variable is positive and the other is zero or vice versa) and are related by a set of linear equations. According to Anitescu-Potra's model [7], for every contact point j, the normal contact velocity v_j and the normal contact force F_j form a complementary pair, i.e., $F_j \geq 0$, $v_j \geq 0$, and $F_j v_j = 0$. This statement is also represented as $0 \leq F_j \perp v_j \geq 0$. Since v_j and F_j are linearly related, $\sum_{j=1}^{n} F_j v_j = \mathbf{F}^T\mathbf{v} = \mathbf{F}^T(\mathbf{AF}+\mathbf{b}) = 0$. where n is the number of contact points, $\mathbf{A} \in \mathbb{R}^{n\times n}$ is a positive semidefinite matrix having information on body mass and inertia and $\mathbf{b} \in \mathbb{R}^n$ is a vector consisting of all external forces and body forces.

If friction need to be considered into the model, a tangential component of force and velocity at every contact point also need to be considered. The magnitude of the friction force at contact point j is linearly related to the normal force by Coulomb's law. But this lead to non-linear complementarity problems which are difficult to solve. Hence the polyhedral approximation of the friction cone is considered. For two dimensional case, the tangent space is a line separating two bodies at contact point [7], [8].

The Anitescu-Potra's model [7] gives a complementarity formulation for rigid body contact problems with dynamic friction that can be solved using Lemke's algorithm [3]. In Anitescu-Potra's model, at each contact point j, the complementary constraints are:

$$0 \leq v_{n_j} \quad \perp \quad p_{n_j} \geq 0 \tag{1}$$

$$\mathbf{0} \leq \boldsymbol{\rho}_j = \lambda_j \mathbf{e}_j + \mathbf{W}_{f_j}^T \mathbf{v} \quad \perp \quad \mathbf{p}_{f_j} \geq \mathbf{0} \tag{2}$$

$$0 \leq \zeta_j = \mu p_{n_j} - \mathbf{e}_j^T \mathbf{p}_{f_j} \quad \perp \quad \lambda_j \geq 0 \tag{3}$$

where $\perp$ indicates the complementary variables, v_{n_j} $(= \mathbf{c}_{n_j}^T \mathbf{v})$ is the normal velocity between the two bodies at contact point j, p_{n_j} is the normal component of the contact impulse, $\mathbf{p}_{f_j}$ is the friction component of the contact impulse, $\mathbf{e}_j \in \mathbb{R}^D$ is a vector of ones, D is the number of edges of the friction pyramid, λ_j is a variable that approximates the magnitude of the sliding velocity, μ is the Coulomb friction coefficient, n_d is the dimension of the configuration space, and $\mathbf{W}_{f_j} \in \mathbb{R}^{n_d \times D}$ is the friction contact wrench matrix that transforms the generalized velocity $\mathbf{v} \in \mathbb{R}^{n_d}$ of the body along the tangential component of the friction cone. For a two-dimensional problem ($n_d = 3$, two translations and one rotation), at each contact point j, there are only two directions $\{\mathbf{t}_j, -\mathbf{t}_j\}$ of the friction component of the contact impulse, each perpendicular to the outward contact normal $\mathbf{c}_{n_j}$. The friction contact wrench matrix $\mathbf{W}_{f_j}$ is $\begin{bmatrix} \mathbf{t}_j & -\mathbf{t}_j \\ \mathbf{r}_j \otimes \mathbf{t}_j & -\mathbf{r}_j \otimes \mathbf{t}_j \end{bmatrix}$, where $\mathbf{r}_j$ is the vector from the body centroid to contact point j, and $\otimes$ is the two-dimensional equivalent of cross-product. In addition to these complementary constraints, we have equations of motion for the system:

$$\tilde{\mathbf{M}}\left(\tilde{\mathbf{v}}^t - \tilde{\mathbf{v}}^{t-1}\right) - \tilde{\mathbf{J}}^T \tilde{\boldsymbol{\nu}}^t - \tilde{\mathbf{W}}_n \tilde{\mathbf{p}}_n^t - \tilde{\mathbf{W}}_f \tilde{\mathbf{p}}_f^t = \left(\tilde{\mathbf{F}}_b + \tilde{\mathbf{F}}_{ext}\right) \boldsymbol{\Delta} t \tag{4}$$

$$\tilde{\mathbf{J}}\tilde{\mathbf{v}}^t = 0 \tag{5}$$

where superscript t indicates value at time t. $\tilde{\mathbf{p}}_n = [p_{n_1}, \ldots, p_{n_K}]^T$ is the concatenated normal contact impulses for all the K contact points in the assembly. Similarly $\tilde{\mathbf{p}}_f = [\mathbf{p}_{f_1}, \ldots, \mathbf{p}_{f_K}]^T$ is the concatenated friction impulses. $\tilde{\mathbf{M}}$ is the concatenated mass matrix, $\tilde{\mathbf{W}}_n = [\mathbf{W}_{n_1}, \ldots, \mathbf{W}_{n_K}]$ is the concatenated normal contact wrench matrix where $\mathbf{W}_{n_j}^T = \begin{bmatrix} \mathbf{c}_{n_j}^T & \mathbf{r}_j \otimes \mathbf{c}_{n_j} \end{bmatrix}$, $\tilde{\mathbf{W}}_f = [\mathbf{W}_{f_1}, \ldots, \mathbf{W}_{f_K}]$ is the concatenated friction wrench matrix, $\tilde{\mathbf{v}}^t$ is the concatenated generalized velocity at time t, Δt is the time step of simulation, $\tilde{\mathbf{F}}_{ext}$ is the concatenated external generalized forces (input torques) for all the bodies in the system and $\tilde{\mathbf{F}}_b$ is the concatenated body forces, $\tilde{\mathbf{J}}$ is the concatenated constraint jacobian matrix of the system. A gear rotating on its bearing is modelled as a gear hinged at its center and the corresponding constraint jacobian matrix becomes,

$\mathbf{J} = \begin{bmatrix} 1 & 0 & 0 \\ 0 & 1 & 0 \end{bmatrix}$. $\tilde{\boldsymbol{\nu}}$ is the concatenated constraint reaction impulses for all the parts. Thus the linear complementarity model [7] for multibody-multiple contact problems can be written as:

$$0 \le \begin{bmatrix} \tilde{\mathbf{A}}_{nn} & \tilde{\mathbf{A}}_{nf} & \mathbf{0} \\ \tilde{\mathbf{A}}_{fn} & \tilde{\mathbf{A}}_{ff} & \tilde{\mathbf{E}} \\ \tilde{\boldsymbol{\mu}} & -\tilde{\mathbf{E}}^T & \mathbf{0} \end{bmatrix} \begin{bmatrix} \tilde{\mathbf{p}}_n^t \\ \tilde{\mathbf{p}}_f^t \\ \tilde{\boldsymbol{\lambda}}^t \end{bmatrix} + \begin{bmatrix} \tilde{\mathbf{b}}_n^t \\ \tilde{\mathbf{b}}_f^t \\ \mathbf{0} \end{bmatrix} = \begin{bmatrix} \tilde{\mathbf{v}}_n^t \\ \tilde{\boldsymbol{\rho}}^t \\ \tilde{\boldsymbol{\zeta}}^t \end{bmatrix} \perp \begin{bmatrix} \tilde{\mathbf{p}}_n^t \\ \tilde{\mathbf{p}}_f^t \\ \tilde{\boldsymbol{\lambda}}^t \end{bmatrix} \ge 0 \tag{6}$$

where $\tilde{\boldsymbol{\lambda}} = [\lambda_1, \ldots, \lambda_K]^T$ are concatenated λs for all the K contact points in the assembly, and $\tilde{\mathbf{A}}_{nn} = \tilde{\mathbf{W}}_n^T \tilde{\mathbf{W}} \tilde{\mathbf{W}}_n$, $\tilde{\mathbf{A}}_{nf} = \tilde{\mathbf{W}}_n^T \tilde{\mathbf{W}} \tilde{\mathbf{W}}_f$, $\tilde{\mathbf{A}}_{ff} = \tilde{\mathbf{W}}_f^T \tilde{\mathbf{W}} \tilde{\mathbf{W}}_f$, $\tilde{\mathbf{A}}_{fn} = \tilde{\mathbf{W}}_f^T \tilde{\mathbf{W}} \tilde{\mathbf{W}}_n$, $\tilde{\mathbf{b}}_f = \tilde{\mathbf{W}}_f^T \tilde{\mathbf{W}}\left(\left(\tilde{\mathbf{F}}_b + \tilde{\mathbf{F}}_{ext}\right)\Delta t + \tilde{\mathbf{M}}\tilde{\mathbf{v}}^{t-1}\right)$, $\tilde{\mathbf{b}}_n = \tilde{\mathbf{W}}_n^T \tilde{\mathbf{W}}\left(\left(\tilde{\mathbf{F}}_b + \tilde{\mathbf{F}}_{ext}\right)\Delta t + \tilde{\mathbf{M}}\tilde{\mathbf{v}}^{t-1}\right)$, $\tilde{\mathbf{W}} = \tilde{\mathbf{M}}^{-1}\left(\tilde{\mathbf{I}} - \tilde{\mathbf{J}}^T \tilde{\mathbf{G}}^{-1} \tilde{\mathbf{J}} \tilde{\mathbf{M}}^{-1}\right)$, $\tilde{\mathbf{G}} = \tilde{\mathbf{J}} \tilde{\mathbf{M}}^{-1} \tilde{\mathbf{J}}^T$, $\tilde{\boldsymbol{\mu}} = diag\,[\mu_1, \ldots, \mu_K]$, $\tilde{\mathbf{E}} = diag\,[\mathbf{e}_1, \ldots, \mathbf{e}_K]$, $\tilde{\mathbf{v}}_n = [v_{n_1}, \ldots, v_{n_K}]^T$, $\tilde{\boldsymbol{\rho}} = [\boldsymbol{\rho}_1, \ldots, \boldsymbol{\rho}_K]^T$, $\tilde{\boldsymbol{\zeta}} = [\zeta_1, \ldots, \zeta_K]^T$.

In this problem, it is considered that gears are running in steady state, i.e., running at constant velocity and transmitting constant torque. This ensures static equilibrium condition at all contact configurations. Hence, the contact configurations are independent of each other as meshing progresses along the line of contact, and doing simultaneous static contact analysis at contact points is equivalent to continuous contact analysis. The steady state input-output torques are shown in Fig. 2. The $\tilde{\mathbf{F}}_{ext}$ in Eq. (4) for sun gear: $[0 \quad 0 \quad -T_1]^T$, for planet gears:

IOP Conf. Series: Materials Science and Engineering **624** (2019) 012033 doi:10.1088/1757-899X/624/1/012033

$[0 \quad 0 \quad 0]^T$ and for ring gear: $[0 \quad 0 \quad -T_2]^T$. T_1 and T_2 are the respective torque values acting on sun and ring gear.

After each time step Δt, we update the positions of gears, and use the LCP (Eq. (1) to Eq. (6)) to find the contact forces at the current time step. PATH solver [9] is used to solve the LCP.

2.3. Calculation of Wear

After obtaining contact forces, contact pressure values are computed using expression [10],

$$p_H = 2F_{nt}/\pi a_H \tag{7}$$

where p_H is the local contact pressure, F_{nt} is the normal load transmitted at a particular point, a_H is the semi-hertzian contact width. Archard's wear theory [5] is applied to obtain the wear depth for each point on the gear tooth flank. Archard's wear equations give relationship between contact pressure, sliding distance and wear.

$$h/S = k_w p_H \tag{8}$$

where h is the wear depth, S is the sliding distance the point is sliding against the interacting surface and k_w is the dimensional wear constant.

3. Results and Discussion

The gear parameters considered for generating gear geometry are given in Table 1.

Table 1. Details of planetary gear train

Parameter	*Sun*	*Planet*	*Ring*
Profile	Involute	Involute	Involute
Type	Input gear	Idler gear	Output gear
Number of Teeth	24	36	96
Module (mm)	2.5	2.5	2.5
Pressure angle ($degree$)	20	20	20
Pitch circle diameter (mm)	60	90	240
Tip circle diameter (mm)	65	95	235
Base circle diameter (mm)	56.38	84.57	225.53
Mass (mm^2)	2827.43	6361.725	45238.93
Polar moment of inertia (mm^4)	1.27e+6	6.44e+6	3.26e+8
Face Width (mm)	15	15	15
Modulus of Elasticity (N/mm^2)	2.06e+5	2.06e+5	2.06e+5
Poisson's ratio	0.3	0.3	0.3
Wear coefficient (mm^2N^{-1})	5e-10	5e-10	5e-10
Torque (Nmm)	3.75e+4	-	1.5e+5
Speed (rpm)	1500	1000	375
Coulomb friction coefficient	0.2	0.2	0.2

In linear complementarity model, multiple static contact analysis are performed for various contact points in the path of contact. Fig. 3a shows the contact force is shared equally between

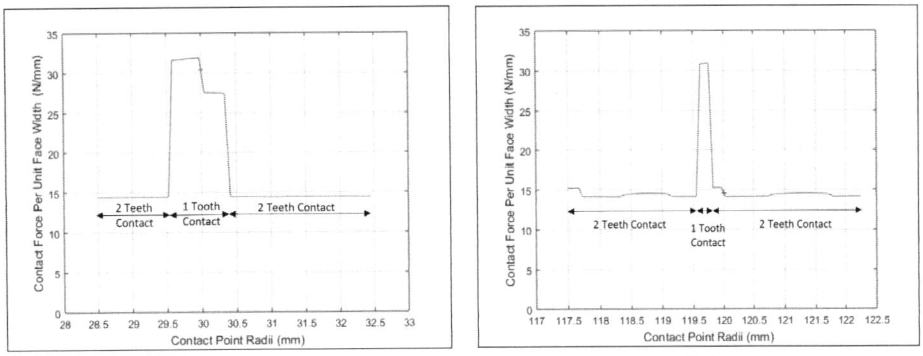

(a) (b)

Figure 3. Load on gear tooth per unit face width vs contact point radii, (a): Sun gear (b): Ring gear

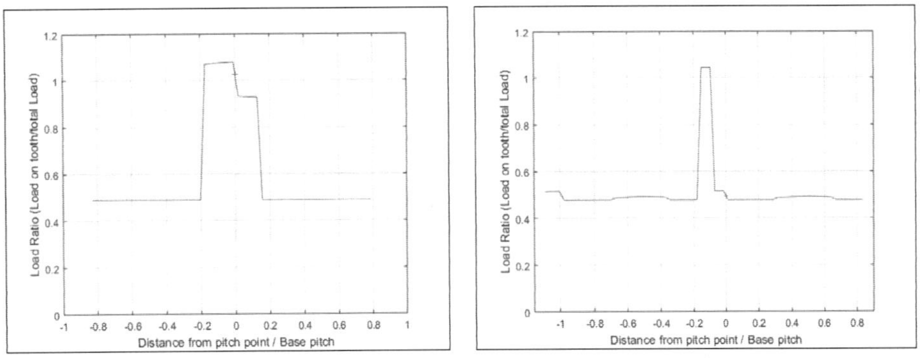

(a) (b)

Figure 4. Load ratio vs distance from pitch point, (a): Sun gear (b): Ring gear

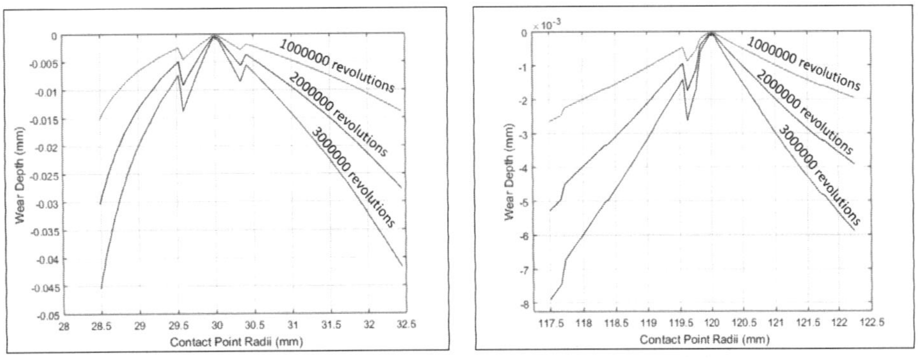

(a) (b)

Figure 5. Wear depth vs contact point radii, (a): Sun gear (b): Ring gear

gear teeth during double pair contact in sun gear. During single pair contact, the single gear tooth takes up all the contact load. A sudden drop in the contact load can be seen at pitch point which is because of the reversal of the friction force direction at pitch point. Fig. 3b shows a single tooth takes up all the contact load during single pair contact in ring gear. The contact force is shared nearly equal between gear teeth during double pair contact. The small variations observed in this region is because of the changing contact force values at the meshing between sun and planet gears and also the reversal of friction force direction at the pitch point in case of sun gear as well as ring gear. Fig. 4 shows the load ratio versus distance from pitch point plots for sun and ring gear. The sliding velocity expressions are given as [5] [11]:

$$v_s = (w_p + w_g)y \ldots\ldots\ldots\ldots [For\ meshing\ of\ external\ gears] \quad (9)$$

$$v_s = (w_p - w_g)y \ldots\ldots [For\ meshing\ of\ external\ pinion\ and\ internal\ gear] \quad (10)$$

where v_s is sliding velocity, w_p and w_g are magnitudes of angular velocities of pinion and gear respectively and y is the distance along line of action between pitch point and arbitrary point chosen. From contact forces and sliding velocities, sliding wear is predicted using Archard's law (Eq. (8)) for different points on the gear tooth profile (Fig. 5). It can be observed that the overall wear depth for ring gear is smaller than that for the sun gear. This is because the sliding velocity between planet gears and ring gear is smaller than that between sun gear and planet gears. Both the wear plots show sudden change in wear due to sudden change in contact load near the region of transition from double pair contact to single pair contact. The wear at pitch point is zero due to pure rolling. The wear predicted goes on increasing as we go away from the pitch point due to increasing value of sliding velocity (Eq. (9) and Eq. (10)). On an average, the computation time required for the analysis of each contact configuration in planetary gear train is 0.2 seconds (with a system having configurations: 16GB RAM, Intel Core i5 processor and Windows 10 OS).

4. Conclusion

In present work, linear complementarity model is extended to a more complicated case of planetary gear train. The reaction forces at bearings are incorporated as bilateral constraints in LCP formulation. The model developed here is meant for rigid bodies. Future work will focus on the extension of linear complementarity model to gears as deformable bodies.

References

[1] Ambarisha V and Parker R Nonlinear dynamics of planetary gears using analytical and finite element models 2017 *Journal of Sound and Vibration* **302** 577-95
[2] Thirumurugan R and Muthuveerappan G Maximum fillet stress analysis based on load sharing in normal contact ratio spur gear drives 2010 *Mechanics Based Design of Structures and Machines* **38** 204-26
[3] Cottle R, Pang J and Stone R 1992 *The Linear Complementarity Problem* (Philadelphia: Society for Industrial and Applied Mathematics)
[4] Pathak M, Gautam M and Rakshit S Contact analysis of a gear pair using linear complementarity 2018 *IFToMM Asian Mechanism and Machine Science Conf.* (Bengaluru, India)
[5] Flodin A and Andersson S Simulation of mild wear in spur gears 1997 *Wear* **207** 16-23
[6] Buckingham E 1963 *Analytical Mechanics of Gears* (New York: Dover)
[7] Anitescu M and Potra F Formulating dynamic multi-rigid-body contact problems with friction as solvable linear complementarity problems 1997 *Nonlinear Dynamics* **14** 231-47
[8] Stewart D and Trinkle J An implicit time-stepping scheme for rigid body dynamics with inelastic collisions and coulomb friction 1996 *International Journal for Numerical Methods in Engineering* **39** 2673-91
[9] Dirkse S and Ferris M The path solver: A non-monotone stabilization scheme for mixed complementarity problems 1995 *Optimization Methods and Software* **5** 123-56
[10] Maitra G 1994 *Handbook of Gear Design* (New Delhi: Tata McGraw-Hill)
[11] Tunalioglu M and Tuc B Theoretical and experimental investigation of wear in internal gears 2014 *Wear* **309** 208-15